T0302223

# The Security Leader's Communication Playbook

# The Security Leader's Communication Playbook

## Bridging the Gap between Security and the Business

Jeffrey W. Brown

CRC Press
Taylor & Francis Group
Boca Raton  London  New York

CRC Press is an imprint of the
Taylor & Francis Group, an **informa** business

First edition published 2022
by CRC Press
6000 Broken Sound Parkway NW, Suite 300, Boca Raton, FL 33487-2742

and by CRC Press
2 Park Square, Milton Park, Abingdon, Oxon, OX14 4RN

ISBN: 9780367570019 (hbk)
ISBN: 9780367723231 (pbk)
ISBN: 9781003100294 (ebk)

Typeset in Times
by Deanta Global Publishing Services, Chennai, India

# Contents

# Preface

When I began writing this book in the Spring of 2020, the great COVID-19 lockdown was well underway. I had just started my new role as the first CISO for the State of Connecticut after working for over 20 years in the financial services industry. I had also met only one person, my boss, in person. Now, I'm almost a year into this role and I've had to forge relationships with my team and stakeholders all by video conferencing, phone calls, email and chat. These are relationships that would normally be formed by face-to-face, in-person meetings. Now, at the beginning of 2021, we are still working remotely for the foreseeable future. While I have managed to meet a handful of people in person under controlled circumstances since then, the future of office work for almost everyone remains unclear. Communication has become more critical than ever. And more difficult than ever.

As a senior security leader, you will be faced with many communication challenges. You will be required to communicate via live presentations, phone calls, Zoom meetings, in-person meetings and in email, just to name a few. You will also face communication headwinds like apathy, language barriers and all the general distractions that all workers in today's workplace face. Is anyone really listening to you? Is your message actually getting through?

This book is not a communication theory book, though it does cover some theory. It is intended to give you "hands on" tools and techniques to help you navigate the complexities of communicating security issues, working with people and getting things done. I will teach you better ways to prepare presentations and get ready for public speaking engagements, how to present to the board of directors and how to communicate with your team and your business stakeholders. I will cover several different communication scenarios and different mediums, but the principles are often the same. Clarity, brevity and active listening should all be high on your list of skills to develop as your career progresses. And when you do reach the top, these are going to be some of the most critical skills to hone and maintain.

Good communication is a process. It is something that you need to consciously think about and practice. In other words, good communication is deliberate. It's important to realize that simply reading this book will not do much to make you a better communicator; you will need to put the tools and techniques into practice to realize any true benefit.

I have seen communication issues derail many technically competent security professionals. By taking the step to work on your communication and soft skills (the two go hand-in-hand), you will hopefully never join their ranks. I am grateful for your time and interest in this subject and sincerely hope you find benefit from the material contained in this book.

# Acknowledgments

There are many people who have helped, encouraged and supported me along the way, both while I was living my life and while I was writing this book. I'd like to take a moment and thank the following people:

- My parents, who knew that I shouldn't buy a word processor but instead a computer that did word processing. Yes, they used to have standalone, single-purpose word processors that couldn't do anything else. This is how I learned computers in the first place.
- Merrill Lynch for hiring me for my first security job in 1997, despite all common sense.
- Brian Redler, for being my first mentor and for being someone who has remained a life-long friend. I learned how to truly be the advocate for my team thanks to you.
- Oscar Gonzalez for a lifetime of advice, wisdom and enthusiasm.
- William Kolbert, for showing me what truly fearless leadership really means. I will always remember my time with you and the many, many lessons you taught me.
- Kostas Georgakopoulos, for being my best hire ever, a true friend and a trusted advisor. It's been a long, strange trip!
- Rocco Grillo, for always standing by me, even when the chips were down.
- Every security team I've managed. This is a continual learning process for all of us. Thank you for your support and for the many who have kept in touch well after our paths have diverged.
- The entire security community, including my many friends at Evanta, ISACA, ISSA, NASCIO and many others. And my fellow State CISOs. I love how we are always so eager to share information and help each other.
- Dan Swanson, for reminding me that I should be writing books like this one and for supporting me along the way.
- CRC Press, for taking a chance on a book that didn't have a lot of clear competition. Thank you for realizing that this was something that needed to be written.
- Gracie, our rescue English Springer Spaniel who came into our lives right when my wife and I needed her the most. I now understand the saying "who rescued who?"
- My wife, Zsuzsa, who put up with me through quarantine while I was working on "the book." You have infinite patience and a wonderful, generous heart.

- My three worst bosses ever. You know who you are. Thank you for teaching me first-hand why soft skills matter and how not to manage a team. I am grateful that our paths crossed and equally grateful that our paths have diverged and that I only carry your lessons forward with me and nothing more.

# Author

**Jeffrey W. Brown**, CISSP-ISSMP, CRISC, CISM, PMP, currently serves as the first Chief Information Security Officer (CISO) for the State of Connecticut. Jeff is a recognized information security and IT risk expert with a strong track record of more than two decades implementing cost-effective controls for global Fortune 500 financial institutions, including Merrill Lynch, Goldman Sachs, Citigroup, GE Capital, BNY Mellon and AIG. Jeff helps senior executives understand and manage cybersecurity risk while keeping a commercial perspective on meeting business objectives.

Prior to his 22 years in cybersecurity, Jeff worked in the publishing industry as a professional editor for HarperCollins before pursuing his passion for technology full time. While at HarperCollins he set up the College division's first web presence and was involved in helping plan distance learning programs, a relatively new concept back in 1995.

Even when he left publishing, Jeff never strayed far from his roots. He is the co-author of the *Web Publisher's Deign Guide for Windows* and *Mission Critical Internet Security*. In addition, he was the technical editor for *Thor's Microsoft Security Bible* and has authored multiple articles for *SC Magazine* and many others. He also worked with the SANS institute, contributing to multiple step-by-step guides and serving as co-editor for their now defunct *Windows Security Digest*.

Active in the information security industry, he is a frequent speaker at many events and conferences. He is the co-chair of Evanta's New York CISO Executive Summit events and works in an advisory capacity with the Cyber Investing Summit and many others. He is a board advisor for iQ4 on their Virtual Cybersecurity Apprenticeship Challenge, which aims to prepare 10,000 students for the workforce and help address the security skills shortage. He is also a former board advisor for a company named NoPassword, which was acquired by LogMeIn in 2019.

In his free time, Jeff enjoys hiking outdoors, playing guitar and enjoying time with his wife and English Springer Spaniel, Gracie. He holds multiple industry certifications, a BA in Journalism and an MS in Publishing from Pace University as well as a certificate in Cybersecurity from Ithaca College, where he also serves as an advisor for the program.

# Introduction

*If I went back to college again, I'd concentrate on two*
*areas: learning to write and to speak before an audience.*
*Nothing in life is more important than the ability to*
*communicate effectively.*
~ Gerald Ford, former president

## WHY A BOOK ON COMMUNICATION JUST FOR SECURITY LEADERS?

According to the National Association of Colleges and Employers (NACE) in their 2020 NACE's Job Outlook 2020 Survey, these are the top 20 attributes that employers like to see on a résumé. Note that this is not specific to a cybersecurity résumé, nor is it even specific to a technical job. These are simply requirements of the modern workplace.

| ATTRIBUTE | %OF RESPONDENTS |
|---|---|
| 1. Problem-solving skills | 91.2 |
| 2. Ability to work in a team | 86.3 |
| 3. Strong work ethic | 80.4 |
| 4. Analytical/quantitative skills | 79.4 |
| 5. Communication skills (written) | 77.5 |
| 6. Leadership | 72.5 |
| 7. Communication skills (verbal) | 69.6 |
| 8. Initiative | 69.6 |
| 9. Detail-oriented | 67.6 |
| 10. Technical skills | 65.7 |
| 11. Flexibility/adaptability | 62.7 |
| 12. Interpersonal skills (relates well to others) | 62.7 |
| 13. Computer skills | 54.9 |
| 14. Organizational ability | 47.1 |
| 15. Strategic planning skills | 45.1 |
| 16. Friendly/outgoing personality | 29.4 |
| 17. Entrepreneurial skills/risk-taker | 24.5 |
| 18. Tactfulness | 24.5 |
| 19. Creativity | 23.5 |
| 20. Fluency in a foreign language | 2.9 |

DOI: 10.1201/9781003100294-101

You might notice from these results that technical and computer skills are number 10 and number 13, respectively. Leadership and variations of communication and other "soft skills" make up most of the rest of the list.

Why the big disconnect? If you were looking for books on cybersecurity, most would focus on technical skills. In fact, almost all of them focus on technical skills. Books on network security, red-teaming and other technical topics abound. Educational institutions that are finally teaching cybersecurity are rewarding skills in engineering and computer science without including communication studies. In the earlier part of your career, you might have been rewarded primarily for your technical knowledge, especially when you were an individual contributor. Once you are no longer an individual contributor, communication and soft skills become far more important. In fact, my perspective is that even when you are a technical or individual contributor, your ability or inability to work on a team may make or break your early success or will at least put some limits on how far your career progresses and how fast. This book is intended to address some of these challenges. Just because we spend our lives communicating doesn't mean that it's easy to do or that we're good at it just because we do it a lot.

IANS Research, a security research and advisory firm based in Boston, set about answering the question of what made the top-performing CISOs the best in class. With their unique access across the security community, they drafted a paper, "5 Attributes of Top-Performing CISOs." In this paper, the following attributes are cited for the top-performing CISOs in any industry.

1) **Business-focused**: These are CISOs solving business problems with security tools, not just people deploying security tools.
2) **Influencers**: They are people who can get the business on board with security objectives without resorting to command-and-control tactics.
3) **Team building**: They are people who drop their ego and empower others. They leverage their entire team and don't have to be the smartest person in the room.
4) **Passion**: They are people who have a driving interest in security. This is, after all, a quickly evolving profession and what you learned yesterday may not have much relevance tomorrow.
5) **Crisis leaders**: Finally, they are people who can be the calm in the middle of the storm. They are approachable and can lead in a crisis.

The first attribute, being business-focused, means that effective security leaders spend less time on the technical details and more time finding ways that security can help solve business problems. To stand out in this category, you need to understand how each security task drives a business outcome. You'll also want to educate your team to think the same way. Being business-friendly is the strongest attribute of an effective security team. Less effective leaders set up teams that operate in silos away from the business and sometimes even away from the rest of the technology team. The "cloak and daggers" approach to keeping security secrets is not effective in the business world.

Security leaders who excel as influencers, the second attribute, are good at getting cross-functional areas of the business on board with security initiatives. Sometimes (believe it or not) security initiatives will face opposition from the business that may not be receptive to security tools and solutions. Let's face it, tools like multifactor authentication (MFA) and other password controls can introduce a level of friction for customers and employees alike. To excel as an influencer takes strong communication skills and a broad understanding of business topics other than security. We will discuss building influence later in this book.

Security leaders who excel at team building, the third attribute, can empathize and connect with subordinates across a wide range. This is where you need to drop the personal ego and really work with and connect to people. Good team builders put their people first and empower their teams. These are the leaders that employees really want to work for, and these teams tend to result in far less turnover than is typical in the industry. I dedicate a whole chapter to working and communicating with your team, because without them your success will be limited.

You would think that having a passion for security, the fourth attribute, might be a given. But in an industry where many are commanding high salaries, there are a good number of people attracted to this profession with only a passing interest in learning about the subject in depth and a strong interest in building a name for themselves, getting a bigger salary or moving on to an even bigger, better role in the future. For the people who are genuinely passionate about the subject, it tends to be contagious for others. Mark Twain once said, "find a job you enjoy doing, and you will never have to work a day in your life." Having passion ultimately makes everything easier because you're enjoying what you do, and this has a knock-off effect for everyone else.

Finally, being a good crisis leader and the calm in the middle of the storm is the final attribute of the best security leaders. These leaders remain calm, tailor their messages to specific audiences, prioritize work based on what the business really needs and have a plan and a process to collaborate with cross-functional leaders. Not just during incidents, but all the time. They bring order to a chaotic list of projects, they help align the resources and they show high amounts of leadership not just in the security group but across the entire organization. These are the leaders who have been invited to have the proverbial "seat at the table" with other senior business executives.

You'll notice that there isn't a lot of deep technical security knowledge outlined in these attributes. In my opinion, the amount of technical knowledge you *need* to have even in the CISO role isn't very deep. That's because the best CISOs aren't technologists working for a business. They are business executives with both technical and business skills who understand how security can help solve business problems.

Many companies are starting to consider if their CISO is really a C-level executive or not. The differentiators are on the soft skills side, not based on the depth of their security knowledge. Your ability to get things done, lead and influence others and sell the security mission on an enterprise scale is the future of this and other security leadership roles. But, as author William Gibson said, "the future is already here—it's just not evenly distributed." It will take some time for organizations to realize that the best CISO is not just the person who knows the most about security. It's the person who knows enough about security and how to work with everyone else to get the right controls in place to protect the organization.

# STRUCTURE OF THIS BOOK

This book is divided into two main sections: foundational communication skills and communication scenarios where you will apply these skills. The foundational skills chapters may seem basic to some, but I recommend spending some time here and not skipping ahead. These are the foundational skills that will help you when the more complicated situations in Part 2 arise. I've also made sure to include some things you may not have heard of before in other communication books, like adopting the right mindset. Strong communication is more of a mental game than many people think it is.

Part 2 dives into the situations you will encounter in the real world, including writing policies and standards, responding to incidents, speaking to the board of directors and communicating effectively with your boss and team. These are the scenarios you will be faced with on a day-to-day basis, and I've included some advice on handling these situations based on the foundational skills you'll learn in Part 1.

My goal in writing this book is not to provide an exhaustive list of every possible communication problem you'll encounter but to provide "just enough" practical skills and techniques for security leaders to get the job done. To a degree, it is deliberately a mile wide and an inch deep across a broad spectrum of situations. These are situations I've either personally been in or I've seen with my peers. Paired with the foundational skills in the first part of this book, I believe that everyone will find something of value somewhere along the way in this book. Another goal of this book is to simply make you think more about communication, your communication style and how making some adjustments could make you more effective.

Finally, I include some samples and templates in the Appendix that I hope you will find useful. Adapt them, modify them and make them your own.

# WHO AM I?

By way of introduction, I have been working in information security for almost two-and-a-half decades at the time of this writing. Many of my roles were at large, global financial institutions, some of which had several hundred thousand employees in more than 100 countries. I consider myself very technical, but I was largely self-taught and have been a life-long learner with computers and with cybersecurity. My path to CISO would be considered non-traditional by most. My formal education was in communications, including publishing and journalism. In the early part of my career, I spent some time doing professional editing, including both print and television work.

While this has raised a lot of curious glances from hiring managers in the past, these skills have served me very well in an industry that is complex, confusing and yet needs to be understood by a large variety of stakeholders. While I currently spend a good amount of time keeping up with technology, this communication foundation has served me far better than a two-decade old computer science degree would have served me. That's not to say that technical skills don't matter though. I have my own virtual computer lab, fearlessly adopt beta software and go deep into a technical level that has never failed to keep up with my peers. I can discuss the merits of certain hashing algorithms over others, why no one should still be using Secure Sockets Layer (SSL) and what's the difference between a red team and a blue team. The real skill comes from not just explaining this from one security person to another. The real skill comes from explaining this from a security person to anyone else who has never even heard these terms.

After all these years in the industry, I have also seen a lot of what works and what doesn't work with communication. Much of the advice offered in this book is from my own personal experience and from my industry peers and from other best practices. I have seen how to resonate with business leaders and how communications can fail. I've made my own mistakes, and plenty of them, along the way. I've given public talks that were strained and presentations that didn't follow any logical format. We're all human and you will make your own share of mistakes along the way. But as philosopher Alain de Button said, "if you're not embarrassed by who you were 12 months ago, you didn't learn enough." The key is to continuously improve and be better than you were before.

Security has changed from a technology issue to a business issue that has worked its way all the way to the board of directors and CEO. The stakes of solid communication skills in the cybersecurity industry have never been higher. While some security people shake their heads about how everyone else will never "get it," we are going to try to tackle this problem head on and turn the lens back on ourselves as the path forward. The only way people are going to "get it" is when we finally learn to explain it in terms that everyone can understand. As Albert Einstein has often been quoted as saying: "If you can't explain it simply, you don't understand it well enough." Fortunately, our problem is a lot easier than explaining the laws of physics.

While some security leaders will no doubt continue to get by with marginal communication skills, the ones who have excelled have done it through a combination of strong communication skills and a strong technical knowledge of the foundations of cybersecurity. It used to be that you needed 90% technology and 10% communication in a security role. But now, I'd hazard to say it's about 30% technical and 70% communication. I know very few of the top CISOs who spend most of their time talking about TCP/IP handshakes and cryptographic hashing algorithms. Sure, it comes up now and then, but most of their time is spent learning about business objectives, gaining buy-in and funding for the program, managing their team, educating people on the less-technical aspects of cybersecurity and helping the board of directors understand their risk and exposure in business terms they can easily understand. The best security leaders are also risk managers. Every issue doesn't carry an equal weight, but if you don't understand the business, you will never be able to put the risks into a context that makes sense. If you're in a top leadership job and not spending time on these things, it may be time to pause and reflect.

World Economic Forum (WEF) released the third edition of its *The Future of Jobs Report 2020* in October 2020. It's an exhaustive document that "aims to shed light on: 1) the pandemic-related disruptions thus far in 2020, contextualized within a longer history of economic cycles, and 2) the expected outlook for technology adoption jobs and skills in the next five years." Some key findings in the document include:

- The pace of technology adoption is expected to remain unabated and may accelerate in some areas.
- Automation, in tandem with the COVID-19 recession, is creating a "double disruption" scenario for workers.
- Although the number of jobs destroyed will be surpassed by the number of "jobs of tomorrow" created, in contrast to previous years, job creation is slowing while job destruction accelerates.
- Skills gaps continue to be high, as in-demand skills across jobs change over the next five years.
- The future of work has already arrived for a large majority of the online, white-collar workforce.
- Online learning and training is on the rise, but it looks different for those in employment versus those who are unemployed.

The document outlines a number of jobs and industries either on the decline or in danger of being replaced by automation. Even some very high-profile jobs were flagged as being on the decline, including accountants and auditors, financial analysts and human resource specialists. The report is not all bleak though; five areas relevant to our industry were shown to have increasing demand both now and over the next five years. These include:

- Risk management specialists
- Strategic advisors
- Management and organization analysts
- Organizational development specialists
- Information security analysts

These disciplines do have a few things in common, including their focus on risk management, strategy, business operations and IT. The future looks a lot like it will be dominated by self-directed, critical thinkers and problem-solvers who communicate clearly and dynamically.

# TARGET AUDIENCE

I would hate to think of this book as being only for CISOs and I deliberately chose to use the word sparingly and not in the title of this book. For starters, CISOs aren't the only ones who need to communicate about cybersecurity. Security is a rapidly growing but still immature industry. There are now many senior security sales roles, consulting roles and other leadership roles that command large salaries and have great influence. A large team at a bank might have several former CISOs reporting to a global CISO. In fact, a lot of financial institutions implement the three lines of defense model, which typically includes a peer-level cyber risk executive. Everyone is talking about cybersecurity now and my goal is to make the conversation easier for everyone, not just a job that constitutes a single role in a single company. CISO is just a title, and there are still huge talent gaps between the good and the great CISOs.

At the end of the day, we all need to have more effective communication about cybersecurity issues. As boards and senior business executives are directly fielding more questions about how they protect corporate information assets, even business leaders need to improve the way they communicate about cybersecurity.

My hope is that this book will be useful for anyone in a technical role who would like to improve their communication skills and for anyone working in cybersecurity. This stated, a few scenarios like presenting to the board of directors will likely apply only to the CISO or equivalent in an organization. Don't let this deter you if you are not in this role yet. If you ever aspire to be in this role, it would be useful to gain some insight into what this communication scenario entails and how it might improve some of your own, day-to-day communication with business executives. To others, it might offer a little insight into why there's such a disconnect at the top of the company and how it might improve in the future. One principle I hope you will take from this book is that there are two sides to any communication. This isn't just about you and how you communicate. It's about connecting with others and getting successful messages across to a variety of stakeholders.

I'm including many "hands-on" suggestions and tools that I hope will be useful to anyone in technology or anyone struggling to be more effective in a technical role that needs to interact with the business. Although some advice in this book may seem basic at first, ask yourself if you really put these techniques into practice. That's what really makes the difference. Knowing what to do and actually doing it are two very different things.

This is technically a "soft skills" book. I hate that term since it is always compared to "hard skills" like measurable technical knowledge that comes from certifications or academic degrees. Yes, hard skills matter. If you don't know what a firewall does or what some of the inherent limitations of tools like Data Leakage Prevention (DLP) are then it probably won't matter much that your communication is clear, because what you're saying will be incorrect or not that useful. But if you have all the knowledge but can't get your point across, you are equally ineffective as someone who has all their facts wrong but communicates misinformation very well.

You'll also notice that I've included some personality material, like the DiSC model. At first, topics like this might not seem to have much to do with communication. But the reality is that people all have their own personalities, quirks and communication preferences and DiSC will help you understand this better and help you tailor your communication strategy so that it's more effective. I've also included a good amount of material on adjusting your mindset. You will never be an effective communicator if you are going into conversations with your own mental baggage and biases.

If you can't interact effectively with the rest of the organization, you are going to fail. A modern security leader needs to understand business objectives and present the information security program as a response to business needs. 20 years ago, the few CISOs that were around might have worked primarily with technologists, but that's not how it works anymore. Today's security leader needs to be effective everywhere from the datacenter to dealing with third parties to interacting with business leaders all the way up to the CEO and the board room.

Your technical skills most likely got you ahead, but your people skills will open your best future opportunities. Your work ethic, your attitude, your communication skills, your emotional intelligence and a whole host of other personal attributes are the soft skills that will be crucial for your career success. Soft skills are undervalued, and often there is little to no training available. Perhaps it is the difficulty in certifying and measuring these skills that makes this a challenging subject. Organizations just expect people know how to behave and get along, communicate, and get stuff done. Teamwork, leadership and communication are supported by soft skills development. The better you are, the better and easier you will lead and influence across the whole organization, not just your team.

In a security leadership role, you will be required to work with business leaders, ensuring that security projects have business value. You will need to collaborate with members of the organization's management team, technologists and end-users. You will need to understand all the technology projects and even many business projects. You will not be able to do this alone; you will require a team of direct reports and support from the entire organization. Leading and managing your team or cross-matrix teams will determine your success or failure in the role.

The good news is you don't need to go get an MBA to learn to connect to the business better. In fact, I know a few CISOs who have an MBA who are still soft skill disasters. They are routinely pushed out the door from one job after another because everything looks good on paper but it falls apart where it counts in the real world. It's not a degree that will make the difference, what you really need is just to take some time and put some deliberate practice into stronger communication and connecting with others.

While I don't talk about a lot of technical security issues in this book, this is really a communication book for people with at least a basic understanding or interest in security concepts like incident response. However, I don't think anyone will find the technical detail overwhelming even if you don't have that understanding.

Finally, I encourage you to read the sections that you don't think apply to you. You're not a consultant or don't work in sales? One thing you will learn from this book is that communication is a two-way process. You will likely deal with salespeople and consultants and it pays to understand their perspectives and approaches to communication. And if you do work in sales or as a consultant, I think you will gain some insight into the challenges faced by a security practitioner. Mutual understanding leads to better connections and thus better communication. Gain some perspective on where the other person is coming from and everything will go smoother.

# WHY SOFT SKILLS ARE HARD, AND COMMUNICATION IS DIFFICULT

Words are powerful. We use words to heal and connect, but also to harm and divide. Living in the information age means that we communicate more and faster than ever before. We constantly communicate with each other digitally or in-person. But despite being much more connected, our use of words and our communication skills seem less effective than ever.

Again, with a misleading name like soft skills, there is a lot of confusion about their importance in the workplace. But what makes the whole concept of soft skills and strong communication so difficult to master? One reason is that these skills are simply hard to measure. Soft skills are not black and white, and they can't be easily tested the way that you can have someone take the Certified Information Systems Security Professional (CISSP) exam and they either pass or fail. "Hard skills" can be tested. You understand a concept or you don't. You know something or you don't. You passed or you failed.

The other challenge with communication and soft skills is that it's not that easy to be consistent. You can't *always* be on. You can't always deliver that great presentation. You can't always craft crystal clear communications. Sometimes we stumble. We choke on our own words in the board meeting. You lose your train of thought during an interview. Practice helps here, as does preparation. These will be two big themes that repeat throughout the book.

Hard skills also don't have to account for personality traits or subtle communication cues. You generally learn a hard skill and then you're done. Once you understand how to enter a firewall rule or what SQL Injection is, you don't have to continue refining and developing that skill. You learn it, you incorporate it, and you move on to the next thing. This isn't the case with soft skills, which need to be acquired, refined and maintained.

Some of the most valued soft skills include:

- **Teamwork and collaboration**: Very few things happen in isolation anymore. You need to be able to work with a team, on a team and potentially even lead a team depending on your role.
- **Oral and written communication**: You need both oral and written communication skills. Almost all communication is a derivative of the written or spoken word. Video chats, SMS texting, email and phone calls are all just variants of written and spoken communication.
- **Analytical reasoning and critical thinking**: Analytical skills are a kind of problem-solving skill. Attention to detail, critical thinking and decision-making ability all play into this area.
- **Complex problem-solving**: Few things are more complex than solving cybersecurity problems. The ability to look at a lot of data and make the many trade-offs that need to be made to be effective is key.
- **Agility and adaptability**: The ability to adapt and change with business priorities has become critical in our fast-paced world. Keeping up with new technology like the cloud, artificial intelligence (AI) and the Internet of Things (IoT) rather than trying to slow down their deployment is important.
- **Ethical decision-making**: Businesses also have compliance and ethics requirements. You cannot go against these requirements. Getting your job done should not be "by any means necessary," it needs to be by ethical means that are in alignment with your company and your own values.

Soft skills affect communication skills and communication is how you interact with the world, your company and everyone else around you. When the focus on hard skills is overemphasized, gaps often develop with employee soft skills. When there are gaps in soft skills, they often manifest themselves in problems like high turnover, an inability to work well with other areas or your termination in the role because no one can work with you. In the consulting and sales world, they can manifest in client retention problems or the inability to make a sale. Gaps with soft skills will also prevent you from leveraging the full power of your team and all their collective knowledge. In other words, soft skills matter, and they matter quite a bit.

# SUMMARY

Hopefully you now understand the value of pursuing, enhancing or developing soft skills and why they matter. In the end, soft skills will let your hard skills shine in a way that wouldn't be possible otherwise.

- Soft skills are becoming the most sought-after skills in today's workforce. Soft skills facilitate interpersonal and relationship-building skills that help people to communicate and collaborate effectively across the entire organization.
- The most effective security leaders spend most of their time not on technical pursuits, but on learning about business objectives, gaining buy-in and funding, managing their team, educating people on the less-technical aspects of cybersecurity and helping the board of directors understand their risk and exposure in business terms they can easily understand.
- Everyone can benefit from making their communication and soft skills better. Security is a topic that is touching every level of the organization. It is not just for the CISO or IT department.
- Soft skills are hard to measure. There are not many effective tests or certifications you can get to prove your abilities.
- Part 1 of this book focuses on the foundational skills needed for the real-world scenarios found in Part 2.
- High turnover and an inability to get projects executed across the enterprise are often a result of soft skills problems.

# REFERENCES AND RECOMMENDED READING

Alex, K. *Soft Skills: Know Yourself & Know the World*. S. Chand & Company Ltd., 2012.
Collette, Ronald D., et al. *CISO Soft Skills: Securing Organizations Impaired by Employee Politics, Apathy, and Intolerant Perspectives*. Auerbach, 2009.

"'The Future Is Already Here – It's Just Not Evenly Distributed' William Gibson, The Economist, December 4, 2003." *Cities & Health*, vol. 4, no. 2, 2020, pp. 152–152. doi:10.1080/23748 834.2020.1807704.

*The Top Attributes Employers Want to See on Resumes, National Association of Colleges and Employers*, www.naceweb.org/about-us/press/2020/the-top-attributes-employers-want-to -see-on-resumes/.

"Why Soft Skills Matter." *MindTools*, www.mindtools.com/community/pages/article/newCDV _34.php.

# PART 1

# Communication Foundational Skills

This section of the book walks you through "just enough" communication foundational skills. While it may be tempting to jump to Part 2 of the book, much of that section has been built on the material found in Part 1. These foundational skills include material I wish I had understood earlier in my career and information that I think you will genuinely find useful.

We will cover all the basic foundations that you need to excel in your role: written, verbal and even visual communication. This section covers foundational skills, acclimating to new companies and communication "superpowers" like learning to say no and negotiating like a boss. There's a lot more and I promise to try and serve you to the best of my abilities. Even as a communication expert, I learned a lot by writing this section and I hope you will also learn a lot by spending some time with me in this journey.

DOI: 10.1201/9781003100294-1

# Foundational Communication Skills

1

*The way we communicate with others and with ourselves*
*ultimately determines the quality of our lives.*
~ Anthony Robbins

The basis of all soft skills is strong communication. To discuss communication, it's helpful to first define the word. Let's use a working definition of communication: "to express oneself in such a way that the message is readily and clearly understood." This sounds simple enough. We have been communicating all our lives. It's typically something that just happens, which is why many people never think about how effective they are at it or try to make improvements. This is a mistake.

The word "communication" comes from the Latin word communis, which means common. Therefore, when we communicate, we are trying to establish "commonness" with someone. In other words, we are trying to share information or an idea with someone and get everyone on the same common ground and create a mutual understanding. This definition is helpful, because obviously the point of communication is to connect with someone and get an idea across to them. The only problem is that often in communication, we use the methods that suit ourselves best. We use our preferred communication styles, jargon and mediums rather than the recipient's. A lot of communication doesn't go well because it is set up that way right from the start.

To communicate effectively, you need to consider factors ranging from people with different experiences than your own, the setting, verbal as well as nonverbal cues and the intended meaning versus the perceived meaning of a message. There's a lot working against successful communication. Distraction, work overload, cultural and language barriers can all interfere with how your message is received.

Communication is ultimately an abstraction of a thing, event or an idea, and not that thing itself. I can tell you what it was like being in downtown New York City on 9/11, but it will not convey the same experience for you as it did for me. Communication attempts to share a concept or experience that exists in your head and tries to get it into someone else's head. No wonder getting communication right is so difficult!

In my two-plus decades working in information security I have watched the role of senior security leader evolve from someone who spends most of their time working with fellow technologists to someone who must communicate with all

DOI: 10.1201/9781003100294-2

## Security Leaders
## Need to communicate
## to all levels of the organization

**FIGURE 1.1**  Good communication is critical for cybersecurity leaders. You will need to communicate with every single employee in a company at some level.

levels of an organization, from the technologist all the way up to the CEO, board of directors and everyone in between. A senior security leader is expected to have a technical background, but the role has also shifted into a risk-savvy business executive capable of leading and influencing across the entire organization (Figure 1.1).

Unfortunately, most people in these roles never receive communication training in their career journey and many struggle with communication challenges. Poor communication skills undermine strong technology skills and will keep you from being fully effective. The more senior you get in your career, the less you are expected to have the same "hands on" technical knowledge as the people who work for you and the more you are expected to be a great communicator and someone who can interact with all levels of the organization. You could write, speak and present all day long, but unless you know how to reach your audience, you don't have the communication skills needed to help provide adequate security to your company and be part of its success.

All leaders need to be good communicators. This is true for CISOs, CEOs and any senior business executive. If you have reached this level, chances are you oversee a significant area that includes people and company resources. In fact, the larger your team and scope, the more you will find that jobs like the CISO role are weighted more towards communication than the technical details of cybersecurity.

History has shown that keeping the cybersecurity function in isolation is not a successful strategy. Information about the security program needs to be shared across all departments and at all levels. Everyone from technical staff to marketing and business staff and all the way up to the CEO and board of directors are responsible for their part in supporting the security program and understanding, at least on a basic level, how security works and operates.

This is what makes communication for a security leader so hard. You live in two worlds: the business world and the technical world. And the technical security world you live in is really a collection of technical fields, including applications, networks, policy, databases and security, which can all be broken down into ever-smaller subdivisions. Each of these areas could take years to master individually.

You'll need to learn how to frame the conversation in terms that executives understand when you're working with business leadership. When you're working with technical peers, you'll need to present specific technical controls that need to be in place. These are very different conversations. Finally, when working with your team, you will need to turn information security into something that is relevant to their role and be a mentor, a teacher, a coach and a leader.

This section of the book lays a lot of the foundation you will need for Part 2. These are important skills that you should not skip over. Even if you think you know them, are you *practicing* them? I will cover topics like active listening, being concise and how to think like a businessperson, so you can better connect with them.

It can take a lot of effort to communicate effectively. But spending time learning and practicing stronger communication pays dividends. Whether you're speaking, listening, writing or reading clear communication will greatly enhance your experience and open new opportunities for learning from and connecting to other people.

# THE SECURITY COMMUNICATION MANIFESTO

*When you wake up in the morning, tell yourself: The*
*people I deal with today will be meddling, ungrateful,*
*arrogant, dishonest, jealous, and surly.*
~ Marcus Aurelius, Meditations

It's helpful to be realistic about business communication. It's bad out there. Really bad. The quote above, from Stoic philosopher and Roman emperor Marcus Aurelius, was one that he used to set his daily mindset and expectations. Rather than wishing that things be different or pretending that they would be ideal, he set his expectations every day that things would likely be difficult. And that was OK and to be expected. You see it, accept it and you keep going anyway.

Using this as a starting point, here are some fundamental truths about communication that you should consider.

1) We are all distracted. Assume that the people you are communicating with have plenty of other things they'd rather be doing. So, get to the point!

2) Security is confusing for non-technical people and their inclination will be to tune out if you don't make what you're saying relevant and interesting to them. So, make it relevant and interesting!

3) Security is mostly bad news. People will want to shoot the messenger; it's human nature. Don't be surprised if people aren't thrilled that you're here to talk with them.

4) No one gets excited when your email arrives in their inbox. They will read your messages quickly and then just as quickly delete them or lose them in a mountain of other messages. There are a hundred others that arrived before you. Make it easy for them and don't be part of the email deluge.

5) Hardly anyone listens, but it's still up to you to find a way to get and maintain their attention.

6) Most people you talk to are working on their response or thinking about other things rather than listening to what you're saying. You are going to need to be engaging enough to make them *want* to pay attention.

7) No one wants to read your policy/standard/guideline. Make it easy to digest anyway. Don't let length or technical jargon be their excuse for not understanding it.

8) The report that you spent hours compiling will likely be scanned quickly or not read at all. You still need to put your best work forward. You are a professional and you always want to prove it by producing quality work.

9) People will avoid difficult conversations like the plague. Guess what? You're going to mostly have difficult conversations as a security leader. Don't worry though, you've got this.

10) You are going to have to repeat yourself between 6 and 20 times for some people to finally "hear" your message. Accept it. You probably do it yourself sometimes.

This may not sound like a rosy picture, and it isn't. But it's reality and you need to work with reality. The techniques and tips outlined in this book will help you overcome some of these obstacles and increase the chances of your message getting through. I'll summarize a bunch of them right now for you in very simple language: *make it easy for your audience, not for yourself.*

# COMMUNICATION SCENARIOS YOU WILL FACE

As a senior security leader, you will be faced with many communication scenarios. Throughout the course of your day, you might find yourself talking to technicians, business executives, your team, financial professionals, vendors and even the board of directors. You oversee a critical function and you have been given a lot of resources (or at least some resources!) to accomplish your mission. You are going to have to regularly report on progress, issues and roadblocks in a way that everyone understands.

You will be expected to speak up in meetings and present to business audiences, and you may even find yourself in public speaking engagements, podcasts or interviews. If you are like most of us, you will also spend a disproportionate amount of time writing, responding to and reading emails. You will probably have a team, and in some cases a big team. You need to communicate with them as well. Being able to communicate clearly and concisely will help you ensure that these resources are lined up, that the company understands the good work you do and the business understands how you help mitigate their cybersecurity risks and enable them to achieve their core mission. Good communication will make acquiring funding for your program easier as well, because you will be speaking in business terms, not technical terms.

Brushing up on communication skills is something that most people don't think about much and something that most people don't work on. It is also a leading reason why many CISOs fail in their role. Security leaders must ensure that controls that align with the goals of the business are put in place. To do this effectively, they need the ability to communicate effectively with a variety of different people and stakeholders.

In a senior security leadership role, you need to cross the divide between technical and business audiences. You need to be able to not only deliver a presentation to a big audience but also communicate effectively in more interpersonal situations. A security leader who can't communicate effectively simply won't last long.

Many security leaders struggle to communicate and collaborate with business leaders, in part because of limited interactions and relationships with them. This problem is exacerbated by the perception of security at the executive level. In a study done by Deloitte Consulting, CISO participants (79%) reported they were "spending time with business leaders who think cyber risk is a technical problem or a compliance exercise." As a result, most CISOs "have to invest a lot of time to get buy-in and support for security initiatives." We're all talking to senior executives more and they are asking more questions as security becomes a bigger topic of concern.

Executive relationships are essential in understanding what's happening in the business and where the greatest risks lie. Since it is virtually impossible to protect all data in an organization, a security leader needs to work with the business to understand which data is critical to the enterprise, where that data resides and the impact should that data be lost or compromised.

Security doesn't have the tight integration and back-and-forth channels established with the business that is enjoyed by functions such as customer service (which regularly provides information on customer demands and trends to other key functions) or finance (which delivers dollars-and-cents data to stakeholders across the organization). Channels for reporting security activity to senior management are only just starting to be established and formalized in some businesses and are completely nonexistent in others. In short, this is a tough topic and it's new to a lot of business leaders.

These days, when a company says it is looking for a security executive, it is seeking someone with the same business and communication skills as any other departmental leader in addition to having the subject matter expertise and technical chops for growing and running a security program. It's not enough to just have the security skills anymore. You need business skills and the ability to communicate at all levels of the organization.

# THE HIGH COST OF POOR COMMUNICATION

We've established now that effective communication is deceptively difficult. That's because not everything can be easily communicated. You don't believe me? How would you explain what it's like to drive a car to someone who has never driven one? Could you explain what the color blue looks like to someone who never had eyesight? Could an astronaut explain what it's like to be on the moon in a way that does the experience any justice? These are just a few examples of subjects that even expert communicators would struggle getting right, if they could do it at all. Some ideas, thoughts and feelings simply can't be transferred from your brain to someone else's easily. And some things, like driving a car, really need to be experienced to be understood. Words, spoken or written, will not do it justice. Fortunately, as a security leader, you will hopefully never have to communicate anything quite this difficult. But that doesn't mean your job will be easy.

Bad communication is rampant, and the costs are high. We see examples of bad communication every day. It's the person who is so boring that you tune them out and do not hear a word they say. It's the person so inflated and aggressive with their ego that they're an immediate turn off. It's the long and rambling presentation that fails to make any memorable points but passes the time. Sadly, poor communication is everywhere. David Grossman reported in *The Cost of Poor Communications* that in a survey of 400 companies with 100,000 employees, each cited an average loss per company of $62.4 million per year because of inadequate communication with and between employees. This doesn't even count customer communications! The same report found that companies with leaders who possessed effective communication skills produced a 47% higher return to shareholders over a five-year period. While it's probably difficult to quantify bad communication that precisely in your organization, we can all agree that bad communication is not exactly optimizing your business.

While I think most firms would acknowledge that internal communication is bad, I find it interesting how few work on formal communication plans or training to make things better. People are not getting their points across. Time is wasted and the general approach seems to be to hope that the situation gets better. But hope, as they say, is not a strategy.

Why does any of this matter for security professionals? If general, day-to-day business communications are this bad, your job is going to be a lot worse. Cybersecurity is complex. There are many obtuse and technical topics that need to be explained to people who may have little to no cyber or even technical background. And more people than ever are either interested in hearing about cybersecurity or have a need to know how the company is being protected. You also need to explain security responsibilities to every employee, such as not clicking on suspicious email attachments. This information needs to be conveyed to a wide variety of people in a way that fits their needs and individual experiences.

# HOW GOOD IS YOUR COMMUNICATION?

If you want to be an expert communicator, you need to be effective with the entire communication process. This means you must be comfortable with different channels of communication—face to face, online, written and so on. So how good are you now? How much room do you have to grow? I imagine that if you believe you have this subject nailed inside out, you probably wouldn't have picked up a copy of this book. But the reality is that we all have room to improve and that we're all probably better with some communication methods than with others.

The following questions may help you baseline where you have some room for improvement:

1) How much do you think about what you say or write before you start to communicate?
2) Do you proactively anticipate where confusion might arise with a communication and address it up front?
3) Have you ever been surprised that someone misunderstood something you said that you thought was obvious?
4) Do you try to understand opposing sides of an argument?
5) Do you double-check your emails and documents for grammar, punctuation and other problems before sending?
6) Do you read any of your writing out loud to make sure it makes sense?
7) Can you read basic body language and other nonverbal cues?
8) Do you consider what's the *best* method to communicate before you set up a meeting, send an email or pick up the phone?
9) Do you believe that your language is inclusive of all your colleagues? Did you ever think that some of the things you say might make people feel left out of the conversation?
10) What do you do if you don't understand something that someone said?
11) How would you go about simplifying a complex issue for a colleague who didn't have the same background as you?
12) How would you persuade someone to see things your way if they weren't naturally inclined to do so?
13) What would you do if you misunderstood an important communication?
14) Is it more important to be a good listener or a good communicator?
15) What new problems or challenges do you see in the digital age of communication?
16) What communication skills are you working on or would like to further develop?
17) What is your favorite communication medium (writing, verbal, text, email)? Do you think that you might default to a mode more often than you should because it's easiest or most comfortable for you?

These are questions to get you started in thinking about communication. There's no right or wrong answers to these questions, but thinking about how you approach communication is the first step towards making your communication more effective. We will explore many of these topics in much more detail later in this book.

# START WITH YOUR MINDSET

*If you want to be more confident and creative, or more*
*extroverted and organized, you can become any or all*
*of those things. If you're timid but want to become a*
*powerful, bold and inspirational leader, you can become*
*that as well.*
~ Benjamin Hardy, Ph.D., from
Personality Isn't Permanent

*The starting point of all achievement is DESIRE. Keep*
*this constantly in mind. Weak desire brings weak results,*
*just as a small fire makes a small amount of heat.*
~ Napoleon Hill, Think and Grow Rich

Throughout this book, I will talk about mindset quite a bit. Mindset is just your attitude about something. In work and in life, I think that mindset is critical. Getting into the right mindset as a first step makes all the steps that follow easier. It helps set the stage for change to take place. Mindset helps you change your thoughts, which influence your actions. Your actions change your behavior, and this is what drives success. This is provided that you're practicing the *right* behaviors, of course.

So, let's start this journey by adopting an appropriate mindset. One mindset you could adopt is that your technical skills will more than compensate for the fact that you can't communicate with anyone or work well on a team. I obviously don't recommend this approach. Another mindset might be that you're simply not very good at some things like written communication or public speaking even though you are pretty good at some other forms of communicating. This is better, but this mentality will hold you back from being the best that you could be in your position. It will also prevent you from taking on communication styles or mediums that would be best suited for your audience rather than using what's best for you.

A more productive mindset would be acknowledging that anyone can get better at any subject with some targeted practice. In his book *Personality Isn't Permanent*, author Ben Hardy comments: "People use the past as the excuse to remain stuck in habits and attitudes that keep them from growing." You weren't born knowing security, right? You learned it. How? With study and practice and time. This is the same way that your soft skills and communication will also improve.

If you don't want to change, you won't. You also won't change if you don't believe it's important enough to give this area your deliberate focus. If you try at all, you will likely fail without having a strong reason for why improvement matters. Yet, if you aspire to the top cybersecurity roles, you need to accept that strong communication skills are part of the job. This means, you should set aside some time to keep these skills sharp, the same way you would do with your technical skills.

In the book *Mindset: The New Psychology of Success*, author Carol S. Dweck defines two different mindsets. If you believe that change is possible and that your ability to improve isn't necessarily capped by any artificial barriers, you have what they call a "growth mindset." On the contrary, if you believe that you aren't good at something and never will be, this is called a "fixed mindset." Fixed mindset people tend to make no effort to improve, because they feel that they are inherently bad at something and won't be able to change. People with a growth mindset believe that intelligence and skills can be developed with deliberate practice.

So why should you care about making any improvements to your communication skills? Everyone can communicate. You technically don't need much in the way of special skills other than a basic understanding of the language and how to speak and write at a very rudimentary level. Yet the most effective business executives are skilled influencers, risk managers and stress managers. They are relationship builders and know how to get the most out of people. They lead. They inspire. These are people that have honed not only their technical edge but also their communication and leadership skills.

Being able to lead, influence and communicate easily will not only make you more effective; it will also make you less stressed when it counts. You'll be able to rally the troops faster. You'll get security defenses implemented faster by working better with your colleagues and other teams. You'll have better support from senior management and hopefully the funding you need to get your program implemented because you have explained in business terms why it's important.

Maybe you think you can't handle all this improvement? Well, you were born a blank slate, just like everyone else. Every skill you know today, including all your cybersecurity skills, was learned at one point in your life. And any skill can be improved with practice and the right kind of deliberate focus.

# WHAT KIND OF LEADER ARE YOU?

> *If you know the enemy and know yourself, you need not fear the result of a hundred battles. If you know yourself but not the enemy, for every victory gained you will also suffer a defeat. If you know neither the enemy nor yourself, you will succumb in every battle.*
> ~ Sun Tzu, The Art of War

It's probably not news to you that people have all kinds of different personalities. People also have unique ways that they communicate with each other. When you understand how people are different based on their communication styles, you will better be able to relate to them and their communication preferences. The first step is to understand your own leadership and communication style using the DiSC assessment method.

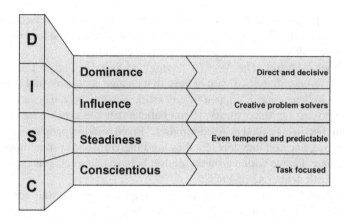

**FIGURE 1.2** The personality DiSC model can help you understand the communication preferences of others and tailor your message to your audience.

The DiSC model is based on the work of psychologist William Moulton Marston back in the 1920s. It remains a popular, straightforward, standardized and relatively easy way to assess behavioral styles and preferences. The tool classifies people's behavior into four types: Dominance, Influence, Steadiness and Conscientiousness. These types are summarized in the table below (see also Figure 1.2).

| BEHAVIOR TYPE | DESCRIPTION | EXAMPLE |
| --- | --- | --- |
| (D) Dominance | Direct and forceful. People who prefer this style value action and achievement and tend to be fast-paced and task-focused. | Arnold Schwarzenegger |
| (I) Influence | Lively and social. People who prefer this style value relationships and enthusiasm and tend to be fast-paced and people-focused. | Bill Clinton |
| (S) Steadiness | Even-tempered and loyal. People who prefer this style value cooperation and predictability and tend to be moderately paced and people-focused. | Jimmy Carter |
| (C) Conscientiousness | Private and analytical. People who prefer this style value accuracy and standards and tend to be moderately paced and task-focused. | Bill Gates |

Most people that know me know that I would fit under the High I, or influencer, category. I'm a big-picture person who likes ideas and strategies. I focus on people and people issues and like to get everyone on board with a vision rather than just telling them that things are going to be a certain way. By the way, there is no "right" category for leaders, and you shouldn't strive to change your category unless it's really causing you a problem. Keep in mind there is a dark side to every category. Some High D leaders are pushy and aggressive, bordering on unapproachable. Some of us High I leaders have our heads so high in the clouds that we lose track of the details. Some High S leaders can't deal with conflict. And High C leaders can get bogged down in details and analysis paralysis.

What type are you? Keep in mind that you may have traits in more than one category. For example, you might be an influencer with some traits of dominance or any other combination. These categories will hopefully give you some insight into your own communication preferences, but what does it mean for the people you work with? Understanding their behavior type can also help you better relate and communicate using their preferences. Below are some communication tips for dealing with different personality types.

# Working with Personality Types

*High D (dominant) leaders key characteristics*
- Direct and decisive
- Likes to be in control of decision-making
- High self-confidence, risk takers
- Bottom-line leaders who like quick results

*Weaknesses of High D leaders*
- May overstep their authority
- May seem pushy and unapproachable
- May not listen to other people's inputs or ideas
- May overextend themselves and their team

*Communicating with High D leaders*
- Be clear and to the point
- Avoid chitchat and frivolous subjects
- Don't go deep into details unless they request it
- Focus on WHAT, not HOW
- Be prepared and organized with an agenda
- Present the facts logically
- Provide alternatives so they can make their own decisions

*High I (influence) leaders key characteristics*
- Creative problem solvers who can think outside the box
- Function best on teams rather than alone
- Enthusiastic, optimistic and persuasive

*Weaknesses of High I leaders*
- Generally not interested in lots of detail, prefer the big picture
- Concerned with people and personalities
- May not handle rejection well
- May get emotional over some subjects

*Communicating with the High I leaders*
- Ask about their ideas on a subject
- Allow time for relating and building rapport

- Don't drive straight into facts and figures
- Help them get organized and put details in writing
- Don't leave decisions up in the air

*High S (steadiness) leaders key characteristics*
- Reliable and dependable
- Even-tempered
- Patient and good listeners
- Will strive hard for consensus
- Sees tasks through to completion

*Weaknesses of High S leaders*
- Likes sticking with routine
- May oppose change
- May be passive and avoid conflict
- May have a hard time saying no, try to please everyone

*Communicating with the High S leaders*
- Build rapport and show sincere interest in them as people
- Draw out their personal goals and objections
- Don't force them to make a quick decision
- Present your case logically, non-threateningly and in writing
- Address specific questions like how you will handle something
- Don't interrupt as they speak, listen carefully
- Be kind and patient, avoid confrontational language or negative body language

*High C (conscientiousness) leaders key characteristics*
- Will think through very detailed processes step by step
- Great operational temperament
- Very thorough and detailed
- Takes pride in accurate work
- Great at analysis and research
- Motivated by research and facts

*Weaknesses of High C leaders*
- May be over critical of others, especially for their work
- May not be open to other people's ideas
- May focus more on tasks than on people
- May prefer to work alone rather than with a team

*Communicating with the High C*
- Provide them with information and the time they need to make a decision
- Approach them in a direct way
- Recognize they may be uncomfortable speaking in large groups
- Build credibility by looking at all sides of an issue
- Don't force a quick decision
- If you disagree with them, prove it with data and facts or testimonials from reliable sources

The DiSC method is a great starting point but not an exact box to force-fit everyone's personality into. Also, some people may vary by their setting—for example, being more dominant at work but a steady personality at home. I hope this provides enough of a framework to realize that there are all kinds of different styles and that one size doesn't fit all when you are communicating.

What style are you? What style is your boss and others that you routinely communicate with? Think about how you might be able to change your approach with certain personality types to better connect with them. If your boss isn't big on details but you insist on peppering them with minutia since that's your preference, this could be a problem. Conversely, if your boss is very detailed and wants to know exactly *how* something will get done, you need to adapt to their needs. A principle that will be repeated throughout this book is that you need to make it easy for the recipient, not for you. Just because you have your style and preferences, it doesn't mean that this will work for everyone. We want successful communication, right? Not easy but ineffective communication.

**Action**: Think about which style you are and try to map your boss and a few other people to their respective styles. Do you think you might be taking the wrong approach based on someone's personality type? Is anyone doing the same with you?

# HOW COMMUNICATION WORKS

*The single biggest problem in communication is the*
*illusion that it has taken place.*
~ George Bernard Shaw

While this is not a communication theory book, I think that providing at least some background on the subject will help you become a better communicator. The following section lays out a general communication framework and will be referenced in subsequent chapters. You'll find from this model that communication is simple. But that doesn't mean it's *easy*.

*The Mathematical Theory of Communication* breaks the communication process into seven basic elements (see Figure 1.3):

1. Source
2. Encoding
3. Channel
4. Decoding
5. Receiver
6. Feedback
7. Context

These elements can be found even in the simplest forms of communications, such as sending a text message, to the most advanced communication scenarios, like a public presentation with a live audience. You can become a better communicator by gaining an understanding of this communications process. It will also help you understand where communication goes wrong. Let's look at each element in detail.

**The Communication Process**

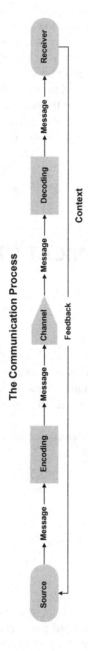

**FIGURE 1.3**   This is a simplified view of the communication process. Understanding how communication works will help you be more effective with your own communication.

# Source

Source means the source of the message or communication. In the context of this book, the source will be you and the message will be what you want to communicate. In this step, you have a clear message and a purpose for communicating it and an audience in mind where you'd like to get this message across.

# Encoding

This stage involves putting your thoughts into a format that you can send, and the receiver can receive. Success in this step begins with knowing your audience and their communication preferences. It also involves getting clear on exactly what you are trying to communicate.

# Channel or Medium

There are countless different channels that you can use to send your message. Verbal communication channels include face-to-face meetings, telephone and videoconferencing, while written communication channels include letters, reports, emails, instant messaging (IM) and social media posts. You might also want to include visual communication like videos, photos, illustrations, or charts and graphs in your message to emphasize your main points. Different channels have different strengths and weaknesses. For example, it's not particularly effective to give a long list of directions verbally, and you'll be better off delivering sensitive feedback in person, rather than via email.

There are many communication channels to master. These include:

- Text message/SMS
- Email
- Phone calls
- Video chats
- Meetings with multiple people
- Virtual meetings
- In person 1-1
- In person with many people
- Webcast/podcast
- Tools such as Slack

A lot of these examples can be summarized as being verbal, written or digital mediums. Digital channels can be verbal, written or both. Careful consideration should be given to how you choose to communicate and through what channels, which I will discuss in the next section.

No matter the medium, there are several ways to make the communication process easier. This foundation section of the book aims to do exactly that: simplify communications regardless of the medium.

# Decoding

Successfully decoding a message is as much a skill as encoding it. To accurately decode a message, you need to take the time to read through it carefully, or to listen actively to it. Confusion is most likely to occur at this stage of the communications process, though that doesn't mean it will always be the decoders' fault. The recipient might lack sufficient background knowledge to understand the message, or they might not understand the specific jargon or technical language that you are using. It's therefore essential that you tackle issues like these at the encoding stage or even the source stage.

# Receiver

No doubt, you'll want your audience to react in a certain way or take a specific action in response to your message. Remember, though, that each person is different, and will interpret messages subjectively. Every receiver brings their own ideas and feelings that influence their understanding of your message, and their response to it.

That means it's your job, as the sender, to take these ideas and feelings into consideration when drawing up your message. To do this effectively, brush up on your emotional intelligence (EQ) and empathy skills, which we will cover later in this book.

# Feedback

Feedback can come in all forms and be written, spoken or nonverbal. Nonverbal feedback (covered later) can be especially useful in gauging if someone has heard and understood your message. Where communication becomes very difficult is while using mediums like email, where feedback is either delayed or nonexistent. Feedback is an important concept to understand, because this really lets us gauge if a message was received and understood.

If you see that people are struggling to understand your message based on feedback, you can try to adapt your message. If you're talking about a complex security subject, it's probably time to dumb it down a bit and remove any technical jargon. You want to aim for everyone in your audience to grasp the message, no matter their background knowledge. Again, the goal is successfully getting your message received and understood.

# Context

The "context" is the situation in which you deliver your message. This might just be the venue, like in a meeting, or it can also factor in other elements like time.

The most important thing to remember is that communication is more complex than it may first appear and that it needs to be a *two-way* street for being effective. Too many people think the message will speak for itself without factoring in listener feedback. They simply speak or send out one-way email communication expecting that everyone would hear and, more importantly, understand their message.

# FORMS OF COMMUNICATION

There are five main forms of communication. These include intrapersonal communication, interpersonal communication, group communication, public communication and mass communication. There are similarities and differences among each form of communication.

## Intrapersonal Communication

Intrapersonal communication is communication with oneself using internal vocalization or thinking. This is literally the thoughts you have and how your own internal processes work. This communication form takes place only inside our heads. Other forms of communication are meant to be received by someone else. So why are we discussing this one? This form of communication is your "self-talk" or your "inner critic" and it matters a lot more than you might think. It's the nagging voice inside your head telling you that your board presentation is going to be a disaster or that you don't even belong in your role in the first place. This communication is influenced by your subconscious mind and can have a lot to do with your own mental health. This self-talk can be encouraging, or it can undermine your every move and every word. Spending time on developing a good mindset is the antidote to counterproductive intrapersonal communication.

## Interpersonal Communication

Interpersonal communication is simply communication between two people. We spend more time engaged in interpersonal communication than the other forms of communication. Interpersonal communication happens every day, but to be a competent interpersonal communicator, it helps to also have conflict management skills and good listening skills. I will cover both subjects later in the book.

## Group Communication

Group communication is communication among three or more people. Group communication is typically more intentional and formal than interpersonal communication. This could be a business meeting or a group hallway chat. Group communication is more affected by social rules, like letting the boss speak first, or in taking care not to interrupt and talk over each other.

## Public Communication

Public communication is typically when a single person conveys information to an audience. Most people are afraid of or dislike public speaking. Public speaking tends to be more intentional, formal and goal-oriented than other forms of communication.

In Western societies, public communication is usually more sender-focused than what is found in interpersonal or group communication. In other words, there is typically a single speaker and an audience that receives the message. We will cover this subject more in depth later in this book.

## Mass Communication

Public communication becomes mass communication when it is transmitted to many people through print or electronic media. This could be in the form of giving an interview to the news to discuss a breach or the latest issues in security. This form tends to be planned the most, since the stakes are much higher if things go wrong.

# CHOOSING THE RIGHT COMMUNICATION MEDIUM

Choosing the right communication medium is also an important part of making sure your message is understood. A lot of people don't give much thought about what medium they are using. In fact, most people just default to the method they are most comfortable using. For introverts, this might be a text or email and for extroverts it might be showing up in person or hosting a meeting.

Using email to send simple messages may seem practical. However, if you want to delegate a complex task, an email might just lead to more questions and many back-and-forth messages. If your communication has any negative emotional content, stay away from email. You wouldn't use email to reprimand an employee. You would want to make sure that you communicate face to face or by phone, so that you can judge the impact of your words through a feedback loop and adjust your message appropriately.

When choosing the right channel for your message, it helps to consider the following:

- The receiver's preferences: put the audience first
- The complexity of the material
- The sensitivity of the subject
- Time constraints
- Geographic constraints
- The need to field questions and receive feedback

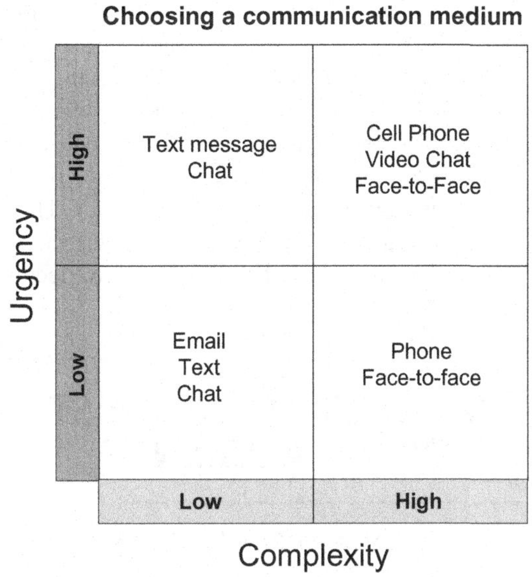

**Choosing a communication medium**

| | | Low | High |
|---|---|---|---|
| **Urgency** | **High** | Text message<br>Chat | Cell Phone<br>Video Chat<br>Face-to-Face |
| | **Low** | Email<br>Text<br>Chat | Phone<br>Face-to-face |
| | | **Low** | **High** |

Complexity

**FIGURE 1.4** Choosing the right communication medium can help your message be understood more clearly. Simple messages work fine with email and chat, but more complex communications are better served by being in person or on video.

The more complex and urgent your message is, the more you should likely default to some sort of in-person (video, phone or in-person meeting) communication. Simple messages that don't need much feedback are fine with mediums like SMS or Slack messaging (Figure 1.4).

Think about which medium you should use before communicating. Should you call? Send an email? Have a face-to-face meeting? It may depend on what would be best for the audience and for the purpose of the message.

As an example, you would never want to fire someone over text (yes, this actually happens). You also wouldn't want to send an email if you know you're going to raise a lot of questions that need to be addressed. You don't want to use email for sensitive communications. Think and strategize before choosing your medium. Some other considerations when choosing the right medium include:

- *Ease of access*: Calling an in-person meeting is sometimes not timely, or practical. It can also be expensive. In this case, you might want to pick a video meeting or choose another medium.
- *Sensitivity*: Private or sensitive issues, including anything with confidential information, are not suited for email or text. Email is a clear-text medium with no feedback loop. It may be archived in your company forever and it can be easily forwarded to many people. You'll probably want to go with face-to-face meetings if possible when you have sensitive material to discuss.
- *Purpose*: If your purpose is to share a report or document with a lot of people, email is probably the perfect way to go.

- *Interaction*: If you anticipate a lot of back-and-forth questions, this would be better suited for in-person meetings, online meetings or phone calls.
- *Need for feedback*: Looking for some feedback and the ability to read non-verbal communication queues? You're limited to in-person and video communication, where you can see in real time how people are reacting. Of the two, meeting in person is by far the strongest method of gauging feedback. Some mediums are especially bad for getting feedback. Think of sending out a newsletter to a wide audience. Maybe someone will comment and maybe they won't. Even if they do, you will not likely receive feedback right away.

# IT ALL STARTS WITH UNDERSTANDING YOUR AUDIENCE

*Speech belongs half to the speaker, half to the listener.*
~ Michel de Montaigne, philosopher

One of the big communication challenges is knowing your audience and being able to tailor your message to their preferences. What do they know, what do they think and what do they expect of you? Communication is not about what you say—it's about what the audience thinks you say. You'll need to be clear about what you personally hope to communicate and what you want people to do or take away as a result of your communication.

Understanding the audience is one of the most neglected communication principles. Yet knowing your audience allows you to tailor the message to what they want to know and package it in a way that they will be most receptive to receiving and understanding.

The most effective way to ensure your message is understood is to take the time and effort to understand your audience. As basic as this sounds, a lot of people don't take the time to really get to know and understand their audience. Understanding your audience can help you craft more effective communications that hit the point better and faster. Some people refer to this as "stakeholder communication analysis."

Regardless of the communication medium, you should always start by considering your audience. Who are they? What do they care about? To really understand your audience, you should have some idea of their background, their priorities and what they care about in terms of the security program. Knowing these factors can help you understand the best way to present information, monitor your use of technical jargon and focus your message on the things the audience want or need to hear. Using a mode of communication that considers the specific needs of its stakeholders will be more effective than simply using your own favorite communication techniques.

Also factor in how much your audience may already know about security or specifically about the security program at your company. If you're talking to information technology professionals, there's a good chance that they may know some of the material. In either case, they can probably tolerate a higher level of technical jargon, though I think this is still something you should always limit.

Poor communication usually happens because we don't know whom we are talking to or why we are talking to them or if this is their preferred communication channel. This is why so many mass email messages go ignored. Mass email is the ultimate example of a communication form that really doesn't care about audience, feedback or relying on anything other than chance that your message is received and understood. For this reason, it should be the last resort for communication that matters.

The more time you can spend understanding your audience, the easier you will find all your communications. And the higher the stakes of the communication, the more time you should spend attempting to get to know your audience beforehand. An example would be presenting to the board of directors, which is covered later in this book.

Once you learn to think like your audience and really put yourself in their position, you will be able to anticipate questions ahead of time and better tailor your message. As you progress through this book, you will begin to understand that tailoring your communications to the audience and choosing the best medium is key to ensuring that your message is received and understood. And that's why we communicate in the first place, to be understood.

Are you writing something? Think like the reader. Who are they? What do they already know? Are you preparing a presentation? Think about what the audience wants or needs to know and what questions they may have (Figure 1.5).

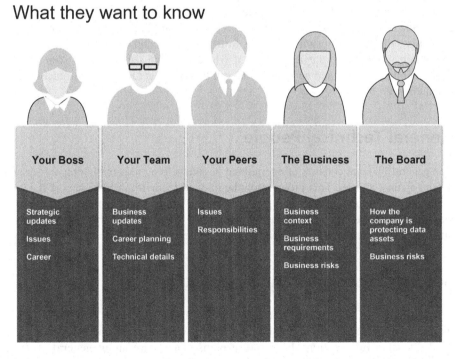

**FIGURE 1.5** Always start by considering your audience and tailoring your message to their needs. Not everyone wants to hear the same information.

A few examples of the types of audiences you'll need to understand and interact with are discussed below.

# C-Level Business Executives and the Board

This audience group wants strategic and possibly some high-level operational information. They want to know where your program is going, how it's going to get there and what value it will add to the bottom line of the business. They want to know key risks that might impact the business. They also want to understand and hear about long-term trends and issues, not necessarily every minor problem that might have an impact right now. They will want an idea of how the security program compares with those of their industry peers.

I will go into further detail on communication with this group later in the book, as this group is one of the most challenging and important groups to get right.

# Non-technical Businesspeople

Many senior security leaders think they want to spend most of their time with senior business leaders. But the reality is that you are going to need to spend time working with all levels of the organization, not just the most senior executives. Non-technical business employees will be able to help you get through legal and procurement processes and other business functions that will be necessary to run your program.

This audience group will not want to hear about technical details and may not care much about your strategic vision or maybe even about security at all. You want to avoid technical jargon and acronyms and you'll want to get to the point quickly with this audience.

# General Technical People

The general IT team consists of operational people helping to actively prepare for and defend against attacks. This group might be on your team or might be part of the general IT organization that is assisting in some way with the security mission. They will want details of specific attacks and vulnerabilities. You can probably provide a good amount of technical detail depending on their interest level or need.

# Security Managers and Architects

This audience group tends to be much more technical as well as more strategic in its information interests. They will appreciate information that can be immediately used, but also information about future plans. They care about general attacker techniques and tactics and specific actions that they can take to help with the security mission.

## Security Operation Center and Your Team

This group will be interested in very detailed technical information, such as indicators of compromise, general trends and threats. You will want to focus on immediately useful information and make it as specific and technical as you can. It's probably OK to use whatever jargon you like with this audience, but make sure they are following you and understand you based on feedback cues. Don't forget this step, not everyone on your team is likely at an identical skill level and will follow everything you say.

This is not intended to be an exhaustive list but is instead intended to give you an idea of the types of groups you will be interacting with in a senior security role. The key message is that you need to take time to consider your audience before communicating. One size will not fit all with communication. But remember, it's the audience that matters, not you. You need to tailor your communication style to match your audience if you want it to be effective. Don't expect everyone to adapt to your own style and preferences.

# DON'T JUST TALK, LISTEN!

*Most people don't listen with the intent to understand;*
*they listen with the intent to reply.*
~ Stephen R. Covey

*My first rule of conversation is this: I never learn a thing*
*while I'm talking. I realize every morning that nothing I*
*say today will teach me anything, so if I'm going to learn*
*a lot today, I'll have to do it by listening.*
~ Larry King

By far the most overlooked communication skill is that of listening. In my experience, the most effective way to build trust with someone is to listen to them and hear their side. When people tell you that security controls are difficult, instead of lecturing them, listen, empathize and help come up with mutually agreeable solutions.

If you look back on our model of how communication works earlier in this book, it is a two-way street and should always consider feedback from the recipient to be effective. Great communication delivers the right message and the right amount of information to the right people at the right time, using the right methods. As a rule, you should be doing a lot more listening than speaking. This might come as a surprise for the security leader eager to prove their value and knowledge.

Why do more listening than talking? We listen to obtain information, to understand, to make better connections with people on a personal level and to learn. Listening is the key to understanding your business and putting their needs first. In Stephen R. Covey's book *The 7 Habits of Highly Effective People*, one of the key principles is *seek first to understand, then to be understood*. In other words, you need to understand the other person's perspective before you can truly connect with them. This applies even if you do not agree with what the other person is saying. You don't need to agree with them, you need to *understand* what they are saying. There's a big difference. Listen to your team, to your peers and to your business partners. You'll be surprised what you can pick up from doing this.

When someone feels that you understand them, they are far more likely to be receptive to hearing your perspective in return. People like to be heard and have their concerns acknowledged. In his book *The Road Less Traveled: A New Psychology of Love, Traditional Values and Spiritual Growth*, author M. Scott Peck states, "You cannot truly listen to anyone and do anything else at the same time." This is a tall order in a world with so many distractions and in a job where you likely have many things competing for your attention. Single-tasking, not multitasking, is the way to go to be a better listener. Learn to use mindfulness and be fully present when you are engaging with someone.

A key skill to practice is active listening. Active listening is a technique that requires the listener to fully concentrate on the conversation. Useful methods like paraphrasing what was just said or asking meaningful questions can demonstrate your level of comprehension of the conversation.

Sounds simple, right? The problem is that it's not. There are many communication blocks that keep us from listening properly. We live in the age of distraction. Phones and notifications are vying for our attention. We are often thinking of what we should say when the speaker stops talking. We daydream. We think about other problems or the next task or how behind you are on email. We worry about the next breach or that we've maybe already been breached. All these things are barriers to listening, but they can also be overcome with active listening skills. Failing to listen can rob you of valuable opportunities to learn and connect.

Why do we listen in the first place? When you listen, you are trying to:

1. Learn or obtain new information
2. Understand another point of view
3. Listen for enjoyment or to build rapport

Without listening properly, you will miss out on all of these opportunities. Active listening takes some practice. But you'll find that when you really listen to what the other person is saying, you can quickly get to the root of problems, build a greater rapport and maybe even learn a thing or two from the conversation.

How do you perform active listening? In general, there are four key components: paying attention, paraphrasing, questioning and acknowledging. Let's look at each of these in detail.

## Pay Attention

Give the speaker your undivided attention. No checking smartphones or thinking about your rebuttal to what they just said. Watch for your own nonverbal communication and body language. Avoid distractions that might take away from the experience. Don't let your mind drift away. Make an effort to pay attention and be present. People need to feel heard, understood and validated. Again, keep in mind this doesn't mean you have to *agree* with what they are saying! You can still have your own point of view.

## Paraphrase, Repeat or Summarize

Paraphrasing means repeating what you have heard using slightly different words (Figure 1.6). Doing this forces you to really think about what has been said. It forces you to take what you have just learned and integrate it. Paraphrasing can help you zero in on key points or concerns in a conversation and make sure that they are crystal clear to both parties. It also demonstrates how you have heard and understood what has been said. You can also repeat or summarize a conversation to demonstrate not only that you heard what was said but also that you *understood* what was said.

## Questioning

Another way of engaging more in a conversation and strengthening active listening is to ask questions. Don't interrupt, of course. But asking clarifying questions helps keep you more engaged in the conversation and can help you connect at a deeper level. Questions demonstrate a basic level of caring about the conversation and what is being communicated.

You'll want to use open-ended questions, which are questions that can't be answered with a simple yes or no. An example of an open-ended question would be, "What is it

**FIGURE 1.6**   While your message may seem clear to you, what people hear and perceive may differ greatly from what you intended.

that you'd like to see accomplished?" or "How did that make you feel?" These questions can't be answered with a simple yes or no and may start a whole new train of thought in the conversation.

## Acknowledging

Acknowledging what the speaker said also demonstrates a high level of engagement in the conversation. Acknowledgement can come in the form of simple nonverbal cues, like nodding in agreement, to making more formal statements that show an understanding of what the speaker said, such as "that must have made you feel very angry."

## The Only Advice on Listening That You Need

Active listening may seem like a lot of effort to have a simple conversation. The good news is that there's only one tip on listening that really you need. If you listen and pay attention and care what the other person is saying, no other technique is necessary. You will be more engaged in the conversation and it will all happen naturally. Assume that you have something to learn from everyone. You won't be disappointed.

**Action**: how good a listener are you? Do you talk more than you listen? Do you make time to listen at all? We all have times where we are multitasking or thinking about something else while someone is trying to communicate with us. How could you be a better listener by adopting some of these techniques?

## Repetition

When you are the one speaking, you need to understand that most people are not good listeners. Less than 2% of professionals ever receive training in listening skills. This means you are likely going to need to repeat yourself before your message can get through. How many times? Some people say three times. Others use the "rule of seven." A study by Microsoft investigated the optimal number of exposures required for people to finally hear audio messages. They concluded between 6 and 20 was best. Yikes. Obviously, you don't want to repeat yourself 20 times in the same conversation, but you should probably factor in that many people won't get what you're saying the first time.

Thomas Smith wrote a guide called *Successful Advertising* in 1885. The saying he used is still being used today.

1) The first time people look at any given ad, they don't even see it.
2) The second time, they don't notice it.
3) The third time, they are aware that it is there.
4) The fourth time, they have a fleeting sense that they've seen it somewhere before.
5) The fifth time, they actually read the ad.
6) The sixth time, they thumb their nose at it.

7) The seventh time, they start to get a little irritated with it.

8) The eighth time, they start to think, "Here's that confounded ad again."

9) The ninth time, they start to wonder if they're missing out on something.

10) The tenth time, they ask their friends and neighbors if they've tried it.

11) The eleventh time, they wonder how the company is paying for all these ads.

12) The twelfth time, they start to think that it must be a good product.

13) The thirteenth time, they start to feel the product has value.

14) The fourteenth time, they start to remember wanting a product exactly like this for a long time.

15) The fifteenth time, they start to yearn for it because they can't afford to buy it.

16) The sixteenth time, they accept the fact that they will buy it sometime in the future.

17) The seventeenth time, they make a note to buy the product.

18) The eighteenth time, they curse their poverty for not allowing them to buy this terrific product.

19) The nineteenth time, they count their money very carefully.

20) The twentieth time, they buy what it is offering.

Again, this was written in1885. If you think about it, people haven't changed that much. But the amount of information we process has dramatically changed. This means that most of your communications will need some reinforcement and some repetition. It also means that consistency is important. If you're going to repeat something, but it changes every time it won't be as effective as being clear, concise and consistent with a single, simple message.

Think of memorable advertising campaign slogans like "just do it" and "good to the last drop." Most people will probably recognize these right away as belonging to Nike and Maxwell House, respectively. Advertising is so effective by creating slogans like this and then engraving them into our memories with repetition and consistency.

# SUMMARY

We covered a lot of material in this chapter. These are the building blocks of communication and the foundation for the rest of the book. Here are a few takeaways:

- There are many communication scenarios you will need to face as a security leader. This may be everything from speaking up in meetings to presenting to business audiences. You may even find yourself in public speaking engagements, podcasts or interviews.
- Poor communication skills will trump excellent technical skills. Cybersecurity has a lot of complexity and a lot of people in a company need to know more about how to protect corporate data.

- We all have room to improve our communication skills. We are likely better with some communication methods than others, but the preferences of the recipient may determine if a communication is successfully received.
- The DiSC model offers a good framework to start understanding different personality types and how this might impact communication. Some people prefer many details and some people like the big picture better. Understanding your personality type and the type of other people you interact with will help you make communication more comfortable for everyone.
- There are five different forms of communication, including intrapersonal, interpersonal, group communication, public communication and mass communication. Each has its own set of challenges.
- Understanding your audience is key. Make sure you are tailoring your communications to their preferences, not your own. Choose mediums that are the most appropriate for what you are trying to communicate. Email is not always the best choice. The need for feedback may make in-person or video meetings the most appropriate method.
- The most overlooked communication skill is listening. Active listening lets you connect better with people and understand their side. Being engaged in communication allows for both sides to connect better.

# REFERENCES AND RECOMMENDED READING

Aurelius, Marcus, and George Long. *The Meditations of Marcus Aurelius.* Shambhala, 2019.
Brown, Steven, and Dorolyn Smith. *Active Listening.* Cambridge University Press, 2007.
Covey, Stephen R. *The 7 Habits of Highly Effective People: Powerful Lessons in Personal Change.* Simon & Schuster, 2020.
Das, Sejuti. "Cybersecurity Mantra—'Train Like You Fight & Fight Like You Train'." *Analytics India Magazine,* 28 Dec. 2020, analyticsindiamag.com/cybersecurity-mantra-train-like-you-fight-fight-like-you-train-says-sudeep-das-ibm-security-systems/.
Grossman, David. *You Can't Not Communicate: Proven Communication Solutions That Power the Fortune 100: How Top Leaders Differentiate Themselves.* Little Brown Dog Pub., 2012.
Hardy, Benjamin Jr. *Personality Isn't Permanent: Break Free from Self -Limiting Beliefs and Rewrite Your Story.* Bantam Press, 2020.
Leal, Bento C. *4 Essential Keys to Effective Communication in Love, Life, Work—Anywhere!: A How-to Guide for Practicing the Empathic Listening, Speaking, and Dialogue Skills to Achieve Relationship Success with the Important People in Your Life.* Bento C. Leal III, 2018.
Peck, M. Scott. *The Road Less Traveled: A New Psychology of Love, Traditional Values and Spiritual Growth.* Rider, 2008.
Rohm, Robert A. *Positive Personality Profiles: Discover Insights into Personalities to Build Better Relationships.* Personality Insights Inc., 2014.
Rohm, Robert A., and Julie Anne Cross. *You've Got Style: Your Personal Guide for Relating to Others.* Motivated Pub. Ventures, 2004.
Shannon, Claude Elwood, and Warren Weaver. *The Mathematical Theory of Communication.* University of Illinois Press, 1999.
Smith, Philip. *Successful Advertising.* Jubilee Edition. Smith's Advertising Agency, 1928.

# People Skills

**2**

*When you're nice to people around you, caring,*
*empathetic, you're always going to get more results.*
~ Mark Cuban

This chapter covers people skills and how to get along better with people. These skills are required to connect better with the people you will ultimately be communicating with. I will cover being a more approachable leader, emotional intelligence (EQ) and adaptability quotient (AQ), which is also sometimes referred to as "adversity quotient."

Whole books have been written about these subjects, so this is intended not to be a comprehensive coverage of these ideas but to get you started on the right path to thinking about the complexity of dealing with other people and how they act, feel and communicate.

## HOW APPROACHABLE ARE YOU?

*The day soldiers stop bringing you their problems is the*
*day you have stopped leading them. They have either lost*
*confidence that you can help them or concluded that you*
*do not care. Either case is a failure of leadership.*
~ Colin Powell

Part of your success as a security leader is based on being approachable to your team and colleagues. Being approachable is also the key to building strong relationships. When you're approachable, people will be able to escalate issues to you before they become full-blown crises because they know you won't react badly to the news. In addition, employees who have approachable managers feel safe contributing ideas and you'll get more out of having a team that supports you. Being approachable is a professional skill. As a leader in your company, your approachability definitely matters.

Too often, some leaders choose to be less approachable to maintain their authority. They either isolate themselves with direct reports only or in some cases they only manage up to upper management and leave the team wondering if they even have a leader. One company I worked with built a state-of-the art security center, but the CISO rarely came to visit, didn't hold skip-level meetings and left a group of over 100

DOI: 10.1201/9781003100294-3

feel like they were on a rudderless ship. In some companies, company culture may even be that employees are reluctant to approach senior management because it isn't appropriate or because there's some artificial hierarchy that they don't feel comfortable opposing. As the leader of your team and a senior executive in the company, you are going to need to break down these barriers to create a level of trust and open communication.

Being approachable means that you use the right body language and the right verbal communication and listening skills. There are also some specific ways to keep yourself approachable. In her book *The Art of Body Language: 8 Ways to Optimize Non-Verbal Communication for Positive Impact*, author Susan C. Young states: "Unfortunately, unapproachable leaders create a tense environment that may prevent their people from bringing their best strengths and talents or challenges and solutions forward." It's clear that in order to get the best out of your team, you need to be an approachable leader.

Managing a team means getting the best out of everyone supporting the security function. But being approachable doesn't just include your team. It includes everyone else in the company as well. Many security leaders have been pushed out of organizations where either their leadership was seen as too "ivory tower" or they were just seen as an overall obstacle to getting things done.

Here are some general guidelines to make sure you stay conscious about your approachability.

## Listen Actively

See the section on active listening earlier in this book. Actively listening to what people are telling you shows that you are paying attention and care about what they are saying. Respond to issues that are being raised to you and show them that you are taking their feedback seriously. Keep eye contact and don't read emails or check your smart phone while the conversation is going.

Being too busy to listen or in too big a rush to get to your next meeting leaves people feeling they shouldn't "bother you" by bringing issues to your attention. If you really are running late for a meeting, politely say so but set a time to follow up and make sure you do it.

## Be Proactive

You are a senior security leader. Some people may not be comfortable approaching you out of hierarchical respect or fear of wasting your time. This means that you may need to make the first move. Be proactive and be inclusive and start the conversation. You need to take the first step and get people engaged. An easy way to do this is by asking open-ended questions. As you'll recall, these are questions that can't easily be answered with a yes or no response. A few questions you might consider are:

- What is the most interesting part of your job?
- What do you like about working here?

- Where would you like to be in five years with your career?
- What's one thing you think we should be working on that's not on our roadmap?

Open-ended questions are great for getting people more engaged and pulling meaningful answers from people who might otherwise be reluctant to go into a deeper conversation. If you ask more closed-ended questions—even things like "how are you doing?"—you might just get one-word answers like "fine." Ask better questions and you'll get better answers.

## Make a Personal Connection

Take an interest in your colleagues and team beyond their projects and operational duties. Getting to know your staff helps build strong relationships. I find that it doesn't take long to find some common ground with people. Are you from the same town? The same state? Do you both like hiking? Reading? Watching the same shows? Find common interests and use them as a way to not talk about business. Not sure what someone's interests are? Ask them! Most people will simply open up and tell you. This doesn't mean you have to be deep into everyone's personal business, but simply asking how their weekend was and being genuinely interested in the answer can help keep the work environment open and friendly and make employees feel liked and respected.

## Keep an Open-Door Policy

If you're in an office, leave your door open when you're not actively in meetings. If you spend most of your time behind closed doors, your colleagues are not going to feel a connection with you. Yes, this means that you will have more distractions and people stopping by, but a large part of your job as a leader is having good relationships and keeping open communication with your team and your colleagues. Manage your time but make some space in the schedule for non-formal chats and conversations.

## Greet Everyone

I'm assuming you know the name of most people on your team, even if you manage a big team. If not, you need to spend some time on this. Remember, some big teams can have 100 people, so it's not that unreasonable that you might not know everyone's name. But a basic habit of greeting everyone by name goes a long way and enforces a "people matter" perspective. Greet anyone you can, and greet them by name. Choosing specific individuals on a regular basis can make people who aren't selected feel alienated or make it seem like you're playing favorites. If you have trouble remembering names, I will cover some tips later in this book.

## Own Your Mistakes

If you're the kind of manager that only likes to hear good news, then don't expect your team to tell you everything that you need to know. This happened famously at General Electric under Jeff Immelt, where a culture of sharing only the good news and being overly optimistic took away from confronting real issues that were facing the business. Let your team come to you with suggestions for improvements, mistakes and potential problems. And don't shoot the messenger. We've all made mistakes; it's how you recover from them that counts. Own your mistakes and make things right and let your team own theirs, but help them through it. You will gain a lot of respect for not throwing other people or even whole teams under the bus by playing the blame game.

Also, admitting to your failures may help team members talk about theirs. Put your pride aside and admit to your failures. This makes you seem more human and will strengthen communication and create a safe environment for learning from mistakes.

## Have a Sense of Humor

While security executives are supposed to have a level of gravitas, the best leaders know when to crack a smile or break the ice with a joke. After all, if you can't laugh about things then what's the point? People are more likely to resonate with a leader who feels like they're part of the team. It's even better if this person also has the power to make everyone's job easier and more successful and do it with a sense of humor.

## Get Feedback

There's an old saying that "perception is reality." Sometimes you may think you're approachable, but you're the only one who thinks so. Get feedback from trusted colleagues and see if you can zero in on the problem if you're struggling in this area. If your organization supports 360-degree reviews, make sure you participate. A 360 review is where performance feedback is solicited from all directions in the organization. Participants in 360 reviews usually include the employee's manager, several peer staff members, reporting staff members and functional managers from the organization. The objective of the feedback is to give you the opportunity to understand how your work is viewed in the total organization and by coworkers in any position. There's a tendency of some leaders to not listen closely enough to feedback from further down the ranks, but this simply creates a career blind spot. If you want to be the most effective that you can be, you need to have the full view of how you're perceived. If your organization doesn't do 360 degree reviews, just ask some trusted team members and peers their thoughts.

# EMOTIONAL INTELLIGENCE

Emotional intelligence, sometimes called EI or EQ, is the ability to understand and manage your own emotions, and those of the people around you. People with a high degree of EQ know what they're feeling, what their emotions mean, and how these emotions can affect other people and what they may be feeling. People who leverage emotional intelligence have a greater ability to influence, persuade and connect with others, which ultimately is all about the way we communicate. EQ is basically the industry-accepted framework for "people skills."

EQ was popularized by psychologist Daniel Goleman, Ph.D., in his popular book *Emotional Intelligence—Why it can matter more than IQ*, which remains the most important book on the subject. *Emotional Intelligence* was named one of the 25 "Most Influential Business Management Books" by TIME Magazine. You may know some people who are academically brilliant and yet are socially inept or otherwise unsuccessful at work. Intellectual ability or your intelligence quotient (IQ) isn't enough on its own to achieve success. This is where EQ comes into play.

EQ measures how well you will work with others and how well you manage relationships. Without having at least some level of EQ, you will find it difficult to build and maintain relationships and communicate your strategic vision in a way that connects with others. People with high emotional intelligence also have a high general level of success. People tend to want to help these people because of the way that these people treat them and make them feel. High levels of EQ affect everything from your performance at work to your mental and physical health. By understanding your emotions and how to control them, you will be better able to express how you and others around you feel.

Emotional intelligence is commonly defined by four attributes (Figure 2.1):

1) **Self-management**: This means that you can control impulsive feelings and behaviors, manage your emotions in healthy ways, take initiative, follow through on commitments and adapt to changing circumstances.
2) **Self-awareness**: This means that you know your strengths and weaknesses. You have self-confidence. You recognize your own emotions and how they affect your thoughts and behavior.
3) **Social awareness**: This means that you have empathy. You can understand the emotions, needs and concerns of other people. You can pick up on emotional cues and feel comfortable socially and recognize dynamics in a group.
4) **Relationship management**: This means that you know how to develop and maintain good relationships, communicate clearly, and inspire and influence others. You work well in a team and manage conflict effectively.

## Assessing Your EQ

There are many books available on EQ and I will provide some recommendations at the end of this chapter. The following sections are meant to provide a broad overview and

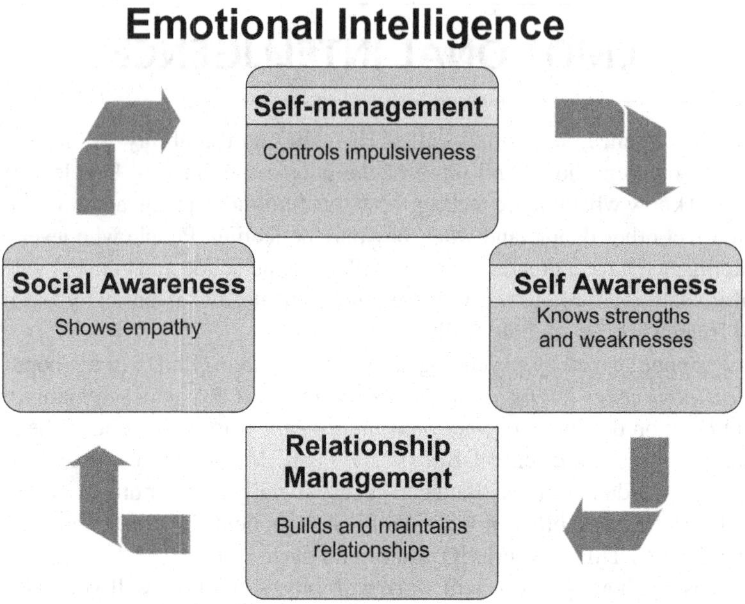

# Emotional Intelligence

**Self-management**

Controls impulsiveness

**Social Awareness**

Shows empathy

**Self Awareness**

Knows strengths
and weaknesses

**Relationship
Management**

Builds and maintains
relationships

**FIGURE 2.1**   Emotional intelligence is an important consideration for how you will connect with and communicate with others. Self-management, self-awareness, relationship management and social awareness will all play a role in how your "people skills" develop.

summary of what you might find should you choose to investigate this subject deeper and to give you just enough information to start thinking about your own EQ abilities.

To give you at least some indication that you may need some work in this area, consider the following questions.

1) Do you become defensive when criticized?
2) How do your emotions impact your behavior?
3) Do you tend to avoid difficult or touchy interactions?
4) Are you calm under most circumstances? How do you act when you're under a lot of stress?
5) Do you struggle to build rapport with people?
6) Can you recognize your emotions as you experience them?
7) What makes you angry?
8) How do you respond when a co-worker challenges you?
9) How do you recover from failure?

These questions are not intended to baseline your EQ or give you a score. They are simply intended to get you thinking about what benefits a high EQ might provide and where your own EQ might need some work. There are several, more formal methods of baselining EQ that are available for free online.

Like all skills, EQ is something that can be improved with some directed effort. Keep your mind open that even your emotional reactions and responses are not set in stone. Whatever has been learned in your life can be changed and unlearned. You can't always control your emotions, but you can control how you respond to them. This is what emotional intelligence is all about.

## Self-Management and How to Improve It

The first pillar of EQ is self-management. Self-management is the ability to control your emotions and impulses. People who score high on this attribute think before they act and stay cool under pressure. They don't shy away from difficult conversations. Some other characteristics of people high on the self-management scale are thoughtfulness, comfort with change, integrity and the ability to say no.

When someone is low on the self-management scale, they can get stressed easily and lose control of their emotions. In other words, they won't be able to act thoughtfully. As you can imagine, it's not easy to make good decisions when you are emotionally stressed. Stress management might just be *the* critical skill for security leaders. We are constantly under attack from both real-world attackers and our businesses that may be resistant to some of the changes that come with implementing a security program. We are over-worked and under-resourced. In the article *9 Reasons Why Cybersecurity Stress Is an Industry Epidemic*, author Jasmine Henry suggests that security professionals are more than twice as likely to report poor work–life balance, more than three times as likely not to take full vacation days and more than five times as likely to worry about job security. Stress in this industry is very real.

If you would like to make improvements in your own self-management capabilities, there are a few key skills to focus on. These include being aware of your feelings, paying attention to your own self-talk and practicing mindfulness.

1) **Be aware of your feelings**. The first step to managing stress and self-awareness is to be able to catch yourself in a difficult or emotional situation. This may take some practice, as it becomes much more difficult when you have already become emotional over a subject. Think of getting into an argument with one of your peers who is blaming your security tools for causing a network outage. By the time the discussion gets heated, it might be hard to see that your own emotions are about to explode. See the meditation and stress resources at the end of this chapter.
2) **Pay attention to your self-talk**. Refer back to intrapersonal communication in Chapter 1. The thoughts you think can definitely have an impact on how you will react in certain situations. Is your boss calling you into his office? This doesn't have to be a worst-case scenario. There's a great saying in the Buddhist community, "don't believe everything you think."
3) **Choose your own adventure**. Remember that you have the ability to choose your response to any situation. Choosing to remain calm and trusting that a situation will work itself out with your leadership is better than yelling at or becoming heated with your team and colleagues. Great leaders choose to be calm under pressure. That doesn't mean that they don't feel the pressure, they just choose their reaction to it.

# Self-Awareness and How to Improve It

*Real wisdom is the ability to understand the incredible*
*extent to which you bullshit yourself every single moment*
*of every day.*
~ Brad Warner, Sit Down and Shut Up: Punk Rock
Commentaries on Buddha, God, Truth, Sex, Death,
and Dogen's Treasury of the Right Dharma Eye

Self-awareness is the habit of paying attention to the way you think, feel and behave. Many problems with communication in the workplace are caused by a leaders' lack of self-awareness rather than by their intent. Many leaders don't think about the impact their behaviors or words might have on people in their organization. This is usually because they didn't intend any negative consequences. But while intentions are nice, what matters the most is the actual impact. A lack of self-awareness can create "blind spots" where leaders don't see what everyone around them sees. This can lead to failure in the role or just general ineffectiveness as everyone struggles to work with and get along with such leaders.

All improvement begins with acknowledging where you are right now and being able to "step out" of your individual situation, almost as if you were an impartial witness. Some people are never able to make this leap, attempting instead to preserve their ego. As a result, they never improve. When you improve your self-awareness, you will be able to pay attention to how you react and behave in certain situations. In other words, what your default response would be under the circumstances. When you become an impartial observer of your default responses, it becomes possible to change them.

There are many benefits to increasing your self-awareness. In fact, self-awareness is the basis and starting point for all self-improvement. Increasing your self-awareness will benefit you in many ways, including building stronger relationships, improving your mood and attitude towards the job as well as your decision-making, communication and ability to connect with other people. The more we know ourselves, the easier it will be to be honest about what we would like and also be respectful of what everyone else wants.

If you're not particularly self-aware or would like to make some improvements in this area, where's the best starting point? At the risk of taking a slightly spiritual turn in the book, the most effective tools are mindfulness and meditation. Meditation helps you take that step back and realize that you are not your thoughts. It can help you gain new perspectives on stressful situations, increase your general self-awareness and help you focus better on the present moment. Meditation has been practiced for thousands of years. It was originally intended as a spiritual development tool, but these days, it is commonly used for relaxation and stress reduction. And who couldn't use more of that?

In his book *Tribe of Mentors*, Tim Ferriss interviewed hundreds of successful people, ranging from Arianna Huffington to Ray Dalio. His aim was to analyze what some of the world's most successful people attributed to their success. Ray Dalio, the founder, chair and co–Chief Investment Officer at Bridgewater Associates (the largest hedge fund in the world), cites meditation in the book as "one of the best or most worthwhile investments" he ever made. In fact, 80% of the people interviewed by Ferriss appeared to have some sort of meditation or mindfulness practice. This makes absolute sense if you see the ability to focus as a prerequisite for being successful. The most successful people in the world take active steps to quieten their mind.

There are plenty of great books, courses and even apps that teach meditation. There are also many forms of meditation, so you can select something that works best for you. There are guided meditations, like Headspace. Mantra or Transcendental Meditation®. And if you are really not the type of person who can sit still for long, you might try some moving form of meditation, like walking meditation, yoga or qigong. The elements of most meditation are basically just focused attention, relaxed breathing, a comfortable and quiet setting and an open attitude.

I won't go into any great detail here, but since I did promise to deliver some hands-on tools in this book, here are a few good starting points for you if you've never tried any form of mediation before.

## *Breath Awareness*

This is great for beginners because everybody breathes, and your breath is always with you. It is one of the few involuntary systems in your body that is also under voluntary control. In breath awareness meditation, you focus all your attention on your breathing. Breathe in deeply and slowly. When your attention wanders, gently return your focus back to your breathing. It's really that simple. But simple does not make it easy. Your mind will invariably drift and you'll start thinking about something else. When you recognize that you have drifted, simply note it as thinking and return your attention to the breath. Don't be overly judgmental if this is hard for you. It's hard for everyone.

## *Mantra Meditation*

You can also repeat a mantra. A mantra is simply a word, syllable or phrase that is repeated in your mind. Repeating your mantra can help guide your thoughts and calm your mind. The universal mantra is the syllable Om (pronounced as aa-uu-eemm), which is said to be the sound of the universe. By giving your mind something to focus on during meditation, you will, with time and practice, become better at being an observer of your thoughts rather than someone who is controlled by them.

OK, I get it. If you're not into the whole meditation thing, at least consider trying qigong, it worked for the Shaolin monks of China. And if none of this works for you, what else can you do to increase your self-awareness?

- Try to pay attention to what bothers you about other people. Believe it or not, sometimes the things that bother us the most in other people are actually qualities we dislike in ourselves. So, when someone bugs you, ask yourself: might this be a reflection of something in myself that I dislike?
- Read quality fiction. Great writers are expert observers of human nature. They notice the tiny details that most of us miss. Good fiction can teach us how to think about people with compassion. And the better we get at observing others, the more likely we are to also look at ourselves.

- Finally, when all else fails ask some trusted friends or colleagues for feedback. But only do this if you are receptive to criticism and open to change. Take any feedback or criticism and try to work on it. In fact, just take ONE thing that you think would help and work on that. When you feel like you've made improvements, pick something new.

Unfortunately, the term "self-awareness" can come across as a bit mysterious. It's actually not. Self-awareness is simply the ability to observe ourselves. It is the ability to notice and pay attention to the patterns in our thoughts and behaviors. It will also give you a very strong grounding for all the stress you are going to face working with security incidents that were probably preventable, businesses that just don't "get it" and bad guys that are dramatically better funded and organized than the good guys. I will leave you with some references for further investigation at the end of the chapter if you're interested in pursuing this topic further.

## Social Awareness and Empathy

Empathy is the ability to understand and share the feelings of someone else. It is also a way to make incredibly strong connections with someone. "Empathy" is a term we use for the ability to understand other people's feelings as if we were having them ourselves. Empathy is often confused with sympathy, but they are not the same thing. Sympathy is a feeling of concern for someone's happiness, perhaps due to a misfortune in their life. Unlike empathy, sympathy doesn't involve shared perspective or emotions.

Empathy involves putting aside your own viewpoint and seeing things from the other person's perspective. You are putting yourself in the other person's shoes and trying to understand how they see things. There are three types of empathy:

- **Cognitive**: I know how you think.
- **Emotional**: I know how you feel.
- **Empathetic concern**: I care about you.

The third type, empathetic concern, is the most sustainable and the strongest method for improving connections and communication. Think about it, I can understand how you think (cognitive empathy) and I can understand how you feel about something (emotional empathy) but if I don't actually care about you, this information can be abused to manipulate you. If you're in a command-and-control environment, this can be especially toxic. Strive to make connections with people, not to use this information to get what you want out of them.

## Relationship Management

Improving relationship management skills is one of the most useful EQ skills for security leaders. Like it or not, a lot of what you are going to accomplish will require other people to take action and support your initiatives. Relationship management can also help when dealing with conflict. When you work with other areas early in the process, you can avoid potential problems down the road.

I think there are three key skills in this area that can help develop or improve your relationship management EQ skills. The first is your ability to listen well and recognize emotions in others. The second is mastering the art of conversation. The third is staying in touch the right way.

We've already covered active listening and empathizing. Both of these skills are critical for relationship building, as they will help you connect with people on a much deeper level. Again, people want to be heard and understood. Giving someone your full attention, minimizing distractions and reflecting thoughtfully can reassure others of your sincerity. Being able to recognize emotions through empathy will also help make this deeper level of connection. The two working in harmony will be a very strong combination to build closer relationships with others.

While some may think that mastering the art of conversation and other things that could be considered "small talk" is frivolous, they are really an important element to building stronger relationships. The ability to make effective small talk will help you establish rapport and make a positive impression. Small talk also leads the way to deeper conversation. In fact, it's not generally possible to get to deeper levels without starting by building trust through more trivial subjects. Small talk is easy enough to master. Any number of subjects from the weather to a recent sports game can help break the ice and get a conversation going. You can then go deeper by actively listening and then asking better questions, which we will cover in Chapter 7. Conversations can introduce you to people who could be future employers, friends or confidants. Without the ability to start basic conversations, you will have a harder time building a social circle and the support from the business that you'll need to implement the security program.

Another way to manage relationships is making sure that you're staying in touch the right way. This means that you shouldn't wait until you need something to reach back to a connection. Keep in touch periodically by just checking in and seeing how someone is doing. Follow ups like this solidify personal connections. In fact, you might reach out and see if there's anything you can do for them. No one likes someone reaching out only when they want something from you.

# WHAT ABOUT AQ?

You may have also heard the term AQ or adaptability quotient. This is also sometimes referred to as adversity quotient. This is a somewhat new buzzword in the business world, but because it is so applicable to security practitioners, I think it is worth mentioning.

Put very simply, AQ is the ability to adapt to and thrive in an environment of change. AQ is an interesting concept. Some people seem to have a strong ability to overcome setbacks quickly and easily and some seem to be crushed by even a light resistance. Having a strong level of resilience obviously has a lot of applicability in the world of cybersecurity, where not everyone is interested in going along with our plans.

The term "adversity quotient" was coined by Paul Stoltz in his pioneering book *Adversity Quotient: Turning Obstacles into Opportunities*. An AQ score measures the ability of a person to deal with adversities in their life. Harvard Business Review called AQ "the new competitive advantage."

What do you think your AQ is? Do you turn obstacles into opportunities? Do you persevere despite setbacks and roadblocks? I think a lot of security practitioners must have a high AQ to continue in this profession. After all, it doesn't really feel like we're winning the war with cybersecurity. But we keep going and we are resilient.

There is an adversity response continuum. Think about where you fall on this scale.

- **Avoiding adversity**: Not dealing with a problem head on or delegating it to someone else.
- **Surviving adversity**: You get through adversity, somehow. You're not looking forward to the next problem.
- **Coping with adversity**: You have sort of accepted the fact that there's adversity, but you are mostly just keeping your head above water while managing it.
- **Managing adversity**: This goes beyond coping. You might even be able to get something positive out of the adversity.
- **Harnessing adversity**: This is the most evolved approach to adversity. It reframes problems as opportunities and gives ways to grow from hardships.

While AQ may seem like an innate personality trait, there are steps you can take to improve how you deal with adversity. These steps are discussed below.

## Reframe the Situation

When something seemingly bad happens, how do you react? Do you get upset? Angry? What if the situation wasn't really bad? There's a wonderful Zen story that illustrates this point perfectly. The story goes that an old farmer's horse ran away. His neighbors told him it was bad luck, to which he replied "maybe." The next morning, the horse came back with three other wild horses. "What good fortune," they said. "Maybe," the farmer replied. The next day, the farmer's son was thrown from one of the horses and broke his leg. "What misfortune," his neighbors said. "Maybe," the farmer replied. The next day, military officials came to the village to draft young men, but they passed by the farmer's son because of his broken leg. What good fortune? Maybe.

Sometimes things are not what they seem and sometimes things need to play out further before you can take a view on them. Don't jump to conclusions. The most powerful way to reframe beliefs is to think that everything happens *for* you rather than *to* you. You've been let go from a job? This is happening for you because there's something even better for you right around the corner.

## Keep a Long-Term Perspective

Even if something seems very difficult in the present, will you really care about it as much a year from now? How about a week from now or maybe even by tomorrow? For sure, no one will care about it 100 years from now. Understand that bad things do happen, and that time heals most problems.

## Own the Outcome

*On any team, in any organization, all responsibility for*
*success and failure rests with the leader. The leader must*
*own everything in his or her world. There is no one else*
*to blame. The leader must acknowledge mistakes and*
*admit failures, take ownership of them, and develop a*
*plan to win.*
~ Jocko Willink, Extreme Ownership: How
U.S. Navy SEALs Lead and Win

When a bad or difficult situation arises, take full ownership. You may not have personally caused or deserved a bad situation, but you do own your own response and reaction. Develop your action plan, own the situation, own the outcome and keep asking yourself what more can be done to make things right. You'll find that even taking basic action will make the situation feel more under control.

Some leaders seem to struggle with this concept. Instead of taking ownership, they try to blame others on their own team or other groups in the company. The simple lesson is that whatever happens under your watch is your fault. A leader owns everything in their world. If subordinates fail, leaders must provide better training and instruction. That's not to say that if a security incident happens that it's directly your "fault," but perhaps you should reflect on how you prioritized the program and if you missed something along the way. In either case, your response to an incident is 100% under your control. Own it.

## Be Willing to Compromise

Life isn't perfect. In fact, life is full of compromise. You won't always get the outcome you want, but the more you can be flexible and willing to compromise, the more you will find a mutually acceptable common ground. Remember, you don't always have to be right, and you don't need to win every confrontation.

This doesn't mean you should cave in or compromise every time there's conflict, but it's more of the ability to remain flexible and seek out win–win situations. There's more on negotiating and compromise in a later chapter.

## Let It Go

In the book *Letting Go: The Pathway of Surrender,* author David Hawkings says, "we hang on to pain. It certainly satisfies our unconscious need for the alleviation of guilt through punishment. We get to feel miserable and rotten." The question then arises, "But for how long?"

Sometimes things hurt. Sometimes things suck. Sometimes it's OK to feel bad about things. But then, let them go and move on. Some people replay a negative event over and over in their mind. They wind up not being able to focus on anything else. Get it out of your system, but then let it go.

## Reflect and Improve

Take some time to reflect every day. Consider taking up journaling, which we discuss later in this book. What did you learn today? What went well or poorly? Where can you improve or how would you do things differently if you had the chance? Take the time to reflect and incorporate improvements.

In his book *Principles*, Ray Dalio uses an equation:

pain + reflection = progress

In other words, taking the time to get through the difficult situation and then reflecting on what went wrong and right can open the path to real growth. Reflect, integrate your findings and grow stronger.

# SUMMARY

People skills are important. Without good people skills, you can't have good communication. The ability to connect with people on a more personal level helps facilitate communication and mutual understanding.

- Being approachable as a security leader is critical to working with other people and your team.
- EQ, or emotional intelligence, offers a good framework for working better with people. The four elements of EQ are self-management, self-awareness, social awareness and relationship management.
- Empathizing and genuinely caring about someone offers the strongest communication connection between people.
- Stress is very real in this industry. Tools like meditation, breathwork and qigong offer some outlets for these feelings.
- AQ, or adaptability quotient (sometimes called adversity quotient), offers a framework for resiliency and managing some of the setbacks you will encounter.

# REFERENCES AND RECOMMENDED READING

## Meditation and Mindfulness Apps

- Headspace
- Calm

- Insight Timer
- Ten Percent Happier Meditation

# MEDITATION BOOKS

Chodron, Pema. *How to Meditate*. JAICO Publishing House, 2016.

Denniston, Denise, and Barry Geller. *The Transcendental Meditation TM Book: How to Enjoy the Rest of Your Life*. Fairfield Press, 1991.

Gunaratana, Henepola. *Mindfulness in Plain English*. Wisdom Publications, 2019.

Hạnh, Nhất, et al. *You Are Here: Discovering the Magic of the Present Moment*. Shambhala Library, 2012.

Kabat-Zinn, Jon. *Wherever You Go, There You Are*. Piatkus, 2004.

Kornfield, Jack. *Meditation for Beginners*. Jaico Pub. House, 2010.

Salzberg, Sharon, and Joseph Goldstein. *Insight Meditation: A Step-by-Step Course on How to Meditate*. Sounds True, 2001.

Skinner, Julian Daizan. *Practical Zen: Meditation and Beyond*. Singing Dragon, 2017.

Suzuki, Shunryū, et al. *Zen Mind, Beginner's Mind*. Shambhala, 2020.

Yates, John, et al. *The Mind Illuminated: A Complete Meditation Guide Integrating Buddhist Wisdom and Brain Science for Greater Mindfulness*. Atria Paperback, an Imprint of Simon & Schuster, Inc., 2019.

# WEB RESOURCES

Best starting point for Qigong, Lee Holden Qigong: https://www.holdenqigong.com

Calm: https://www.calm.com

Best site on introductory mindfulness: https://www.mindful.org

Easy introduction to meditation: https://www.headspace.com

Many other resources: https://www.soundstrue.com

# CHAPTER REFERENCES

Bradberry, Travis, and Jean Greaves. *Emotional Intelligence 2.0*. TalentSmart, 2009.

Ferriss, Timothy. *Tribe of Mentors: Short Life Advice from the Best in the World*. Houghton Mifflin Harcourt Publishing Company, 2018.

Goleman, Daniel. *Emotional Intelligence*. Bloomsbury, 2020.

Hawkins, David R. *Letting Go: The Pathway of Surrender*. Hay House, Inc., 2018.

Stoltz, Paul. *Adversity Quotient: Turning Obstacles into Opportunities*. Wiley, 1997.

Stoltz, Paul. "When Adversity Strikes, What Do You Do?" *Harvard Business Review*, 23 July 2014, hbr.org/2010/07/when-adversity-strikes-what-do.

Young, Susan. *The Art of Body Language: 8 Ways to Optimize Non-Verbal Communication for Positive Impact (The Art of First Impressions for Positive Impact) (Volume 3)*. ReNew You Ventures, 2017.

# The Language of Business Risk

# 3

This chapter focuses on working with business leaders. These are the non-technical people whom you will need to communicate with on a regular basis. These are the people who generate the revenue, run the business and ultimately fund your program. They also own business risk, which is why you're going to need to work closely with them.

Cybersecurity risk is fundamentally a business risk. CISOs don't own information. The security group doesn't own information. IT doesn't own information. The business owns information and therefore they own the risk associated with the loss, alteration or exposure of their information. While you may have been hired to protect business information, ultimately it is the business that owns the risk to the information as well as the budget to implement security controls around their most important information. At the end of the day, it is their risk and their budget that pays for the security group and drives what kind of controls should be in place. Your job is to help the business understand their risk so that they support you in putting the right security controls in place to protect their information, commensurate with risk.

What is business risk? Strictly speaking, business risk is the exposure a company has to events or threats that could lower profits or lead to operational failure. Anything that threatens a company's ability to achieve its financial goals can be considered a business risk, including reputation damage. It is not possible for a company to completely avoid all risks, so risk management strategies are employed to manage risks to reasonable levels. Risk management strategies include avoiding risk, mitigating risk, transferring risk (like insurance) and accepting risk.

While many security leaders over-emphasize avoiding or eliminating risk, this is not always the best business strategy. In fact, many businesses put thought into the amount of risk they can tolerate by forming a "risk appetite" statement that is approved by senior management and the board. This statement would typically discuss acceptable levels of risk, knowing that avoiding all risk is not practical. It might detail that reputational risks won't be tolerated, but moderate risk with the credit markets might be acceptable to maximize profit. The balance often comes down to customers and regulatory expectations as well as how peer companies are handling cyber risk. It's a balance; perfect security doesn't exist.

It may come as a surprise to new security leaders that a business might accept some risks without doing anything to address them. This is especially important since you will likely discover quite a few new risks under your watch and maybe even inherit a lot of old ones. Not every risk identified will result in a well-funded initiative by the business to address it. While the sense of apathy for information security has greatly diminished over the years, that doesn't mean that every issue you point out will get

DOI: 10.1201/9781003100294-4

attention, funding and resources. This is actually a good thing, because it would result in chasing more problems than anyone could possibly go after. Instead, it's best to take a risk-based approach to security. This means that risks are identified, ranked in order of potential impact and likelihood and then the focus is put on the most meaningful risks in the context of the business. It is up to you, as the senior security leader, to make sure that the *right* programs get funding to address the *most important* business risks.

The best way for you to understand the most important business risks is to first understand your business. What business or businesses are you in? Do you know where the biggest revenue streams are? Are you in the services industry? Manufacturing? Do you understand how the company runs marketing and in what mediums? You need to start thinking like a business owner and focus on protecting the things most important to keeping the company profitable, safe and out of hot regulatory water.

What's the best way to get started? Ask. Talk to your business leaders and division heads, I have found that almost all are willing to take some time to explain their area to you. This should always be followed by you asking how you can help them.

You can also gather quite a lot about a public company from their investor relations page and SEC filings, especially the 10-K report, which will typically even include a line or a few paragraphs disclosing the companies perceived cybersecurity risks. The Securities and Exchange Commission (SEC) encourages registrants to describe the nature of information technology–dependent business activities and the effect and financial costs of cyber risk on those activities. These statements will typically be brief and surrounded by other operational risks. They will often read like:

### Information Security, Cybersecurity and Data Privacy Risks

If our efforts to provide information security, cybersecurity and data privacy are unsuccessful or if we are unable to meet increasingly demanding regulatory requirements, we may face additional costly government enforcement actions and private litigation, and our reputation and results of operations could suffer.

We regularly receive and store information about our guests, team members, vendors, and other third parties. We have programs in place to detect, contain, and respond to data security incidents. However, because the techniques used to obtain unauthorized access, disable or degrade service, or sabotage systems change frequently and may be difficult to detect for long periods of time, we may be unable to anticipate these techniques or implement adequate preventive measures. In addition, hardware, software, or applications we develop or procure from third parties may contain defects in design or manufacture or other problems that could unexpectedly compromise information security, cybersecurity, and data privacy. Unauthorized parties may also attempt to gain access to our systems or facilities, or those of third parties with whom we do business, through fraud, trickery, or other forms of deceiving our team members, contractors, and vendors.

Prior to 2013, all data security incidents we encountered were insignificant. Our 2013 data breach was significant and went undetected for several weeks. Both we and our vendors have had data security incidents since the 2013 data breach; however, to date these other incidents have not been material to our results of operations. Based on the prominence and notoriety of the 2013 data breach, even minor additional data security incidents could draw greater scrutiny. If we, our vendors, or other third parties with whom we do business experience additional significant data security incidents or fail to detect and appropriately respond to significant incidents, we could be exposed to additional government enforcement actions and private litigation. In addition, our guests could lose confidence in our ability to protect their information, discontinue using our RedCards or loyalty programs, or stop shopping with us altogether, which could adversely affect our reputation, sales, and results of operations.

The legal and regulatory environment regarding information security, cybersecurity, and data privacy is increasingly demanding and has enhanced requirements for using and treating personal data. Complying with new data protection requirements, such as those imposed by the recently effective California data privacy laws, may cause us to incur substantial costs, require changes to our business practices, limit our ability to obtain data used to provide a differentiated guest experience, and expose us to further litigation and regulatory risks, each of which could adversely affect our results of operations.

This excerpt comes from the March 2020 10-K annual report filing for Target, who suffered a large data security breach in 2013 that resulted in over $18 million in settlement fees and affected more than 41 million of the customer payment card accounts. The text is a good example of how to translate cybersecurity issues to real business risk. Target is now especially sensitive to the disruption that a cyber incident can have on the business.

**Action**: If you are working in a public company, have you read your company's most recent 10-K filing and do you know what's in it? Can you explain to someone how your business makes revenue and where their most important income streams are in simple terms? You can find a company's public filings at https://www.sec.gov/edgar.

# GETTING THE BUSINESS TO OWN RISK

*As cybersecurity leaders, we have to create our message of influence because security is a culture and you need the business to take place and be part of that security culture.*
~ Britney Hommertzheim

How do we get business leaders to understand their role in risk ownership if they don't already feel that sense of ownership? Security is fundamentally just one of several risks that need to be considered in the course of doing business. Other operational, market and credit risks also get the attention of the board. If you're the only one being kept up at night worrying about security, then the ownership message probably isn't clear. Worse yet, efforts to protect your firm's most important digital assets may also be off course. In a world of ever-expanding threats, more security professionals are being asked to identify which threats really need to be addressed and which can be accepted as a part of doing business. This shouldn't be up to you; it needs to be a dialogue with the business.

Taking a risk-based approach to security helps you better focus limited staff and budget and make sure you're protecting the most critical business assets. This approach varies from a "check the box" compliance program that scrambles to address every threat without having a deeper understanding of the actual risk to the business and the underlying business processes. In a risk-based security program, compliance becomes just another factor in understanding your company's overall risk profile. But moving from a compliance-based program to a risk-based program is not without its challenges.

Good risk management can influence strong business decisions and leverage resources more efficiently by focusing mitigation efforts on the most important risks. Understanding business security risk involves looking at the problem through multiple lenses. There are business process risks, information risks, technology risks and disruption risks. No single role in the organization has the complete risk picture, which means that a risk-based approach must be collaborative. Security leaders must take on the role of a facilitator to help the business understand and manage their security risks. They must build and maintain relationships across the enterprise to get this discussion started.

You also need to develop a stronger understanding of risk assessment and risk management concepts. If your company has an enterprise risk management or operational risk management department, work with these groups to understand what frameworks and functions may already be in place. If not, there are several industry standard frameworks that can help, such as COBIT, FAIR and NIST.

If risk management isn't mature in your organization, getting the risk conversation started involves gathering key stakeholders, including business leaders, the chief compliance officer, legal, privacy, internal audit, human resources (HR) and the CIO. This collective group can work collaboratively to identify risks and decide how best to manage them. This group can also serve as a steering committee for the security program. You will also need to build the capability to monitor risks proactively over time. An organization's risk tolerance will change as the business and threat landscape changes.

It is impossible and undesirable to apply every security control to every asset. Taking a risk-based approach will help focus your efforts and ensure that the most critical assets are getting the most attention. Then, maybe everyone can sleep a little easier at night.

## Risk Exceptions, Risk Acceptance

There will always be scenarios where a security issue can't be addressed right away or sometimes be addressed at all. Open security issues can lead to breaches, audit issues or regulatory problems. Since they represent a known issue, they should be documented along with the action plan.

I'll make a distinction between a risk exception and a risk acceptance. A risk exception is when an issue has been identified and can be fixed, but it can't be fixed right away. This might be because someone needs to change the application code or because a system is so old that it can no longer be patched, and it needs to be sunset. In this latter case, there may be a project to migrate to a supported platform, but there's still some time to go before this is complete.

A risk acceptance is when there's an issue that either can't be fixed or the business simply doesn't support spending money on to get fixed. An example might be a mainframe system not meeting modern security requirements, but also being too costly to migrate from in the near future.

Both scenarios should be documented and approved by senior management. Obviously, in the case of a risk acceptance, you may need a higher level of authority to make sure that the most senior people in the business understand and accept the risk. The business owns this risk, not the security team. And therefore, the business needs to acknowledge that they own the risk and that someone senior enough in the organization is willing to sign off on it.

Paperwork doesn't make the issue go away. While documenting these issues is important, it does not reduce risk in any way. For this reason, both risk acceptances and risk exceptions should be reviewed periodically. I suggest no less than annually for a risk acceptance. Risk exceptions should be tracked closely to make sure that remediation efforts stay on track. A risk exception should have a program of work to close the issue, so you want to make sure you're monitoring progress and that issues are still a priority.

## Handling Risk Exceptions

When a "fixable" issue has been identified, you need to identify a senior issue owner. This would be someone that could ensure there are enough budget and other resources to fix the problem. Typically, this could be the CIO, CTO or equivalent department head. Issues that are going to take some time to fix, say longer than three months, should be formally documented. If an issue can be fixed quickly in a few weeks or a month or two, you may choose not to wrap as much formality around the paperwork and focus on fixing the problem instead.

You'll want the action plan to be reasonable and plausible. Migrating 30 servers to a new operating system isn't going to happen in a week, but it's probably not acceptable for it to take two years either. There's a negotiation and balance that needs to be hit here, but any plan that is defined may need to be defended to auditors, regulators, the board or other stakeholders. Make sure the plan makes sense.

## Handling Risk Acceptances

By its nature, a risk acceptance will not be fixed any time soon and possibly ever. This means that you really want to document the issue and make sure that senior management signs off that they are accepting it. Now let's be clear for a moment. You can advise the business not to accept the risk. You can plead with them not to accept it. But if you have done your job and they truly *understand* the risk and accept it anyway, then you are not in a position to say that they can't do that. You can question if it's been approved at the right level of the organization. You can suggest some compensating security controls. But if everyone understands the issue and they still have no appetite to fix it, you'll want to get some paperwork in place documenting who knew about the issues, what was discussed, the risk it represents and who made the final approval over moving forward without fixing the problem.

You'll want a sign-off on this document to make people think. You'll want this to ultimately read somewhat like a contract. You'll want it to read in a way that might give someone some pause before signing it. Consider this language:

> I understand that compliance with Company XXX's information security policies and standards is expected from all operating units. I believe that the security controls required by our information security policies and guidance from the cybersecurity group cannot be complied with due to the reasons documented above. I understand and accept the risks documented in this form and certify …

I've included a sample risk acceptance document in the appendix that can be used interchangeably as a risk exception document with a few modifications.

# THE SECURITY LEADER AS A BUSINESS ENABLER

I hate to disappoint you, but the top security jobs probably spend more time communicating and managing stakeholders than doing any of the "fun" stuff that got you into security in the first place. The more senior you get, and the more you become a security leader, the less time you will spend on packet captures, forensic images and cryptography. Instead, you are going to be communicating to all levels of the organization, helping them understand and own business risk, educating them on security risk and gaining funding and support for the program. So, if you're going to spend this much time communicating, you might as well get good at it.

Senior business audiences don't always care about the technical aspects of security. Some may not care about security at all, seeing it as a necessary (or even unnecessary) nuisance. Business priorities include profits, market share, business risk and organizational reputation. They don't necessarily match with the objectives you are trying to drive. I once heard a senior business leader refer to their security team as "those annoying password people." This is not the reputation to strive for in your organization.

The language of most for-profit businesses is financial. This means that one of the easier ways for a businessperson to understand the concept of security is to see it as a form of insurance. Insurance usually works with historical damage data to determine premiums, but unfortunately, this kind of actuarial data are not mature or reliable for information security incidents. This means it will be difficult to quantify the benefit from most information security expenditures.

Don't expect your business leaders to become fluent in security. While there is a greater interest in the subject these days, there is also a lack of time and technical training for most business leaders. Don't expect them to adapt to you, you need to adapt to them. They aren't going to learn the language of cybersecurity, which is why you need to learn the language of business.

At the same time, you also need to adopt a "business-first" attitude. Can you articulate how your programs are supporting the business in their strategic efforts? You should define your operations around business issues that have security solutions. Don't focus on how to implement new security tools. Instead focus on how your tools help improve and secure the business.

It's also worth spending some time to understand your role as security leadership. I don't care if you are a senior leader or the CISO, you very likely don't own any business data and don't get the final say on taking business risks with that data. This may come as a shock as a new CISO and is a big reason that many new CISOs fail. They throw security mandates at an unreceptive business that either doesn't understand their security risk or that does understand it but is moving forward anyway. Very few CISOs in the industry have enough authority to completely halt the business based on an issue. That's not to say there aren't circumstances where you should escalate to the highest levels of the organization.

In the old paradigm, many security professionals interpreted their role as having the right to say yes or no to requests that came through the security team. In some firms, the security group has been jokingly called the "department of NO" because they only say no to security requests. Unfortunately, this led to a lot of people finding ways around security controls. The security group was never directly engaged in the first place, because everyone knew their answer already. This paradigm not only doesn't work today but never worked in the first place. The security group should strive more to be the department of KNOW. They need to know a lot of information about a lot of things and they need to be able to help figure out the most secure way of meeting business requirements. Instead of shooting down everyone else's ideas, they need to put skin in the game to help define secure solutions that meet business requirements.

Now this doesn't mean that one business unit in a large company can put the whole firm at risk by having a "business requirement" for unfettered bi-directional access to the Internet. In these scenarios, your role is to make sure that the most senior executives in the firm understand your position on this and support the stance that no one business unit can accept a risk for the whole firm. These situations do come up, but thankfully not often and probably wouldn't come up with today's increased awareness around cybersecurity issues.

When a business still wants to move forward with something that you consider overly risky, make sure that your senior management agrees with your stance. If they don't agree, make sure they really understand the risk. Did you really do everything you could to explain the risk clearly? Does senior management really understand the likelihood and impact if the risk is accepted? If they do and want to move ahead anyway, make sure the risk is documented and formally signed off by senior management. If this is happening routinely, it's probably time to find a new role.

# START WITH YOUR OBJECTIVE

When you're interacting with business management, you should consider that most interactions will involve a goal or objective that you need addressed. For example, is there some action you are trying to get? Is there a timeframe for that action? Are you looking for support on an initiative? Approval? Funding? What's your goal? You should be clear and upfront about this. You don't want objectives and goals to be lost in the conversation. Security people have a lot of work to do and wasting time is not going to help. The same goes for business executives, so it's always best to have a clear plan and strategy. While active listening and understanding your business partner's view is high on the list of any interaction, ultimately you are being paid to get things done and move the security program forward. Make sure you are making the most of your time and keeping your eye on the key goals and decision points needed to move the program forward.

Sometimes, however, an interaction is just that, casual and without goals. And that's fine. It's not always about productivity. You need to make sure that you have a rapport with key business leaders in your organization, because the odds are that at some point you will either need to either work closely with them on an issue or need their direct support for an element of your program.

Some general goals of business communication include:

- **Informing**: A situation where you are looking to create awareness about an issue or other topic.
- **Requesting help or support**: A situation where you are looking for either action or support.
- **Giving or receiving a position statement**: A situation where you are making your view on something known for the future or trying to understand their position.
- **Advising and consulting**: A situation where you are looking for advice or input or you are advising them about something.

Think about the objective you have going into your communication. If you want the business to accept your ideas, or change their opinion about something, how are you going to get them to do it? You can inspire, motivate or negotiate with them to take action. But you can't do any of those things until you know what you want the ultimate outcome to be and which tradeoffs you can accept.

# BLUF YOUR WAY THROUGH COMMUNICATIONS

*Many attempts to communicate are nullified by saying*
*too much.*
~ Robert Greenleaf

BLUF, or "bottom line up front," is one of my favorite mechanisms for communication to senior management and honestly to just about everyone else as well (Figure 3.1). This term originates in the military and is intended to help you get straight to the point. Contrast this with providing tons of background information that builds up to an

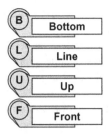

**FIGURE 3.1**   Putting the point that you're trying to make at the beginning using the BLUF technique can increase your odds of being understood.

amazing conclusion, ten or twenty minutes from now. By then you've lost your audience. The more senior the executive, the less they are going to understand technical security issues and the more unrelated subjects they will have already on their mind, like running the business.

Let's think about the life of a C-level executive for a minute. They are typically in charge of large areas with many people supporting them and a corresponding number of challenges, crises and all the problems that come from having this much responsibility. They also know their business, not cybersecurity and technology. They are interrupt-driven, meaning they can be called away at a moment's notice for the latest crisis. Don't assume you will have undivided attention even if you have a 30-minute meeting scheduled.

BLUF helps you get straight to the point and then helps steer how much detail a listener is interested in hearing. BLUF works for all communication mediums, including writing and speaking. It's basically always a good idea. The goal is to get to the most important details first, so that you don't delay the point you're trying to make. I've had C-level managers pulled away to emergencies in the middle of presentations and you simply lose the moment. If you are working up to making a point, you are going to miss your opportunity as they are pulled away and you have to reschedule.

BLUF is a fantastic communication tool in business and in life. No matter your audience, they will always appreciate you quickly getting to the point. Receiving too much information actually interferes with the audience's ability to process it all. Resist the urge to bury people in a mountain of evidence that supports your point. Instead, stick to your strongest points. Pick one or two points that really make your case and keep the other evidence on hand in case anyone's interested in digging deeper. But let them decide how much detail they want.

While I have seen some people physically write out BLUF or put "bottom line up front" in their written communications, I feel this is a little artificial unless this is a standard format in your organization. I think it's enough to just get to the point in the first sentence or use the word "summary" or "executive summary." Either of those is a little more business-friendly, especially since many people probably won't know what BLUF stands for.

The BLUF technique can and should be applied to almost all of your communications in a business setting. Get to the point quickly and provide just enough detail for your audience. Remember that the quicker you get to the point, the more likely your point is to be heard. Attention spans are short and everyone's busy. You have a limited window to make your point.

Even with longer documents or business reports, you should structure it so that key points aren't buried. You can always include supporting information or have it at the ready. For example, I like to add supporting details in the appendix of a PowerPoint presentation, but have the main slides be in the 3–4 range. That way, if someone wants to know how I drew certain conclusions or wants more detail on a particular subject, the information available and I can pivot into a more detailed discussion on demand. In either case, I let the level of interest drive the amount of detail provided.

# FEAR, UNCERTAINTY, AND DOUBT (AKA "FUD")

It's also worth spending some time talking about fear, uncertainty and doubt, also known as FUD. This term has been around for some time and even predates the information security industry. A lot of security people are tempted to use FUD as a means to get funding for projects or to justify elements of the security program. FUD tactics typically focus on overly negative scenarios and news to exaggerate negative consequences. The problem is that FUD is actually a disinformation tactic that paints an overly pessimistic picture of worst-case scenarios. The idea is to use outlier or extreme cases to gain support of funding for unlikely security events.

While true that some cybersecurity incidents can be catastrophic for a company, these are typically very small companies. Even Ashley Madison is still operating after a hack leaked their customer database and it made the news for what felt like months. The same goes with TJ Maxx, Target and many other high-profile companies that suffered a cyberattack, saw their stock fall, and then watched it rebound stronger than ever. The fact of the matter is that a lot of these attacks happen, we clean them up and we move on. Don't think for a minute that your business executives aren't aware of that fact.

Instead of using FUD, have a meaningful conversation around the most likely risks to your businesses and how they can take steps to manage them. Risk management is a big subject, but in general, you need to focus on the most likely threats with the biggest impact that represent the greatest risk to your company. Being the person who cries wolf, or the person saying that the sky is falling is not going to get you very far with your business colleagues.

I once asked a Chief Operating Officer (COO) what kept him up at night. It wasn't a security incident; it was something more like Amazon entering their industry and undercutting their margins by 60%. In this scenario, a competing business doing the same thing cheaper and better would cause the company to bleed a slow and painful death if they couldn't respond fast enough, which he knew they couldn't. A security incident generally won't compete with this kind of existential problem, so try to keep some perspective.

It's notable that one item that sometimes falls into the FUD category is compliance. Yes, compliance issues are business requirements. They are requirements from regulators or legal sources and compliance items need to be done with little debate. While compliance efforts can certainly be supported by security efforts, you will still be better off highlighting the benefits of cybersecurity, rather going deep into breach horror stories or possible giant compliance violation fines. Forceful compliance discussions should be a last resort, not an opening tactic. Trying to force security programs through compliance requirements fails to highlight the benefits of a good security program and makes it seem more like eating brussels sprouts. It's good for them, but that doesn't mean they're going to like it.

# A NOTE ON THE USE OF TECHNICAL JARGON

*Developing excellent COMMUNICATION skills is absolutely essential to be an effective engineer. The engineer must be able to share knowledge and ideas to transmit a sense of urgency and enthusiasm to others. If an engineer can't get a message across clearly and motivate others to act on it, then having a brilliant idea/ design doesn't even matter.*
~ Gilbert Amelio, President and CEO of National Semiconductor Corp

Avoid technical jargon and acronyms wherever possible, especially when you're working with business management. No one will be impressed by how many acronyms you know or what level of technical detail you have learned over the years. Your knowledge doesn't matter if the listener has no idea what you're saying. What matters is that your message is heard and understood by your audience. As basic as that sounds, a lot of people miss this mark and instead try to use jargon to make issues seem more complex or to make themselves seem smarter. All security issues can be dumbed down for everyone to understand.

We spend a lot of time learning security concepts and it is a field that uses many acronyms and a lot of technical jargon. Save this stuff for talks with your fellow security professionals, not for the C-level management or your business users. Not only will senior executives be unimpressed, but they will also have no idea what you're saying and tune out of the conversation altogether. Always ask yourself if there is a simpler way to deliver your message without losing its intended meaning.

I'm not advocating leaving out necessary technical detail, but make sure your language is as clear as possible and that the technical detail really is relevant. If you find it necessary to use technical terms or acronyms, just make sure that everyone understands your meaning. Sometimes, it's easiest to simply add a definition to the technical term as you speak. For example, "Our competitor was breached by a technical attack called SQL injection. This is where an attacker uses an application in a way that wasn't intended to expose or change stored customer information." Don't make assumptions that people understand what SQL injection is or why they should care about it. Also notice, I didn't start talking about back-end databases or what SQL is or stands for. It's

Technical talk is not always understood by your audience

**FIGURE 3.2**   Many people do not readily understand technical jargon. Avoid using acronyms and technical terms so that everyone can understand what you're saying.

not relevant to the conversation and you don't need to "educate" business users on this level of technical detail to get your point across (Figure 3.2).

This takes us back to knowing your audience. The same conversation with your team might be worded very differently. You might even gain a bit of credibility if you can "talk tech" with the technical people by going deep into the details. Unfortunately, as you move up the management chain, technical jargon gets less and less effective and you need to adopt other tactics.

What is considered technical jargon? If you're not sure, you probably shouldn't use it. If you couldn't explain it to your mom, you probably shouldn't use it. Technical jargon, by definition, is terminology that can only be understood by those with a technical background. Use easy-to-grasp words that can be understood by everyone. Jargon is a type of shorthand that is used to simplify communication among certain groups. It typically involves the use of words and phrases that are otherwise meaningless, especially when taken out of context. Some other concepts to keep in mind include:

- Shorter sentences are easier to understand than longer, convoluted sentences.
- Use simple words. Use plain, direct conversational language.
- Think about how many technical details are necessary to provide. It's easy to jump into the technical weeds, but this isn't helpful if your audience doesn't understand what you're saying. Remove everything that doesn't support your story and speak/write for the general audience even if a few of them do have a technical background.

# "SELLING" INFORMATION SECURITY

This brings up the subject of how you should sell a security program if you're not using FUD tactics. Like it or not, you are often going to be faced with the need to "sell" what the security team is doing and spending money on to senior management. No function in a business has a blank check and security is no exception. Security is a cost center. As much as people talk about the return on investment (ROI) of security, it is fundamentally a cost center by nature. Security efforts are rarely directly connected to revenue generation or real cost savings. This means that business leaders will typically see security as more of a necessary evil than an investment. Or at best, security plays a supporting role to other efforts that do generate real revenue.

The best way to sell security is by creating a business case for your projects and programs. It's OK to view tools as investments for improving security and you should be able to explain exactly why a capital spend is necessary and what value or risk reduction is gained from the spend. What business risks is your program addressing? You need to be able to explain this. You can't just go in looking for funding with arguments like we need this new product because everyone else has it and we don't.

You may find that some business executives have a sense of apathy about information security. Breaches like the SolarWinds issue in 2020 highlight that no one is safe from or immune to cyberattack, even if you do the right things. Therefore, they may feel like they have already been breached and that the attackers are already inside. Others may cite how breaches like those in Target, Home Depot and others happen and then life moves on with little lasting impact.

Thankfully, while an apathetic view on cybersecurity does exist, the overall view of the importance of security is steadily becoming positive. This said, some security leaders still struggle to make a clear and defensible business case for investing in a security program. To do this, you are going to need to be able to explain the benefits of information security tools and programs in business terms. You need to put yourself in the position of a business manager (using empathy) and ask yourself:

1. Why should I care about this?
2. Why would I spend money on this?
3. What would happen if I didn't do this at all?
4. What benefit will this give me?
5. What risk does this address, and do I care about that risk?
6. If I do care about that risk do I think it's still likely to happen and would have a big impact if we failed to address it?
7. Why do I need to spend money on this, again?

If you can't answer these questions plainly and simply, then you have some work to do refining your story. And you definitely want to answer these questions yourself before someone else asks you. Being caught without the answers to basic questions like this undermines your own credibility and will make you seem like you don't have any business sense. Don't be the person who simply says, "we need to buy DLP." Put some time and thought into what the benefits of a program will be and how you can personalize this in a way that doesn't even require the use of the term "DLP."

# UNDERSTANDING BUSINESS DRIVERS

> *We will bankrupt ourselves in the vain search for*
> *absolute security.*
> ~ Dwight D. Eisenhower

If you are going to communicate the value of information security to business leaders, you need to understand what drives a business. Business drivers speak to the bottom line and get senior executives to understand the importance of taking an action, including funding a program or aligning resources.

Understanding business drivers will help you better connect to what makes your business tick and how to better connect with leaders and executives who may not have security on top of their mind as they deal with the day-to-day challenges of running the business. Spending time understanding these drivers will also help you to tailor your own business cases in a manner that speaks to the bottom line.

While not exhaustive, a list of typical business drivers is discussed below.

## Revenue Generation

Everyone understands that a for-profit business generally operates to make money. This means that most activities that support or enhance a business's ability to make money often get the most attention and funding. While security by itself does not typically fall under this category, it can help enhance efforts that do. For example, if a business is scaling a process by moving it to the cloud, security controls can be perceived as an enabler that supports this effort. If you can explain how programs are tied to revenue growth and market differentiation, it will be significantly easier to sell these programs to senior management.

## Customer Request

Customer requests typically originate directly from large customers asking for new features or functionality. It's not typical for security features to be requested by customers, but this may change over time. For example, the ability to implement multifactor authentication (MFA) may exist in a competitor's platform and raise questions from customers on why you aren't supporting stronger authentication as well.

## Regulatory or Legal Requirements

Depending on your industry, regulatory and legal requirements can also serve as a major business driver. Public companies have requirements from the Securities and Exchange Commission, and large financial companies may have up to hundreds of global regulators. Other industries that tend to be highly regulated include public utilities and healthcare. In addition, you may fall under many legal requirements at both the state and the federal level as well as falling under some global requirements like the General Data Protection Regulation (GDPR) for companies that do business in the European Union. Regulatory and legal requirements are not negotiable. They must be done. On the plus side, they often have security requirements baked in and some programs can be sponsored by simply being a regulatory or legal requirement. As I stated before, I don't like using compliance to *force* a security program through, but sometimes an effort is already underway and tying a security program to it makes perfect sense.

Security also provides some legal liability mitigation. Being able to demonstrate due care by implementing industry standard security controls can help limit future liability claims in the event of an incident. There has been a recent rise in cyber negligence cases for corporations that failed to put reasonable security controls into place. Just don't use this as an excuse for FUD tactics we discussed earlier in this chapter.

## Risk Reduction

Security's sweet spot tends to be risk reduction as a business driver. Businesses must manage their risk if they expect to continue operating. In a financial institution, risk can fall into multiple categories, including credit and market risk. Security tends to be rolled up best as an operational risk, meaning that it is something that needs to be managed in the day-to-day operations of the business. Cybersecurity risk is really just a subset of operational risk, which covers end-to-end business operations risk. Risk management options include accepting, avoiding, transferring, mitigating or accepting risk.

## Reputation and Brand Protection

Product brands and corporate reputations have inherent value. Most companies want to protect their brand name and reputation, especially in trust-based businesses, like management consulting and financial services. Security incidents can have a negative impact on both brand and reputation. This can also be used as a differentiator by establishing an image of trustworthiness and strong governance. For example, a financial institution that has been hacked multiple times (or even only once in some cases) might see an outflow of customers. Trust is an important part of doing business, and managing security risk can help protect the firm's reputation as a safe place to do business.

Product brands have an inherent value and are exposed to competitive or malicious damage to the resources (intellectual property and knowledge) associated with the respective brands.

## Availability and Resiliency

Resiliency means keeping both front- and back-end business processes running to service customers. Most business leaders are starting to understand how disruptive a cybersecurity incident can be, and they are taking it more seriously. Companies take an average of 280 days to identify and contain a breach, according to a recent IBM study. This lengthy amount of time can cost the business millions of dollars, especially in the case of a ransomware outbreak.

## Data Integrity

The integrity of data, especially financial data, is a clear driver for some businesses. At a minimum, firms want to ensure that no one has tampered with their financial books and records. Changes or alterations to this kind of information have a direct impact on the business operations and the bottom line. Integrity is a strong driver in many financial institutions. In a hospital setting, patient records would be equally important. Get to know your business and what is important to them. A lot of security professionals are so focused on data access control and encryption that they sometimes don't spend enough time ensuring that records aren't altered by unauthorized changes.

## Previous Experience

Any recent security incidents that impacted the enterprise can also serve as a reminder to avoid these issues in the future. Worm attacks, ransomware and phishing campaigns are big disruptors for business. Has anything similar recently happened at your company? Even if the event has been wrapped up, you will likely still be dealing with fallout from people who still have it fresh in their memory.

## Changes in the Business Environment

Responding to security implications of change in the business environment (for example, geopolitical risk) can also be a business driver. Monitoring a changing environment means keeping abreast of security investment trends, best practices and approaches, executive focus, and the use of information security as a business differentiator.

Having a strong understanding of what drives business decisions will help you make your own case for security. If you are trying to purchase a tool or product that doesn't fall in the list of one or more of business drivers, you might have a tough time selling the need to senior management. Make sure you're not just trying to buy the latest security toys because everyone else has them.

# THE SEVEN Cs OF COMMUNICATION

It's especially important to be crisp and clear with your communication when you are communicating with non-technologists. If you'd like to make some improvements in this area, I find that a framework called the "Seven Cs of Communication" can help.

There are several variations of the Seven Cs of Communication, but the first one was thought to be provided by Cutlip and Center in 1952 in *Effective Public Relations*. This is an interesting framework to apply to any communication, regardless of the medium.

As shown in Figure 3.3 the modern version of the Seven Cs is generally represented as:

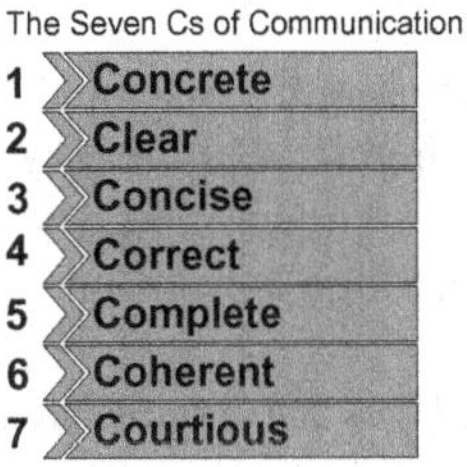

The Seven Cs of Communication

| | |
|---|---|
| 1 | Concrete |
| 2 | Clear |
| 3 | Concise |
| 4 | Correct |
| 5 | Complete |
| 6 | Coherent |
| 7 | Courtious |

**FIGURE 3.3**   The Seven Cs of Communication can serve as a checklist to make sure that your message is concrete, clear, concise, correct, complete, coherent and courteous.

1) **Concrete**: The message must have meaning and relevance for the receiver and not be vague. Messages should be tailored to the audience and be something that they want or need to know. The core message needs to be related to something the audience cares about—don't talk technical details to the CFO, for example.

2) **Clear**: While the content should be timely and relevant, the message must also be clear enough to be understood. Complex messages must be made simple to understand, based on the skill level of your audience. Remove or explain acronyms and avoid technical jargon. A jumbled message will confuse the audience and leave them unsure as to what they should understand or do. One topic per message should be the goal.

3) **Concise**: The message should be concise and to the point. Use BLUF. Avoid lengthy sentences and try to convey the subject matter in the least possible words.

4) **Correct**: Obviously in security, what you are saying needs to be credible. Are you passing along something that is fact-checked? Are you propagating security FUD? Make sure you are factually accurate. Along with this comes general grammar and spell-checking for written communication, which we will cover later in this book.

5) **Complete**: Your message should be complete and include all the relevant information needed to inform the recipient. Complete information should anticipate questions that might arise and address them proactively.

6) **Coherent**: Take into consideration the receiver's opinions, knowledge, mindset, background, etc., in order to have an effective communication. Does your message make sense? Does it flow logically from one sentence to the next?

7) **Courteous**: Be polite and considerate of the recipient. Communications shouldn't require effort from the recipient. Don't expect them to adapt to your information; present it in ways that mesh with their current capabilities.

The Seven Cs represents a good checklist to make sure that you are putting the audience first and making it likely that they will receive and understand your message. Making life easier for your audience is what effective communication is really all about. Some other considerations include:

- **Structure**: Every good communication should have three critical elements: an introduction—a short, high-level summary of the problem and the solution/recommendation; the argument—which will flesh out and fill in the details of your message; and the conclusion—which will reiterate the message and the next steps.
- **Consistency**: Mixed messages and inconsistency lead to distrust. Keep all messages consistent both within single communications as well as across the entire program whenever you are revisiting a subject.
- **Medium**: There are myriad ways to deliver a message and successful communicators match the medium to the audience, time frame, culture and venue. This is especially important with a business audience. Don't be the one persona that calls in to a meeting that everyone else attends in person.
- **Open and close strong**: People tend to remember the first and the last things you tell them. Make sure you have a strong opening and a strong close.

# SUMMARY

We covered a lot in this chapter. When it comes to business communication, you want to learn the business and teach the security elements. This means that you need to do the heavy lifting for business communication and help them understand you.

- Do you understand how your business operates? Make sure you learn as much as you can about your business by studying financials, speaking to executives and spending time with investor relations reports like the 10-K.
- Security doesn't own the risk; the business owns the risk. Security is just one of many business risks that senior executives need to understand and manage. Risk is a dialogue across multiple disciplines. It is impossible and undesirable to apply every security control to every asset. Taking a risk-based

approach will help focus your efforts and ensure that the most critical assets are getting the most attention.

- Make sure you understand your own goals in working with business leaders. Giving information, requesting help or advising on security issues are all valid reasons and can help you keep conversations productive.
- Bottom line up front is a great tool for focusing communications so they will have the most impact and get right to the point.
- Fear is never a good way to sell a security program, instead focus on the benefits that a sound information security strategy brings.
- Technical jargon should always be minimized, but especially when you're working with non-technical people. Keep it simple.
- Make sure you understand business drivers and how to tie security initiatives back to them. Why should the business care about or support your initiative?
- Use the Seven Cs of Communication: concrete, clear, concise, correct, complete, coherent and courteous can help serve as a checklist that you are being effective with the way you position your communication.

# REFERENCES AND RECOMMENDED READING

Driver, Janine, and Mariska van Aalst. *You Say More Than You Think: The 7-Day Plan for Using the New Body Language to Get What You Want.* Three Rivers Press, 2011.

Ekman, Paul. *Emotions Revealed: Recognizing Faces and Feelings to Improve Communication and Emotional Life.* St. Martin's Griffin, 2007.

"How to Read Body Language Quickly In 8 Easy Steps." *Body Language Project.com*, 25 January 2014, bodylanguageproject.com/.

Navarro, Joe. *The Dictionary of Body Language: A Field Guide to Human Behavior.* William Morrow, an Imprint of HarperCollinsPublishers, 2018.

Navarro, Joe, and Marvin Karlins. *What Every BODY Is Saying: An Ex-FBI Agent's Guide to Speed-Reading People.* Harper Collins, 2015.

Pease, Allan, and Barbara Pease. *The Definitive Book of Body Language: How to Read Oher's Attitudes by Their Gestures.* Harlequin, 2017.

Stanley, Andy. *Louder Than Words.* Multnomah Publishers, 2004.

Asplund, Jan-Erik. "BLUF: The Military Standard That Can Make Your Writing More Powerful." *Animalz*, 17 July 2020, www.animalz.co/blog/bottom-line-up-front/.

"Cost of a Data Breach Study." *IBM*, 2020, www.ibm.com/security/digital-assets/cost-data-breach-report/.

Cutlip, Scott M., et al. *Effective Public Relations.* Pearson Prentice Hall, 2006.

Schneier, Bruce. *Beyond Fear: Thinking Sensibly about Security in an Uncertain World.* Springer Science+Business Media, 2013.

# Company Culture

# 4

Company culture has a big influence on communication. I'll define culture as a set of customs, methods and traditions followed by a given company. Basically, it's the way a company conducts business. Cultural differences are not just about being from a different country. Culture can vary widely from firm to firm, but it also has a big implication on your communication strategies. If you work in an Agile or DevOps environment, a lot of your communication is probably in-person and verbal. If you are part of a 300,000-person global firm, your communication might revolve a lot more around video conferences and phone calls. When starting with a new company, it's important to try to get a feel for the culture before trying to change it. I've seen a lot of CISOs come in and think that they'll change the culture to fit their personal operating style and it has yet to do anything except chase the best talent out the door and massively disrupt the team. Eventually, the CISO gets shown the door too. While you might have been hired to be a change agent, make sure you're not overdoing it and are taking it at a measured pace and creating a positive work environment.

Company culture is important and not only varies from company to company but can also vary from department to department within the same company. In fact, cultural differences go all the way down to the individual level. Someone born in the 1960s in a city might have a very different perspective than someone born in the 1980s in the suburbs. Learning to respect and tailor your communications to respect cultural differences can make a world of difference.

How you personally adapt to corporate culture may be the deciding factor on your success at a given company. Here are a few tips on how to adjust and make the transition as painless as possible.

## ORIENT YOURSELF USING THE OODA LOOP

The first step in adapting to company culture is to get an understanding of what that culture looks like. Don't assume on day 1 that you have a good feel for culture; it can really take some time. Observe how people behave. Talk to people and set up 1-1 meetings with them that are more informal in nature. Take in as much information as possible. An interesting tool to consider for orienting is called the OODA loop (see Figure 4.1: OODA loop). The OODA loop is a cycle of observing, orienting, deciding

DOI: 10.1201/9781003100294-5

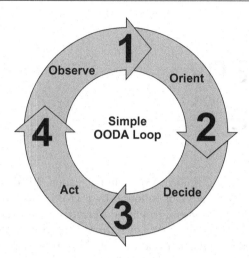

**FIGURE 4.1**    The OODA loop has been around for a long time in military circles. This technique can help you orient yourself and reduce uncertainty when used properly.

and then acting. It was developed by military strategist and United States Air Force Colonel John Boyd. It is an incredible tool when you take a little time to understand it.

During the observing phase, you want to factor in how company decisions are made, how performance is managed and how the company generally conducts business. I once interviewed at a company that didn't believe in managing consistently poor performers out the door because their culture prided themselves on having never done a layoff before. You can imagine that the people caught on to that fact and took the path of least resistance by just keeping their jobs and not doing a lot more. Make sure this is really the kind of company you'll succeed at before signing up for the role. You'll also want to understand how the company makes decisions. Is there a bias for action or analysis and consensus? Do managers want to see a lot of supporting data or just progress and results? Don't jump right into decision-making, taking action trying to prove yourself; learn to observe as your first step.

The next phase of the OODA loop is orienting. Now that you have taken in information, it's time to start considering what you know and what conclusions you can draw. During the orientation phase, analyze all the observations you've made. Analyze, evaluate and prioritize what you have learned. These are the fundamentals, or the first principles, of the situation. It can include threats, opportunities and other useful information.

The next phase of the OODA loop is deciding. In this phase, you want to move from raw data analysis to actionable insight. This step is about choosing from many options and trying to determine your *best* option. Each option should be informed by what you learned in the orientation phase. What will deliver your optimal outcome? What is the most likely strategy to succeed?

The fourth phase of the OODA loop is acting. At this point, you have a good hypothesis and a decision on how to move forward. In this last step, you are executing your decision and determining if your hypothesis was right. What is learned about the validity of the hypothesis is repurposed throughout the entirety of the next cycle of the OODA loop. Ideally, future cycles will be both more accurate and faster. Remember, this is not the *last* step of the process, OODA is a continuous loop, so the next step is to go back to the beginning and observe the results from your actions and start all over again. This is where a lot of people misunderstand the tool.

The OODA loop is often oversimplified and may not seem that useful at first. But it is a learning system and an amazing method for dealing with uncertainty. The OODA loop can help you deal with changing and challenging circumstances and still come out on top. Remember, Boyd developed this concept to explain how to direct one's energies and defeat an adversary and survive in war. This should be more than ample to simply get your bearings in a new job. Some key elements to remember with the OODA loop that may not seem that obvious at first include:

1) In the observation phase, you might be observing incomplete or imperfect information. You might even have too much information to process. This is where judgment is more important than an abundance of information. It's the ability to pull a pattern out of data that's more important than just lots of data or what you perceive as enough data. Don't get stuck continually looking for all the data; typically you'll see the pattern pretty quickly, at which point you can move on to the next phase.

2) When circumstances change, you need to re-orient. Instead of continuing to view things the way we think they should be, we need to re-assess how things actually are right now. This continuous reorientation makes the OODA loop a great tool for dealing with change.

3) We must continuously refine and adapt in the face of new observations. Remember that OODA is a *loop*, not an end-to-end process. I've seen a lot of depictions of OODA that seem to end with "act" and then it looks like everything stops. When you hit the act step, you immediately go back to the observe step to see what impact your actions have had and start the whole process all over again. Loop and then iterate over and over again.

4) Uncertainty is everywhere. Even as we gain more knowledge and perspective of one area, we may grow in uncertainty with other areas. For example, the more you may learn about security leadership, the more dated you've probably become on some of your technical skills. Don't ever expect perfect certainty across every aspect of whatever it is you're trying to observe.

5) Mental models grow outdated. The more we use the same models while the world changes around us, the more we grow stale and ineffective. This is why the loop in OODA is required, to keep the model fresh and always act on new information by reorienting.

6) According to Boyd, the ability to orient effectively and quickly is what separated the winners from the losers in a battle. Challenge your beliefs, keep an open mind, challenge your assumptions and question everything that you think you already know.

7) When we decide and then act, we are doing what we can with our best hypothesis and the data we know at the time. The uncertainty principle means we're only making an educated guess at what we think is right. We go back to observing our results and going through a reorientation if our hypothesis was wrong and again go back to trying to regain our bearings.

Hopefully, you now see the power of this seemingly simple process. Getting oriented quickly will help you become more effective in managing all the uncertainty that we're faced with every day in our difficult roles. This simple framework can give you a powerful way to help figure out which end is up and get back to the business of running the security program and making an impact.

# "FITTING IN" TO THE COMPANY CULTURE

Starting a new role or working with your existing company culture can be tricky. How you communicate, how fast you try to move and how much you try to influence or change culture can sometimes cause a lot of friction. It's not always best to muscle through this friction by moving faster and ignoring the feedback loop. The more you can adapt to the basic culture that's already in place, the more likely you will gain trust and begin to influence change.

## How Do People Communicate?

When you start a new job, look at how people primarily communicate with one another. Is it through email and instant messages? In-person meetings? Are these scheduled interactions or often ad hoc and unplanned? Realize that you may not like the answer to these questions. For example, if you are used to sending emails when company culture is stand-up scrum meetings, don't expect to stay in your comfort zone and still be an effective communicator. You're going to need to adapt. You will need to embrace the company's way of doing business and let this be a strong input to your choice of communication mediums.

## Change the Culture or Adapt to It?

Changing or even influencing company culture is a big job and is not very likely to be a successful strategy unless it is taken slowly over a long period of time. Culture does change, but it changes meeting-by-meeting, day-by-day and person-by-person. Real and lasting culture changes take years. Changing the entire company culture is usually far out of the scope of what you are expected to do as a security leader. At most, you are expected to influence a culture of security through education and awareness. Don't think you're going to make flat organizations hierarchical or introduce massive change that would best come from the CEO. Unless you have been expressly directed by senior management to help change the culture in specific ways, you should likely assume that your role is to fit into the existing culture and influence positive change gradually over time. Don't be the egomaniac who comes in and tries to change how everyone operates and replaces the culture that may have taken more than 50 years to develop. Instead, you need to adapt the best you can and then influence change over time.

# Take It Slowly

You will likely want to come into a new organization and make an immediate impact. If you are in the top security role, there might be press releases announcing your arrival and you will feel that you are under some pressure to show immediate value and make big changes. I recommend putting together a 100-day plan, but part of that plan needs to factor in that you are probably not going to make gigantic cultural shifts in 100 days. This is a journey, not a destination. If you do aim for big changes, make sure they are positive changes, like getting the team enhanced training or removing some obstacle that has been a headwind.

# Be Open and Be Flexible

Don't assume that you need to make any changes at all to company culture. You may have minimal influence over changing the culture outside of your own team, and even that might be very disruptive and difficult to manage. Don't assume that you can't succeed with the culture exactly as it is right now, unless it's truly dysfunctional in your own team and needs to be addressed to be effective. Also, don't use the culture as a reason for your own success or failure. Culture isn't an excuse or a roadblock; it's simply the way the business works, and you've been hired because someone believes you are going to work with that culture, not replace it. Moving from industries, like the public to the private sector, can be especially challenging since the chain of command that exists in the public sector isn't always present in the private sector. The private sector, and to a degree even the public sector, works best when you lead by influence. This is true even if you do have a chain of command in the public sector. See the section on "Building Influence" later in this chapter for more ideas on leading by influence.

# The Change Agent Mandate

Most companies are resistant to outsiders coming in and driving change, but what if you have been told that you're being brought in as a change agent to shake things up and challenge the status quo? Be careful with this situation. A lot of people fail with this mandate because they either have limited support to *really* shake things up or your management may get cold feet as the HR complaints increase and you'll be left alone to deal with the repercussions. If you have been given the change-agent mandate, make sure you fully understand the parameters. Change what? Your team? The entire company and the way they conduct business? Yes, you want to help build and influence a culture of security across the whole firm, but you also need to be realistic about what you can do and the support you'll get from senior management when the going gets tough. Some people mistakenly think they'll change the culture by cleaning house and firing everyone from the previous regime. This way of thinking only replaces the problem with the need to create a new culture with the people brought in to replace them. I remember watching this unfold at one of my previous firms and the only thing everyone had in common was that they knew the head of the program. They didn't, however, know each other and they spent a lot of time jockeying for position and competing with each other.

Determine what you can successfully challenge in the existing culture, and then when and how you should do it. Security is hard enough without trying to take on the CEO role and make broad, dramatic changes across the whole organization and the way everyone conducts business. Leave this to the CEO.

# BUILDING INFLUENCE

Effective security leaders should make it a point to regularly interact with executives outside of corporate IT. They should seek to nurture meaningful relationships with executives outside the context of executing projects, collaborating on risk issues and influencing enterprise-level decisions. These people see these networking opportunities as essential and core to their success.

As a security leader, you are going to need to influence a large variety of stakeholders that do not report to you. You are going to need to influence business leaders to change course on their own projects and potentially delay their own efforts to work on yours. You'll need to advocate where your company's budget is spent and how much of it should go to security. You'll need to do this all through influence and your communication skills. Some of your non-IT stakeholders include the board of directors, internal and external audit, the Chief Financial Officer (CFO), Chief Data Officer (CDO), marketing, business leaders and even vendors.

Influential people can inspire others to take action and can provide a great deal of motivation for the organization. Without influence, you will not be seen as a leader in your organization. In fact, you may not even be seen as an effective manager without at least some level of influence. You get things done through influence and you may not get much done if you haven't mastered this skill.

Everyone starts out on an even playing field when they start with a new organization. People gain influence in many ways. While they might inherit some perceived influence by the nature of their role, it will quickly evaporate without some genuine skill or credibility. The good news is that this is something that can be developed and improved with a little work.

In 1959, social psychologists John French and Bertram Raven identified five bases of power:

- **Expert power**: This is gained from having the knowledge, experience and ability to solve problems and to make good decisions.
- **Referent power**: It is defined as the ability of a leader to influence based on respect for that leader.
- **Legitimate power**: This power comes from the leader's position, which gives them the right to ask others to do something. An example might be your boss or the CEO.

- **Reward power**: It is the power to control salary increases, promotions and other rewards that team members value.
- **Coercive power**: It is the power from having the authority to punish others. An example might be law enforcement.

In 1965, Raven added a sixth type of power: Informational. This is power that results from controlling the information that others need to accomplish something. Personally, this still seems like a variation on expert power, but it's interesting to think about information as its own class. Knowledge is, indeed, power.

A lot of security professionals depend on expert power to accomplish their tasks. After all, security is hard, and you are the expert. Right? The problem with relying too much on expertise like this is that you can become too isolated to create a lasting impact. People can wind up deferring everything to you (since *you're* the expert) and leaving you to handle things rather than establishing more collaborative and productive relationships. This is particularly the case if your expertise is not strategic for your organization, which is usually the case with security. Some people might only see you as relevant to a specific part of the organization, like IT, rather than as an integral part of the business. You will literally become the "security person" with this approach and everyone assumes that when it comes to security you have it covered for everyone else and that keeping the company secure is squarely on your shoulders.

When you think about successful leaders, they typically lean more on their influencing skills than on raw power. You can classify this as referent power or leading by influence. Influential people shape a positive environment and motivate others to do great work without the use of overt power. High-level influencers often have a great deal of charisma and people want to work for them (refer back to the DiSC model earlier in this book).

Charismatic influencers tend to be confident and lightly assertive. For lack of a better term, they have executive presence. They are engaging and likable and have an air of authority about them even if they may not even be specifically in charge of what they're influencing. They somehow just seem trustworthy and worthy of respect.

So, if you weren't born naturally charismatic, how can you learn to build influence? Fortunately, there are ways.

- **Prove yourself**: Drawing on expert power, you want to be good at what you do in your organization. You want to be the expert, but that alone is not enough, you'll need to know how to get things done in your organization and establish a track record of accomplishments without steamrolling over other departments and colleagues. Again, beware of relying exclusively on expert power.
- **Respect the culture**: You'll want to live your company's values and respect the company culture. In other words, you'll want to "walk the talk."
- **Build trust and integrity**: Be a person of your word and tell the truth even when people may not like hearing the truth. Be impeccable with your word. If you make a promise, follow through. Build trust with your peers and with everyone around you. No one is going to be open to your influence if they don't trust you or think that you have integrity.

- **Be consistent and reliable**: Inconsistency is a great way to ruin your reputation. Great leaders are not unpredictable people, where you never know what their reaction is going to be to an event or circumstance. Consistency will help people realize they can depend on you to perform and get things done. Set consistent standards for your team as well, as they are ultimately a reflection of your leadership.
- **Be assertive, but don't be a jerk**: You will often need to push to get what you need in terms of staffing, funding and priority across multiple programs of work when you're working in cybersecurity. Be assertive, but not aggressive. Make your case and make sure you've been heard. Just don't go overboard and be too pushy.
- **Be flexible**: While this may be at odds with being consistent and assertive, you also don't want to be the stubborn one that always takes the "my way or the highway" approach. Life is about compromise and negotiations. Don't be the immovable object and strive to strike a balance.
- **Default to action**: People with influence get things done; they don't wait to be told what to do or just talk about what needs to be done. If you see a problem, take ownership of it and see it through to resolution. If you want to build a higher level of influence, take on problems outside the security group. Volunteer to be a mentor or tackle a common company problem as a side project. Your actions will speak louder than your words.
- **Be visible**: It doesn't matter how much you get done if no one knows about it. While I'm not personally big on bragging or attracting a lot of attention, the fact is that you are marketing the strength of your team and you as a leader every day. You'll also want to get in front of other senior leaders in the organization by attending some of the same meetings. See if you can get security on the agenda as a standing update. If no one knows you're doing great work, you might as well not be doing great work.
- **Stay relevant**: Make your contributions central to the organization's success. Figure out how your role contributes to the mission of the business and ultimately the bottom line. Be able to articulate this.
- **Listen**: Influence is a two-way process. Take in ideas and feedback from those around you. Listening to your peers, your team and other colleagues and then creating a partnership is a great path to building influence and gaining support. People want to be heard and understood and they will be more receptive to your own ideas if they think theirs are being validated.

Remember that you are trying to become more respected in the workplace and increase your influence. You are not looking to just boss people around and push them into working on your projects. Get things done, but get things done the right way.

I don't want to create the impression that expert power is not important. In fact, in the security field, it is critical to your success. People will quickly dismiss you if they realize you don't know what you're talking about or your facts are wrong. If you have expert power, your team is likely to follow you, and you'll find it easier to motivate them. But this is where expert power usually turns into referent power. Expert power is for doctors and lawyers; we defer completely to them because they have studied the subject more than we have and we will let them lead because our own expertise might be negligible on the subject. Referent power from your own team means that I may understand that you're an expert, but I might also be very knowledgeable. I want to help you anyway, because you've earned my respect. This is true leading by influence.

The security field is always changing. Remember that you will need to constantly develop these skills and knowledge to keep hold of expert power. There are many ways to increase not only your expertise but the *perception* of your expertise. Having expertise is not enough if no one is aware of it or if they think that you don't have the right background.

As someone who was educated in liberal arts, I found that I often had to really prove my technical expertise earlier in my career. Interviews were very pointed until I got to a much later point in my career where my accomplishments spoke for themselves. Here are some ways you can increase your expertise or perceived expertise (and thus your own expert power):

- **Attain certification or educational credentials**. I still have a few colleagues who argue that certification isn't important. I don't buy it. While there are definitely limitations to what a certification like Certified Information Systems Security Professional (CISSP) proves in terms of skills and knowledge, it has become the entry fee for most job applicants. Yes, even the CISO should have this credential in my opinion. I received my CISSP in 1999 and went on to obtain several other certifications. I was proving knowledge to myself as much as to everyone else. There are now many competing certifications you can consider and if you work better with a formal education background then there are now many college programs and master's degrees in cyber as well. Don't get the wrong idea on my stance with certifications though. Experience trumps certification every time and certification alone will never be enough. The two together are a very strong combination.
- **Stay current**. Almost as important as getting educated in your field is *staying* educated. Cybersecurity changes fast and you need to keep up with what's current. Having a firm grasp of up-to-date facts is essential if you want to build and maintain your expert power. Read relevant books, blogs, journals, etc. There is so much available online now. Carve out some time to keep on top of the latest news and happenings.
- **Share your expertise**. Be generous with your knowledge. Take advantage of mentoring opportunities, speaking opportunities or even publishing opportunities. Give back generously. The more you invest in helping the people around you, the more your own professional value will grow. You can also keep your own skills sharp this way. There's no better way to master something than by trying to teach it to others.
- **Keep your ego in check**. Unfortunately, cybersecurity has become a place where egos grow out of control quickly. There are now million-dollar CISOs, but the last time I checked they weren't exactly winning the cybersecurity war. Again, expert power can alienate and leave you alone on an island when you push it to the extreme. Don't rely on it as your only source of strength. Rampant egos also lead to blind spots, a close-minded attitude and a profound

lack of self-awareness. So, if you want to be a successful jerk because you're paid triple what anyone on your team makes, this might be the way to go. But if you want to be a better leader? Worry more about your whole team and your contribution more than about yourself, your ego and your paycheck.

- **Maintain your credibility**. Don't make careless comments about subjects you don't fully understand. It's a lot easier to lose credibility than it is to gain it. One ill-informed comment can put all your other comments into question. If you don't know the answer to something or don't fully understand a subject, resist the urge to weigh in with an opinion or defer to someone better positioned to answer.

# DISAGREEMENTS: DON'T BE THE DEPARTMENT OF NO

I don't think I need to tell you that if you work in cybersecurity, you are going to have many disagreements. People will challenge you that something really is as big a risk as you say it is. They will push back on work that is falling on their shoulders and some may just be outright hostile to the security function because it seems to just get in the way of progress. Scott Adam's Dilbert cartoons had a character named Mordac, "the *preventor* of information services." Trust me, you don't want to hold this kind of reputation. It's totally counterproductive.

I've found that many of these disagreements are just communication breakdowns. Make sure you understand the root cause of a problem before taking a view. And please, don't just ignore problems and differences. You didn't get into security to be a doormat, did you? You'll want to take problems on directly. But make sure you really understand the problem on the opposing side. I have neutralized many arguments simply by using active listening and making sure I heard the concern on the "opposing side" (a terrible way to think of things) of the discussion. Even in complex situations, there are usually elements that are negotiable and elements that are non-negotiable.

I was recently given a request from a business that wanted to put up their own Internet connection and go around our security infrastructure. Knowing full well that they didn't have good technical staff, this request got much scrutiny and made its way to my desk for review. People were rightfully concerned that this would represent a giant security exposure if it was implemented. Knowing that this business wasn't very technical, I knew there was something deeper to the request, which in this context didn't make a lot of sense. After really digging into the issue, it turned out that what they really wanted this new connection for was to act as a backup connection because our network had experienced a few storage area network (SAN) outages. While I could focus on the security risk of this new implementation request, which would indeed be high, the reality is that the secondary network connection wouldn't have addressed stability issues with the SAN at all. It would have been two paths to the same destination and the solution didn't solve the problem. There was no need to argue this part of the request and be confrontational. Suddenly the problem transformed from confrontational to explanatory. I could validate the concern using empathy, but also explain in simple terms that this wouldn't fix their problem. It was a very different conversation than it might have been without making sure I really understood the problem and instead focused only on trying to prevent their risky solution.

Make sure you don't put too much emphasis on a disagreement until you understand the request behind the request. And don't be too fast to stand in the way until you really know that you need to do so. In all honesty, if the business had insisted on setting up independent networking infrastructure without firewalls and an approved security stack, it would be worthy of standing in front of it and making sure that top management understood why I was declining the request. This implementation could have put everyone at risk by bridging trusted and untrusted networks, so in my view no single business can accept the risk of putting every other business at risk. Had they truly insisted on this implementation, I would have insisted that they remove themselves from our network. This would be a much more heated conversation, I'm sure. But it never had to happen, because I took the time to understand and ask questions like "wait, what are you trying to do …?"

A better way to build trust is to lead with empathy and to use your active listening skills. People want to be heard and to be understood. They want their issues and concerns validated. *Then* we can validate what is and isn't a reasonable solution for everyone. It's better to be the department of KNOW than to be known as the department of NO. If you become too difficult to work with, you can count on people finding ways to work around you instead.

This also applies with trying to force security requirements on other groups or individuals. It's human nature that people don't like being told what to do. It can feel like our freedom is being taken away. Continually enforcing that something is a compliance or security requirement and therefore *must* be done is not an effective way to work with other teams and gain their cooperation. Yes, security requirements are important and some things, like having a firewall, should be considered non-negotiable. But instead of using the brute force method to force your way on everyone else, consider simply asking for a commitment to help achieve security objectives that benefit everyone. Trying to force people to see or do it your way creates tension and can backfire. Lead by influence instead and help people *want* to support you by asking for their commitment.

# BUILDING A SECURITY CULTURE

We've already established that you don't want to change company culture by yourself. But you do want to gradually build a "security culture" in your organization over time. Your organization's culture is critical to a successful security program. While you can't change a culture overnight, you can influence it and gradually move it in the right direction. A good security culture is where people understand their part in the security program. A good security culture doesn't assume that the security group is taking care of everything. A good security culture is where people recognize potential issues and raise concerns openly.

A bad security culture focuses more on blame and punishment. I have heard of instances where employees are actually being fired for failing phishing simulations. While some might think that this is appropriate, this creates a fear-based culture. It's the kind of culture where people might not be open about reporting potential issues. Think of it this way, if you accidentally clicked on something at a company that was known for firing people who did this, are you going to tell anyone? People need to feel comfortable that they can share a mistake. In this phishing scenario, you are also blaming the victim for the crime. It's more productive to raise awareness and educate people to protect themselves and the company in a more positive manner.

There are three elements I like to consider for building a strong security culture (training and awareness). These elements are providing continual awareness, keeping the security team approachable and using positive reinforcement for the right behavior. These three elements help encourage open dialogue and a punitive-free environment where everyone can share problems and concerns openly.

There's a whole chapter on training and awareness in this book, since it's a critical communication topic. Building out a formal awareness program that does more than just annual courses and a few phishing simulations is important. No one is going to retain information that isn't periodically reinforced in a positive manner. A good awareness program starts with easy-to-read, easy-to-digest security policies and progresses through online training, lunch-and-learn sessions, newsletters and just about every communication medium out there to keep the message front-and-center in the organization. Use every opportunity as a teaching opportunity. If someone requests access to a website that's blocked by policy, educate them on why it's blocked and what the risks are of losing sensitive data or encountering malware.

Keeping the security team approachable is the next element in building a strong security culture. This means educating everyone on the team that they need to adopt a business-friendly mindset, use simple language with the business and not treat security like it's too complicated for everyone else to understand. It also means making sure that no one is making anyone else feel stupid for locking out their account for the 200th time or reporting a phish that's actually a legitimate update from HR.

Similar to keeping the team approachable is rewarding good behavior when you see it. This means that when someone does report that legit communication as a phish, you should still reward their suspicion, help them understand *why* the communication isn't really a phish and thank them for doing the right thing and reporting something suspicious in the first place. This enforces the right behavior and next time when the phish may be real, it will be reported.

Building a strong security culture takes time and the right mindset. Avoid punishing employees and use the three elements of a good security culture: providing continual awareness, keeping the security team approachable and using positive reinforcement for the right behavior as ways to gradually create an environment where people learn and embrace their role in securing the enterprise. Remember that this takes time. Using every opportunity as a learning opportunity will help you move faster and make continual progress.

# A WORD ON ETHICS

*I'm not upset that you lied to me, I'm upset that from now*
*on I can't believe you.*
~ Friedrich Nietzsche

As a security professional, you need to be always candid and truthful. The granddaddy security certification, Certified Information Systems Security Professional, is an independent information security certification granted by the International Information System Security Certification Consortium, also known as (ISC)². The (ISC)² felt that being truthful and ethical was important enough to mention it twice in their code of ethics:

- Tell the truth; make all stakeholders aware of your actions on a timely basis.
- Give prudent advice; avoid raising unnecessary alarm or giving unwarranted comfort. Take care to be truthful, objective, cautious, and within your competence.

Your word must be impeccable regardless of whether it is spoken or written. You are a trusted advisor to the business, and you need to be always factual and candid. If you have lost the trust of your business, you have lost everything and will not succeed in your role. While the truth doesn't always come out, I wouldn't bet my career on it staying quiet. Trust can take an eternity to establish and a moment to destroy. Think and act carefully and always act truthfully.

# SUMMARY

- Company culture can have a big impact on communication. Some firms depend on video more and some may be in-person more. You'll need to work with the culture to facilitate your own communication strategies.
- The OODA loop (observing, orienting, deciding and acting) can help you orient to a new culture quickly and help you stay oriented as things change.
- It is easier to adapt to a culture than try to change it. Resist the temptation to change culture without a clear mandate. Even then, be careful.
- Take it slow and be flexible. Adapting and influencing cultural issues takes time.
- Build influence in your organization to be more effective with your communication using the five sources of power: expert power, referent power, legitimate power, reward power and coercive power.
- Be ethical in everything you do. Once you lose your company or team's trust, it can be impossible to come back from this.

- Don't be the department of NO. It's better to be the department of KNOW than be known as the department of NO.
- Be impeccable with your word and in your actions. Trust can be destroyed in an instant by not being truthful.

# REFERENCES AND RECOMMENDED READING

Carnegie, Dale. *How to Win Friends and Influence People*. Diamond Pocket Books, 2019.
"Code of Ethics: Complaint Procedures: Committee Members." *Code of Ethics | Complaint Procedures | Committee Members*, www.isc2.org/ethics/.
Coram, Robert. *Boyd: The Fighter Pilot Who Changed the Art of War*. Back Bay Books/Little, Brown, 2004.
Hayes, Ryan. *The OODA Loop*. North Carolina Justice Academy, 2013.
Schein, Edgar H. *Corporate Culture Survival Guide*. Wiley & Sons Canada, Limited, John, 2019.

# Better Business Writing

**5**

*The first draft of anything is shit.*
~ Ernest Hemingway

---

## WHY GOOD WRITING MATTERS

---

Writing is thinking. To write well is to think clearly. That's why it's so hard.
*~ David McCullough*

I'd hazard a guess that not too many people went into the cybersecurity field to become a writer. Yet writing skills can be critical to your success and to the success of your program. During the lifecycle of a security program there might be requirements to write, read, edit or review:

1) Policies and standards
2) About a trillion emails
3) God knows how many corporate chats
4) Roles and responsibilities documents
5) Training and awareness material
6) Program strategy documents
7) Program charters
8) Security project portfolios
9) Incident reports
10) Contract language

A lot of people don't enjoy writing or the writing process. They don't try to improve their writing or grammar skills and they struggle through even simple emails that fail to make the proper points or get timely responses. Worse yet, bad grammar and incorrect punctuation can distract from your message and undermine your credibility. Words shape the impression you make on others. Poor writing can wind up with points that are never made, wrong interpretations or even creating the wrong impression about your education.

In the new remote world, clear writing takes on a new and more urgent importance. We are relying more than ever on the written word to communicate our thoughts, ideas

DOI: 10.1201/9781003100294-6

and strategies. The good news is that clear writing makes remote work more efficient and can be more inclusive for your colleagues. Good writing saves time and can create an efficient flow of information in the organization. This is especially true if a team is dispersed across multiple time zones. In global organizations, it is not always possible to schedule meetings that everyone can attend due to time zone limitations. Writing up a good meeting summary can be inclusive and save a lot of unnecessary follow up with members who couldn't be there. Best of all, good writing can eliminate the need for a meeting in the first place. Amen to that.

There is only one way to get more comfortable at writing, and it is to write. In fact, write a lot. Writing, like anything, takes practice. Write a little every day and you will see improvement and get more comfortable with the process over time. I get that your focus is security and that you may not want to be the next Stephen King or Tom Clancy, but learning to write clearly and succinctly will serve you well in any career path, but especially with cybersecurity.

Being a professional writer is a terrible job for most. It doesn't pay well. It's hard, and often thankless work. But the reality is that being a better business writer will pay dividends throughout your career. It is something that will only help you. Better business writing will save you time, help prevent communication and interpretation problems and will make your cyber knowledge shine. Writing is an important skill in our technology-dependent world and is even more crucial now that people are working remotely and the amount of face-to-face communication that we do has diminished. Our reliance on digital communication channels, such as Slack, also means that we need to learn how to write effectively across a wide range of communication channels.

Writing effectively can help you to produce better technical reports such as incident summaries, emails and memos that are concise and drive results and business presentations that don't leave everyone fast asleep. Communicating ideas through writing is an important skill. No matter if it is PowerPoint presentations, emails or blog posts, good writing translates to good business. Poor writing, on the other hand, can make you look unprofessional, ineffective and irrelevant in a world that communicates quite a bit using email. You don't want to have people ask you to clarify your message or become frustrated while they are trying to understand whatever point you're trying to make.

Business writing, meaning the kind of writing you use in an office setting, is different from other forms of writing, like fiction writing. Internal memos are usually more about bullet points, not long paragraphs and page-after-page of text. I'm sorry to remind you that most business documents will be skimmed through, read fast or deleted on the spot if someone feels they are not relevant to them personally. Fiction, on the other hand, tends to be something the reader chooses to read and isn't quite the chore that business reading can feel like. The reality is that if you're writing a memo, formal report or even just an email, people don't have the time or patience to spend a lot of time trying to figure out what you mean by your communication. We are in an information overload, so people will skim, skip and delete. Don't take it personally and I know that you do it too, every now and then.

This means that you really need to concentrate on strong subject lines, short paragraphs, like six sentences or so. And short paragraphs should consist of short sentences like no more than 15 words or so. Finally, it means that the first sentence in the first paragraph could determine how much of your document gets read. This is called a "topic sentence" and you want to always make sure to set the point in the first sentence in a paragraph and that it compels people to read the rest of the paragraph.

Don't worry, you don't need to start counting words or anything like that, this chapter will help you think about writing in a way that will make the process feel more natural. We will review recommendations to make improvements in your writing and your ability to communicate through this often difficult medium. If you remember back from the beginning of this book, there is not a strong feedback loop that comes from writing. While you can solicit feedback or reply to an email to provide feedback, it is not timely and, in some cases, it can create additional confusion since now the sender needs to also understand the response. There's a lot of room for error.

# THE WRITING PROCESS

Writing is a process. When I write anything of substance, I generally think of it in four phases (see Figure 5.1):

1. Plan
2. Write
3. Edit
4. Publish

Writing doesn't have to be a linear process either. For example, as I write this book at any given time, I am researching some chapters, writing others and then refining and editing others. Publishing, of course, is sort of standalone at the end when everything else is finished. If I am having trouble writing and getting the flow of something moving, sometimes I will shift into editing things I have already written or into basic research looking for new material to incorporate in my writing. But I'm always in a distinct mode and mindset for each phase. If you bounce around from one to the other without changing your mental focus from, for example, writing to editing, I'm willing to bet you won't have much work to show for it when you're done. You can put in a lot of time with writing without making a lot of progress. Or you can have a lot of progress in terms of words on the page, but they may not be very good and will need significant rewriting.

The Writing Process

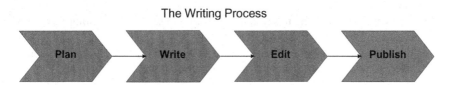

**FIGURE 5.1** Writing doesn't have to be a complicated process, but it helps to organize the process in a simple framework. Planning and editing may take up even more time than writing text.

Writing anything from a corporate memo, to a policy, to even simple emails will generally fall into this simple, four-step pattern. Consider for a moment writing a security policy or standard. You start with researching what is needed, what this policy is about and what points need to be made in it, including which security controls should be prescribed. Then you capture these ideas on paper, or more likely on a computer. Next up you need to read and edit what you have written and fix problems. And the odds are extremely good that your first draft will need some work. In the case of writing a policy, you had better get some significant feedback from stakeholders as well. This is all part of the editing process. We will talk about policy writing, specifically, in the second half of this book. Finally, you need to publish somewhere so that people know that the document exists in the first place. In the case of a security policy, publishing may include an elaborate communication plan as well as an actual project around the implementation of the policy if all of the security controls are not currently in place.

Even writing a simple email tends to fall into this four-step pattern. You think about what you want to say, you say it, you make sure it all makes sense and then you hit send. If you're skipping the edit step (making sure it all makes sense) and just hitting send, you're missing the most important step. As the excellent guidebook, *The Elements of Style*, by William Strunk and E.B. White, puts it:

> A sentence should contain no unnecessary words, a paragraph no unnecessary sentences, for the same reason that a drawing should have no unnecessary lines and a machine no unnecessary parts.

Your first draft of almost any communication isn't going to be perfect. Even professional writers don't get everything right on the first pass.

## Have a Plan, Have a Point

The first step to better business writing is to figure out your objective, not to just sit down and start writing. While it's tempting to get right to work, you really should take a moment and plan. In fact, you should probably spend about 40% of your time planning what you are trying to communicate!

Planning should help you answer several important questions:

1) What am I trying to say?
2) What point am I trying to make?
3) Who is my audience?
4) Given what I am trying to say and who I want to say it to, is writing really the right communication channel?
5) Do I have all the information to start writing, or do I need to gather some background information and do some research first?

Just asking these basic questions can help you get in the right frame of mind, even if writing does turn out to be the most appropriate channel for the task. Still, resist the urge to just start writing without planning. Think hard about your points and maybe jot them down as a sort of outline. In a long document, you may want to put a formal outline together before getting to the mechanics of writing.

Writing a book, one of the most difficult writing tasks, generally starts with a proposal of what the book is about and why it should be written at all. This forces you to think about questions like whether the book really has a strong market, since there may be a lot of other competing titles. It forces you to think if your book is different enough from the rest of the competition. It would then be followed with an outline of chapters or a full table of contents before you ever sit down and start writing the main text. This process really makes you think about what you're doing and why you're doing it before you start wasting effort with nose-to-the-grindstone writing.

Don't get me wrong, sometimes when you're over-procrastinating the right thing to do is to just get started on doing *something*. The blank page is not a writer's friend. Defaulting to taking action is always a good step, but not having at least some high-level concept of what you're doing and why you're doing it sets you up for writing something for which even you may not understand the purpose. We've all received one of those messages in our inbox, now haven't we?

# Writing

*Less mental clutter means more mental resources available for deep thinking ... to produce at your peak level you need to work for extended periods with full concentration on a single task free from distraction. Put another way, the type of work that optimizes your performance is deep work.*
~ Cal Newport, Deep Work: Rules for Focused Success in a Distracted World

Once you're through the planning phase, it's time to start writing. Again, I tend to work in different frames of mind, so during the writing phase it is best to use the input from your planning phase and just start writing. You're in a bit more of a creative zone now, so don't worry too much about fixing and editing; that's for the next step. You want to just start getting as much material on the page as is necessary to make your points. Write without judgment and let it flow. Capture ideas, not sentences. You can fix all that later.

Writing anything of substance takes some concentration. I recommend putting yourself in a distraction-free zone. Turn off cell phones and maybe put on some music that doesn't have lyrics, which can be very distracting when you're trying to write something important. I keep a good pair of noise-canceling headphones on hand to really go deep into the zone. If you want a good way to just put some atmospheric music without lyrics, check out the Endel app (http://endel.io). This app uses AI to generate soundscapes to help relax and focus. It even has an Amazon Alexa skill, so you can play it straight through an Echo device if headphones aren't your thing. Otherwise, you might try one of the white noise soundtracks from Spotify or Apple Music. The main point is to drown out distractions and focus on the task at hand.

Here's a power user tip: put your phone on airplane mode. It's a quick way to silence all distractions quickly and then turn them back on just as quickly. Good news for Mac users, there's also a "do not disturb" mode for your computer. Failing that, at least exit email or turn off the notification feature while you're doing "deep work" like writing.

# Editing

*I didn't have time to write a short letter, so I wrote a long one instead.*
*~ Mark Twain*

*Examine every word you put on paper. You'll find a surprising number*
*that don't serve any purpose.*
~ William Zinsser, On Writing Well: The Classic Guide to Writing Nonfiction

In my life before security and when I was a much younger man, I spent some time working in a large publishing house as a professional editor. There are tons of tips on refining and editing your writing, and I won't try to capture them all here. My objective isn't to teach you to be an editor, but to help you to understand when your writing might still need some work and how to approach fixing it. I will share some tips that will give you just enough information to refine and edit your own writing, covering everything from a short email to a lengthy book. I hope to give you just enough tools and advice to get you started, as well as some resources at the end of the chapter if you would like to dig deeper into the subject.

It's also worth noting that proofreading and editing are two separate things, or at least two separate ways of looking at a written document. A proofreader looks for misspellings, wrong punctuation and other inconsistencies. Editors correct issues at the core of writing like sentence construction and language clarity. A thorough editing and a good proofreading improve the readability, clarity and tone of the text. And yes, it's totally possible to edit *and* proofread your own documents. But I would do this in two different passes, because it's really two different perspectives and two different frames of mind.

The need for multiple passes means you will need to spend a good amount of time in the editing phase. Some people are surprised when they realize that most of your time isn't spent writing, it's spent in the planning and editing phase. It makes sense when you think about it though; even fiction writers need to think about what they're going to say and how they are going to develop characters. Typing words on the page takes a fraction of time compared to all the planning and thinking it takes to figure out *what* words will go on the page.

My first tip for people who don't aspire to do editing professionally is to simply read your work out loud, even if it's just a simple email. Did you write at least somewhat similar to the way that you speak? Did you avoid flowery language and technical jargon? Ask yourself these simple questions:

1. Do I use the bottom line up front (BLUF) technique and get right to the point? Always lead with your main message and give just enough detail to make your point. Edit the rest mercilessly.
2. In anything longer than a short email, make sure your topic sentence—the first line of each paragraph—gives the reader a sense of what's coming in the

rest of the paragraph. Make sure everything in a paragraph is all related or you probably need to start a new topic in a new paragraph. Write your topic sentence with the assumption that it's the only sentence someone might read in the whole paragraph. You'll probably be right.

3. Use active, not passive voice. *A plan is being defined by the team* is not as effective as *the team is defining a plan*. More on active voice later in this chapter.

4. Does your document flow? Are your ideas organized in a logical order? You don't want a lot of random thoughts jumbled together. One sentence should flow to the next and one paragraph should flow to the next. If you did a quick outline in the planning phase, this should not be an issue.

5. Do you know how many editing passes it takes on average to produce a quality piece of writing from a moderate size email to a full-blown book? Six passes. How many are you doing? That stated, you can keep doing editing passes forever. When you find yourself changing things back to the way they were when you started, it's probably time to stop.

Good writing is a continuous improvement exercise. Writing, even simple writing like email, can always improve by some deliberate thinking and some critical editing. Read your own words. Even read them out loud. Do they make sense to you? Could they be made more succinct or clearer? If so, then you're not done yet. Take the extra time to edit, revise and simplify and the reader (even the reader of a simple email) will thank you for being clear and getting to the point. Remember our audience-first principle: write for the audience and make it easy for them.

## Publishing

Now for the easy part. Click send on that email. Send your manuscript to the publisher. Publish your policy on your company's website. If you're doing more than sending an email, there might be a few more steps here, of course. For example, if you're publishing a security policy there may be an entire awareness campaign that needs to happen next. If you are publishing a "living document," like a policy, you'll also have to revise it one day as they do grow stale. At that point, the whole cycle starts all over again.

# GET BETTER AT WRITING THE EASY WAY

If you could spend just five minutes a day centering yourself, planning your day and setting yourself up for success, would you do it? What if that same exercise improved your mindset and helped you become a better writer at the same time?

While I tend to write a lot in both my professional and personal life, the one thing that keeps me on point every day is keeping a journal. It's a critical part of my morning routine that I consider non-negotiable even on weekends. While some people insist that hand-written journals are the way to go, I personally use the Day One journaling app on IOS and the Mac (https://dayoneapp.com/). This allows me to practice writing every day in the same way that I will do most of my other writing: at a keyboard or computer. Journaling allows you to practice your writing voice without sharing it with anyone. If you're a Windows user, you will need to find another app. I hear Diarium (https://timopartl.com) is good, but I have not personally tried it. You could just as easily use notepad though; the app is not as important as the process.

I won't go into a lot of detail about the benefits of keeping a journal, but they are numerous. A few nominal benefits include everything from reducing stress, helping you think through difficult issues and plan for the future, boosting your mood or just keeping track of the important elements of the day. Using an app like Day One, I can see where I was on vacation two years ago and look up important facts or events and how I felt about them at the time. Journaling is the ultimate tool for self-awareness, which is also a key component to building stronger emotional intelligence (EQ).

Here are six ways you can think about journaling:

1) **For reflection**. This is one of my favorite reasons to journal. Generally, you're looking at what happened, why it mattered and how you might do it differently in the future. This is a great method for continuous improvement. It will help you make better decisions in the future and forces you to really deal with some situations that maybe should have been handled differently. You can also capture what went right and how you can keep that up in the future.

2) **For emotional release**. Didn't get that big pay raise? Went through an incident that wasn't very pleasant? Journaling can help you confront and release some of the emotions that may have built up from the event.

3) **For goal setting**. Having written goals puts you in the top 3% of the population, according to some studies. Re-write them and manage them daily and you'll be in the top 1%. Be as specific as possible and put some dates and metrics to them. You'll be less likely to abandon your goals if the steps are clearly written out for you. These don't all have to be business goals either. We all have personal and private goals, bucket lists, etc. Writing them down can be the first step in making them happen.

4) **For ideas**. While this one works best for professional writers, it can also work for the stream-of-conscious idea generation. Brainstorm every place you want to visit before you die and write it down in the journal. Great, now you have a giant list of potential future vacation ideas. Same goes for bucket lists, books to read or movies to watch on Netflix.

5) **For memorization or deeper understanding**. Do you want to know the best way to learn anything? Teach someone else. Journaling can help you explain how a process works, step by step. Writing it down will help you learn the subject deeper. On the other hand, just writing down what happened on a particular day can serve as a future point of reference. You don't remember when you had your last physical? A searchable journal like Day One can not

only find this for you (presuming you wrote it down in the first place) but it can also show you what you were doing and thinking a year, two years or more ago on this day.

6) **Journaling for gratitude**. Not everyone will resonate with this one, as it may come off as a little bit new age. But not everyone appreciates what they already have in life because they are so focused on the next thing. I make it a point to find 3–5 things every day to be grateful for. It could be as simple as that cup of coffee in your hand or just having gratitude for being alive on planet Earth at this time in history. It also helps set the tone for the day since I do this first thing in the morning. Again, everything starts with your mindset and starting with a positive mindset sets you up for success for the rest of the day.

Some people have trouble getting started with journaling. The sight of a blank page just doesn't help getting the ideas of the day flowing. This is why it helps to either have templates with some standard questions or use journal prompts. Here are a few examples of good journal prompts you can use to get the creative juices flowing.

1) What kind of day are you having, and why?
2) What are three things you're grateful for today?
3) What do you like to do? How does it make you feel?
4) What's something you're good at? What makes you good at it?
5) What is something I could do to make today great that is actually under my control?
6) What would you change about yourself or your life? Is there a way for you to change it?
7) What keeps you up at night worrying? Are your worries realistic? Is there anything you can do about them?
8) What's the most important thing I need to do or remember today?
9) Do you have a philosophy of life? If so, what is it? If not, what is your method for making important decisions?
10) What are your most prized possessions?
11) What is something someone else has that you envy? Describe it and your feelings about it.
12) What is a book, movie, song or television program that has influenced you, and how?
13) What is a mistake people often make about you?
14) What's your favorite: season, color, place or food? Describe it.

By journaling you can get better at writing while also going deep into self-improvement territory. Talk about a win–win process. Incidentally, there's a simplified version of this called "bullet journaling," which focuses more on capturing bullet items for the day. I don't care for this as much because it's better suited for fractions of sentences and simple idea capture. This method is still useful, but probably won't do much to help you become a better writer.

# GENERAL WRITING TIPS

*Words are the source of misunderstandings.*
~ Antoine de Saint-Exupéry, The Little Prince

One of the problems with written communication is that things may not be perceived the way you intend. Written words do not convey emotion easily and can be open to misinterpretation. Remember, writing does not have a strong feedback loop. English is also a deceptively difficult language because words can have implied meanings that are sometimes open to multiple interpretations.

With any written communication, you should do some high-level spot checks:

- Review your style. Is it business-friendly?
- Avoid jargon, slang and acronyms.
- Check your grammar and punctuation.
- Check also for tone. If you think the message could be misunderstood, it probably will be misunderstood.
- Consider any cultural context. If there's potential for miscommunication or misunderstanding due to cultural or language barriers, address these issues in advance.
- Think about if the use of pictures, charts and diagrams would help. There's nothing like a visual to help people quickly understand your message.

Writing regularly will help you train yourself to communicate more effectively and hopefully more concisely. It doesn't even matter if you plan on publishing your writing or if it's simply a day-to-day journal entry; the act of writing regularly will help you become a better communicator.

# A HANDFUL OF SIMPLE GRAMMAR RULES

*The difference between the right word and the almost right word is the difference between lightning and a lightning bug.*
~ Mark Twain

Don't worry, this isn't a grammar book, and I won't spend time on dangling participles and compound adjectives. I will, however, share a few pointers that I think will enhance your writing. If you are interested in learning more about grammar in a way that won't put you to sleep, I recommend grabbing a copy of *The Elements of Style* by William Strunk and E. B. White. If you want even less reading and a fun presentation on the go, try the *Grammar Girl Quick and Dirty Tips for Better Writing* podcast. I will include some other resources at the end of this chapter.

# The Only Grammar Tip You Really Need

The most important writing tip I can share is that if something doesn't sound right to you, then there's a problem. If it doesn't sound right to you, it probably won't work for your audience either. Read your text out loud, does it sound OK? If not, there is something wrong with one or more of your sentences. Try to pinpoint exactly which word or phrase is off if it doesn't feel right and fix it. Good, or at least better, writing can be as simple and easy as this. Learning grammar rules and what the problem is called can come later.

Now that we have this easy tip out of the way, here are some helpful grammar and style tips that can improve your writing right away. Again, this is not a grammar book and it is just scratching the surface of a big topic. I'm really aiming to provide you with a "least you need to know" approach so you can get back to the business of running the security program.

# Watch Your Tone

Without a strong feedback loop, writing can't let you gauge body language and other cues that your message is being misinterpreted. You might get no feedback at all. This means it's critical to spend more time up front to make sure that your message is clear. Part of making sure your message is clear is minding your tone. Consider the following example of a very short sentence asking for a report.

> Get me the report first thing tomorrow morning.

If this simple email came from your boss, these eight simple words might convey a bit of terseness or impatience. In fact, many people might start asking themselves questions. "Am I late with this?", "Is my boss mad at me?", "Did I forget to do something?". Perhaps these questions are all valid, but it's also possible that the person who wrote this was simply in a hurry and didn't think about how the message might be received. Consider a lighter example, basically saying the same thing:

> If you can get me the report first thing in the morning, I'll have it ready to go for my 10 AM meeting. Thanks!

Notice, I kept both messages deliberately short. No one wants to hear a big story about why I need the report. But on the other hand, the tiny bit of extra context gives the recipient more information on why the report is needed by a certain time. Finally, the addition of "Thanks!" conveys a much more pleasant tone compared to the first draft which sounds like an angry boss giving orders to lazy or forgetful employees.

**Action**: Are you leaving your emails or other writing "open to interpretation"? Do you read your writing before sending it to the recipients? Think about how your audience might receive your message and what problems they might have understanding what you're saying.

# Use the Active Voice

Which one of these sentences sounds better to you?

**Passive voice**: The files were stolen by the attacker.
**Active voice**: The attacker stole the files.

The active voice not only makes your writing seem more dynamic but also makes it clear. Consider this active/passive example:

**Active voice**: The dog chased the squirrel.
**Passive voice**: The squirrel was chased by the dog.

In the second example, the reader is forced to think for a minute: "wait, who's chasing who?" The active voice example also conjures up images of the dog running after the squirrel, whereas the passive voice doesn't conjure a strong mental image because you needed to think more about it.

Without getting too deep into mechanics, active voice is constructed as: subject, verb and object. The passive voice is generally constructed as: object, verb and subject. You'd be surprised how much passive voice creeps into people's writing. It doesn't do anybody any favors. It's not technically incorrect grammar, for the record. But using the active voice is just a clearer way of writing. In fact, the passive voice example above might make more sense in a story where the squirrel was the main subject, not the dog.

The squirrel was on his way home when he ran into the Rex, the neighbor's dog. The next thing you know, the squirrel was chased by the dog.

Using the active voice conveys a strong, clear tone and the passive voice is subtler and weaker. Some people seem to think that passive voice may sound fancier or more elaborate. If you don't learn anything from this book, you will learn that clear communication counts way more than being fancy or elaborate. There are no bonus points for flowery prose.

For more on this subject, see https://www.grammarly.com/blog/active-vs-passive-voice/.

# Avoid Run-On Sentences

Have you ever read a really, really long sentence that seems to meander from topic to topic never really making a point but going on anyway in a way that seems to lose more and more clarity with each subsequent word? This was a not-so-subtle example of a run-on sentence. Keep your sentences short. There's a reason for keeping them short. Because short works. Short sentences are easily understood and remembered. Of course, you don't want to keep all of your sentences 3–4 words long, as that can also get tedious. Try to strike a balance.

The fix for run-on sentences is simple. Add more periods and break up thoughts into different sentences. Consider the first sentence in this paragraph rewritten and edited for clarity:

Have you ever read a long sentence that meanders from topic-to-topic without making a point? Long sentences lose clarity with every subsequent word.

Hopefully, you found this version a little easier to read and understand.

# Proper Punctuation

I see this mistake quite a bit. If you were in a group of CISOs, there wouldn't be an apostrophe among you. However, you might find one if you're on the CISO's team. Note the use of the apostrophe in the last two sentences. I've seen a lot of really smart people make mistakes like this even on presentations being discussed with the board of directors. The apostrophe in "CISO's" shows a possessive form of the word, meaning something that belongs to the CISO. Without the apostrophe, the word is plural, meaning that there is more than one CISO.

Before you think that punctuation like this doesn't matter and most people wouldn't care, think of the potential miscommunication with the following:

- **Call me Jeff**. This seems straightforward. My name is Jeff and that's what you can call me.
- **Call me, Jeff**. Now someone is asking me to call them?

See how a single comma or one misplaced word can completely change the meaning of the sentence? In a real-world example, a magazine once published a blurb on the cover that featured chef Rachel Ray who "finds inspiration in cooking her family and her dog." Yikes. Let's try that again with proper punctuation this time: she "finds inspiration in cooking, her family, and her dog."

Punctuation matters.

# Avoid Technical Jargon and Be Careful with Acronyms

As I mentioned several times already, avoid technical jargon and acronyms that aren't explained. There are countless opportunities to use both in security. Strive to use neither. I get that you are the expert and it's tempting to show off how smart you are on your particular subject. But unless you're writing for other security professionals, you can't assume that everyone knows the difference between cross-site scripting and cross-site request forgery. You are going to need to simplify and keep things clear for everyone.

Most companies have their own set of acronyms and can't or won't eliminate them. I've even worked at companies that had acronyms that conflicted with other acronyms and you needed to know the context to know which interpretation was right. You can still use acronyms—just make sure that in your first use that you expand the meaning. For example, in the first reference you would say: "Certified Information Systems Security Professional (CISSP)" and then every subsequent reference in the same document or email can simply say "CISSP."

# Use Grammar Checkers, but Don't Depend on Them

If your grammar and punctuation skills could use some refining, you can look at tools like Grammarly (http://grammarly.com), which can help you revise your writing for both clarity and correctness. It's probably worth the price if you're making a lot of basic mistakes. While word processors like Microsoft Word do have some basic grammar checking abilities, a tool like Grammarly might be worth the subscription price if you're looking to make more tangible improvements in your writing. You might give it a shot and see if it works for you. Unlike the built-in grammar checkers, Grammarly can also give you readability scores, check for plagiarism and offer suggestions on tone. For better or worse, bad grammar will create an impression of a lack of education and can undermine an otherwise good communication, be it written or spoken.

If you're going with the Microsoft Word built-in grammar checker, make sure it is turned on (it is by default). Just don't trust every change it suggests. English is a complex language, and a machine doesn't fully understand context, so it can definitely steer you in the wrong direction. If you're not sure about a suggestion a tool makes, reject it or look it up and confirm it.

If you do use tools like spell check and grammar check, use them to make yourself better at both topics. A lot of people become dependent on these tools and then when they need to write freehand, it's atrocious. The best way to leverage these tools is to let them flag what they are going to flag and then figure out the correction yourself instead of accepting the suggested revisions. This is how you'll go from always spelling "acceptible" wrong to the correct spelling, "acceptable." Tools are great, but there is no substitute for the human mind. Learn and grow from your mistakes, don't depend on machines to come in and clean up after sloppy work. I personally turn off autocorrect as well, as it drives me crazy when it gets something wrong repeatedly.

Obviously, when you are simply texting or sending emails you don't always need a lot of writing, rewriting and editing. Shooting a text message to your friend on meeting up for lunch doesn't deserve the same attention as your next board update or incident report. Still, when you get in the habit of good writing, you will likely not tolerate your own poor writing even in trivial texts or instant messages.

---

# MAKING PEACE WITH EMAIL

---

> *If you keep interrupting your evening to check and respond to e-mail, or put aside a few hours after dinner to catch up on an approaching deadline, you're robbing your directed attention centers of the uninterrupted rest they need for restoration. Even if these work dashes consume only a small amount of time, they prevent you from reaching the levels of deeper relaxation in which attention restoration can occur.*
> ~ Cal Newport, Deep Work: Rules for
> Focused Success in a Distracted World

It's worth going a little deeper into the most notorious written communication tool of all: Email. Like it or not, you are probably going to spend an awful lot of your time writing, reading and responding to emails. Everyone complains about email. Some tech companies are even trying to get rid of it entirely by replacing it with Slack or Microsoft Teams. Unfortunately, email is still the most reliable way to contact most people. Almost everyone has an email address, but these other platforms can't be counted on for universal access to people. Therefore, it's worth spending some time to think about how to make this more efficient and effective. Poorly composed emails result in confusion, miscommunication and even large delays in response or a lack of any response at all. On the recipient end, you can become buried in the deluge and get very little else done all day but email.

While it's a bit of a stereotype, a lot of technologists really are socially awkward and introverted, myself included. Defaulting to writing email or text messages is a favorite for many of us. It is also a way to extend the workday and deal better with global time zones. A well-placed email late at the end of the day Eastern time will arrive first thing in Asia. Email helps move projects and drive business. It is unquestionably a convenient medium to choose for communications. It has many great aspects, and many drawbacks.

Email is sort of an asynchronous communication method. Synchronous communication occurs when messages are exchanged in real time. It requires the people involved to be present at the same time and space. Examples of synchronous communication are SMS, video meetings and chat. Asynchronous communication, like email, happens when messages can be exchanged independently of time. In other words, people can respond to email at their convenience. The benefit of email is that it does offer a communication mechanism that doesn't demand an immediate response.

In his outstanding book, *Deep Work: Rules for Focused Success in a Distracted World*, author Cal Newport introduces the concept of "deep work." Deep work is defined as:

> Professional activity performed in a state of distraction-free concentration that push your cognitive capabilities to their limit. These efforts create new value, improve your skill, and are hard to replicate.

If we didn't have email, it might be much harder to get deep work done, as you would constantly be interrupted by synchronous communications like chat or phone calls that demand a response.

Some people treat email as a synchronous communication mechanism, literally watching their inbox all day and responding to email in near real time. Others spend hours at the end of the day trying to catch up with all the emails that have piled up over the day. Master email and you can regain hours and spend your time focusing on deep work and value-add activities that will drive the security program forward. Do email poorly, and you will create confusion, waste time on frivolous activities, and get nothing important done.

So how can you do this, and why is writing an email different than writing anything else? Email is intended to elicit a response from the reader. Emails inform, instruct, educate or help collaborate. According to Forbes, office workers receive an average of 200 messages a day and spend about two-and-a-half hours reading and replying to emails. For many people I know, the numbers are actually much worse.

The email you send, like it or not, will be surrounded in some cases by hundreds of other emails vying for the recipient's attention. Unfortunately, email is a necessary part of life in most corporations, and it is not likely to go away anytime soon.

One of the particularly tricky challenges of email is the feedback loop we discussed earlier in this book. Email tends to be either a delayed feedback loop or potentially a NO feedback loop, making it a one-way communication channel. Without feedback, you can't tell if your message has been read, understood and actioned. If you do get feedback, it could be the next day or much later for those of us with jam-packed inboxes. In short, one of our most used communication mechanisms is also one of our least effective in terms of feedback channels.

Here are a few rules to help you tame the email beast.

1. **My number one email tip is, don't use email**. Sure, I get that email is a necessary evil in today's business environment. You can't *literally* stop using email in most cases. But you also don't have to be part of the problem. Start by asking yourself if a quick phone call, text/instant message, or meeting or dropping by in person might not be more effective and much faster. New research shows that verbal communication helps us feel more connected. Don't forget that email doesn't easily convey emotion and is extremely prone to interpretation errors. Email can also be forwarded on to any number of people. Don't put anything in writing that you wouldn't want to work its way to the boss's inbox. Many firms archive email for seven years or more and consider it part of the firm's books and records. This means they are discoverable in the event of litigation. Think carefully about what you put in an email.

2. **Turn off email notifications while you're working**. Email is a productivity destroyer. Some executives get hundreds of emails a day and each one is typically asking you to do something that is likely more their priority than your priority. I see stressed out executives reading and replying to an email and then they switch gears as soon as a message bubble pops up from someone else. The cycle continues all day. If you can get away with it, set aside time 2–3 times a day to check email and spend the rest of the day getting actual work done. Some people might argue that this only increases the time for any feedback loop in the form of replies. Sure, this is true. But would you rather have a fast but unthoughtful response or a well-thought-out response that was crafted without a lot of distraction? Again, if it's an emergency pick up the phone. Otherwise, it can probably wait.

3. **Have a clear purpose**. Don't send an email without having a clearly defined purpose. As basic as this sounds, I receive many emails every day that seem to fail in this basic objective. I scratch my head asking basic questions like whether the sender wants something from me, is just informing me of something or if it was really intended for me at all. If you are sending emails without a clear purpose, you are part of the problem. If you remember earlier in the chapter, don't just start by writing. Start by planning!

4. **Think like a journalist**. Think like a journalist and write a good headline in the form of the subject line. The subject line of an email offers an

opportunity that many people miss. A compelling email subject can be the difference between having your email opened today or never. Think of tagging your subject line with things like "Time Sensitive," "Question," "Action needed" or "Please reply by XXX." For informational emails, use "FYI" or something similar. Better yet, you could attempt to tell recipients as much as possible in as few words as you can manage. Try communicating the whole email in one sentence, just like the headline in a newspaper.

5. **Check your tone**. In fact, check everything. Again, email does not easily convey emotion. It's a busy world, so people consistently misread emails because they read them too fast. Your choice of words or even simple brevity can easily be misinterpreted in the absence of other visual and auditory cues. Read your message proactively before sending it for any potential interpretation issues.

6. **Send your message at the right time**. Don't send important messages last thing on a Friday afternoon just because that's when you finally got around to finishing it. These messages will likely work their way down everyone's inbox as the weekend goes by and new messages start to arrive. Also factor in regions. If you're sending an email from New York to employees in Singapore, it turns out that 8 PM EST is the best time to send it. This is 8 AM in Singapore and you have a 16% higher chance of getting a response. There's a great tool for these kinds of calculations at https://www.yesware.com/best-time-to-send

7. **Deliver ONE key message in each email**. Keep it simple and show some empathy for the recipient. Try to keep to a single point in your email and, you guessed it, bottom line up front. Shorter emails are better. Shorter sentences are better. Keep it simple. Make your point and go back and edit them down to the essentials. I've personally passed on emails that would print out in multiple pages with the intention to read it "later." Guess when later comes? Never. Of course, you can't make it so simple that there's not enough context to understand the rest of the message, so strive to just keep things short, simple and clear.

8. **Limit attachments**. Just like messages that are too long, emails that come with seven attachments just scream to me to be deferred to "later." At a minimum, try to at least catalog what each attachment is in the body of the email as a bulleted or numbered list. Yes, you can also zip the attachments into a single file. But getting seven attachments is still getting seven attachments. Think hard about if this is the best way to accomplish your objective. And when you are including an attachment, remember to actually attach it in the first place.

9. **Don't be part of the email problem**. If someone ends their email to you with a "thanks" you really don't need to send "you're welcome" back. All of us have too many emails to read. Bad emails result in more bad emails. Let long threads die. How about even adding "no need to reply" to emails where you really don't need a reply? You'd be amazed how this is a relief for both you and the recipient. In either case, every email you send should try to get the subject to the finish line in as few replies as possible. Looking to schedule

a meeting? If you can see free/busy time you shouldn't be sending an email, then it should just be a meeting invitation and skip on the email altogether.

10. **Filter your mail**. Take advantage of email rules; use unsubscribe buttons and aggressive SPAM filtering to try to reduce the number of emails you get in the first place. I created a rule that looks for the word "unsubscribe" in it to move these to their own folder. Bingo, seemingly hundreds of vendor emails and spam are now in their own folder away from my important email. Other words that seem to work well in the security world include "webcast" and "summit."

11. **Skip on the emojis**. Emojis work for friends but not so much for formal emails in a business setting. This stated, I'm OK with them in corporate chats like Slack or Teams. Emojis are a poor-man's feedback loop, they can at least indicate when something is just a joke or not. Just don't put emojis in your next board presentation or your email to the CEO, OK?

12. **Mass emails are almost always a bad idea**. If you're sending an email to dozens of recipients, you are probably not using the right medium. Mass emails are the epitome of not understanding or even caring who your audience are, if they've understood your message or if it was even read at all. If you're sending long mass emails you can almost guarantee most won't be read. Add three attachments to that mass email and see how effective your response rate is. One notable exception is security awareness training, since this really is intended to go to every recipient. But I also guarantee that at least some of those people hit the delete button before reading more than five words.

13. **Keep your emails professional**. We used to have a saying at General Electric: "Digital is forever." Don't forget that anything you send in email can be forwarded, stored, discovered during litigation, etc. Don't put anything in an email that you wouldn't say to someone's face. Remember that the email format doesn't convey emotion very well. This means that messages are subject to interpretation. Keep it professional and for God's sake pay attention to the difference between "reply" and "reply to all."

14. **Blind Carbon Copy (BCC)**. BCC is your friend. Sending an email to a lot of people from different companies? Using BCC will keep addresses private. Sending an email to a wide distribution using BCC also avoids the "reply to all" problem (see the previous item). By the way, the term "carbon copy" derives from carbon paper, which comes from the typewriter era. Since there were no copy machines, you could type something once and have it reproduced on a second sheet of paper with a piece of carbon paper between them. When the typewriter keys hit the first piece of paper, it would make a carbon imprint on the second sheet. We've come a long way.

15. **Use stock answers for unimportant messages**. I save a lot of time using a product called Text Expander (https://textexpander.com). This application, and others like it, can help you craft stock replies or signatures with a few keystrokes. If you've ever felt like you are writing the same emails repeatedly, then this can be a real time-saver. Also note that the Mac operating system and IOS both have a basic version of this built right in called text replacement under the settings section for the keyboard. If these don't work for you, how

about just using some template answers that can be cut and paste. Just read the response before hitting send, no one likes to receive a mail starting out with "Dear, First Name."

16. **Keep your emails short**. And I don't mean "shorter" emails, I mean short emails of a sentence or two whenever possible. Do you enjoy receiving long emails? I thought not, so don't send them either. There are very few situations that call for long emails and there are plenty of benefits to writing short emails. I know what I do when I get a long email with a lot of detail: I put it in a folder to read "later." And that's exactly what most people are probably going to do with your long email. Try to make the maximum size of your emails two paragraphs. Anything more, and you are probably using the wrong communication channel. Set up a meeting or pick up the phone.

17. **Don't just forward messages**. If you're forwarding messages to additional recipients, try to add a little more value than just "FYI." I personally hate receiving a long email thread that I now have to read in reverse to figure out why I would care about this FYI message that was just sent. Why not provide a short summary of the thread along with any expectations like any actions to take.

Writing short emails is a skill. However, if you keep your emails concise, people will appreciate it and you will also spend less time *writing* emails. We have less time than ever. We read emails on mobile phones, Apple Watches (OK, maybe I'm the only one who does that) and other tiny screens. No one wants to plow through your lengthy email to try and figure out what point you're making or what action needs to be taken. Before you send your next email, imagine your recipient reading your email on their smartphone on the subway. Did you get your point across in a few lines? Or do you imagine them squinting while they scroll down, line after line, trying to figure out what this message is all about? Or do they simply move on to someone who had the courtesy of writing them a shorter email.

An added benefit of this is you could save hours a day by not adding unnecessary details and typing out long emails. Strive for a few sentences at most. There are very few things you can't get across with a good subject line and a few lines of supporting text.

Maybe this isn't working for you? How about gamifying it? How could you say the same thing in fewer words? What's the absolute smallest number of words you could use? How could you cut five more words out from there? Consider the following request:

*Hi,*

Please see the attached report recapping the incident that happened recently at Division B in our company. I am interested in your feedback and any changes/additions you may have. I would like to get these changes in by the end of the week, so please try to have any feedback to me before then. This was a tough report to put together, so please let me know if it all makes sense to you. Thanks in advance!

Now, that's not horrible, but consider how this could have been a little more "reader friendly" and a lot more to the point.

Please review the attached and send me your feedback by Friday. Thanks!

Now some of you might want to go to extremes and cut it even further.

Review and comment by Friday.

The problem with that last one is that it comes off sounding too directive and cold. Also, review what? Oh, the attachment. Finally, I also think the addition of "thanks" is important. You are asking someone to do something and even if this person is on your team, you are requesting their feedback and should end on a positive note by thanking them for their time.

## Managing Your Inbox

I'm a fan of Inbox Zero, but I confess I don't always get there. In the Inbox Zero system, you work your inbox down to zero emails by the end of each day. Yes, zero. Inbox Zero is a rigorous approach to email management aimed at keeping the inbox empty or near-empty. It was developed by productivity expert Merlin Mann and later adapted by productivity guru, David Allen, of *Getting Things Done* (GTD) fame.

To borrow what I found to be the most useful part of the *Getting Things Done* methodology, ask yourself if an email is actionable or not. If it's not, you probably want to file it away for future reference or delete it. If it is actionable, do it now if it takes two minutes or less, delegate it to someone more appropriate, or defer it for later if it's going to take some time to get through. Don't forget to come back to it though—this is where I personally fail with this system. A big bunch of lengthy emails that require a big bunch of time is not really how I like to spend my day. This is known as the four Ds of email management: do it, delete it, delegate it or defer it (Figure 5.3).

In a vastly simplified version of GTD that I've adjusted over the years, I have three email folders: @action, the built-in default archive folder and a follow-up folder for me to keep track of something that can't be actioned right now (see Figure 5.2).

Simplified Getting Things Done (GTD) inbox

1  Your Inbox: things that need to be processed

2  @ Action: things that will take more than 2 minutes

3  @ Follow Up: things that need follow up

4  Archive: your default archive for messages

**FIGURE 5.2** Adopting an email workflow can help you speed through your inbox much faster. Workflows like Getting Things Done (GTD) have become very popular to strive towards "inbox zero," where the aim is to get to zero messages in your inbox every day.

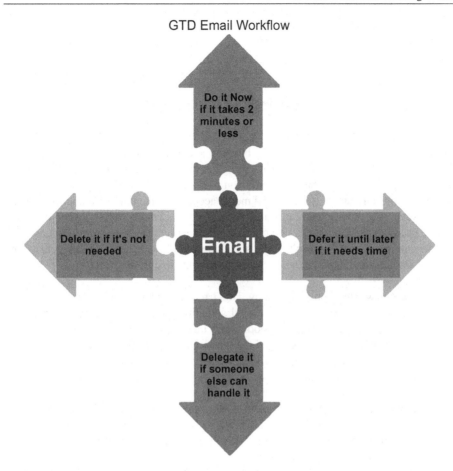

GTD Email Workflow

Do it Now if it takes 2 minutes or less

Delete it if it's not needed

**Email**

Defer it until later if it needs time

Delegate it if someone else can handle it

**FIGURE 5.3** Managing your email will allow you to focus on more productive activities. This is a simplified view of the Getting Things Done email workflow.

If something can be done in two minutes I do it on the spot. If it's going to take some time, I get it out of my inbox and into my @action folder. If it's useful for reference, I file it away in archive. I don't use subfolders in the archive because the search function is more than adequate and ultimately saves me a lot of time tagging and filing since you'll probably look things up a lot less often than you think you will. Finally, I use a follow-up folder to remind me about items that may be either on hold or delegated to my team.

You don't need to use this system. Take some time and personalize your own email workflow to suit your own needs. Taking a little extra time creating things like email filtering rules will ultimately save you a lot of time in the long run and will pay dividends in terms of the productivity gained. I'm not shy with the delete button either, as even the delete folder can be recovered or searched if the need arises. Another helpful rule of thumb is to open and look at an email only once before deciding what to do with it. That way, the only emails you'll see twice are the ones that make it to the @action folder. Most people don't have an email workflow. You're probably one of them if it says you have 4,000 unread messages and about 100,000 overall messages sitting in your inbox.

If you'd like to learn more about GTD methodology, go to https://gettingthingsdone
.com. It might just change your life.

# SOCIAL MEDIA

Most of the companies I worked for blocked social media sites like Facebook and
Twitter due to regulatory requirements and other concerns. Still, I believe a lot of companies do allow access to these sites and you may very well manage a personal account
as well. While somewhat useful, social media accounts come with a number of security
and privacy risks. Attackers can mine social media accounts to get information that
can be used in targeted attacks like obtaining the answers to password reset questions.

If your organization doesn't have a social media policy that takes cybersecurity and
privacy into account, here are some tips and elements to consider.

- Have a written company policy. Don't rely on tribal knowledge or general
  understanding of policy. You need something written down and approved
  by senior management. A good social media policy includes what can be
  posted and who can post it and from which accounts. Note that this does not
  necessarily have to come from the security group and could come from HR,
  marketing or public relations.
- Use a dedicated email address. I don't recommend tying the account to an
  individual for a variety of reasons. One of these reasons is so that the specific
  person in charge of the account isn't directly targeted by attackers.
- Establish a primary contact or working team that can represent security, marketing and legal/privacy issues.
- Have a policy on employee personal accounts. If employees are going to participate in corporate social media campaigns, you may not want them doing
  this from their personal accounts. If they do have a personal account, you
  may not want them talking about company business on the same account
  where they are tracking friends and social interests.
- Consider a policy of zero trust and require that all posts be vetted by the
  social media team or communications for content prior to publishing. This
  obviously doesn't apply to employee personal accounts.
- Establish a review cycle for the policy. Annually should be the minimum, but
  you may want to do it more frequently.

If you are only concerned about your personal social media account, there are a few
other things to keep in mind:

- **There is no privacy**. There is no privacy on social media. Facebook, Twitter
  and LinkedIn allow you to tweak all kinds of settings in the name of privacy.
  Don't expect any of them to work. Anything you say on social media is,
  by definition, pretty much public. It's also much more routine for a hiring

company to do background checks that include reviewing your social media accounts. People have lost jobs over posts that they thought were private or that they thought the company would never see.

- **You don't represent your company.** Unless you've been hired to manage your company's web presence (doubtful if you are reading this book) then don't represent views online that may reflect badly on your company. Don't post anything online that you wouldn't want your boss or the CEO to see.
- **Park your ego.** If you are trying to make a name for yourself online, don't get involved in Twitter flame wars and controversial subjects. Try to focus on serving the community rather than growing your personal brand. If you focus on serving, your personal brand will follow without any effort.

# READING

> *If you want to be a writer, you must do two things above all others: read a lot and write a lot.*
> ~ Stephen King

The final tip I can provide for becoming a better writer is to read extensively. Read both fiction and nonfiction. Read web articles, magazines and whatever else you can find. Reading lets us hear other styles and voices, which ultimately helps us improve and become a better communicator. Reading can also help you recognize when your own writing doesn't look correct. It helps you understand the flow of sentences better and can teach you techniques and ultimately help you develop your own voice. All this, plus the added benefit that you'll likely learn quite a bit along the way.

A quick note on reading fiction. I used to read only fiction when I was a kid (I was an especially big horror fan). Later in college it was a mix of both fiction and nonfiction (I have a minor in English Literature). Now, later in life, I have been spending most of my time with nonfiction, but before you dismiss reading fiction as trivial or a waste of time, consider this: fiction helps you develop imagination and empathy. It can help you understand how people behave and think. It can also be fun. Nonfiction, on the other hand, helps you learn lessons from people who have been there before you. I largely built my foundation in information security knowledge by reading and taking that knowledge and applying it in the real world. Biographies can inspire us to reach new heights. Both fiction and nonfiction have their place, and both should find their way to your bookshelf.

# SUMMARY

Strong professional writing is an essential skill for anyone who wants to get ahead in the business world. One of the best ways to improve your writing is to focus on what the reader wants to know, rather than on what you want to say. Make it easy on them by giving them bite-size information that they are most likely to want to know. The rest of getting better at writing is practice, practice, practice. Like anything in life, everything gets easier with practice.

- Good writing matters and you may need to do a lot more of it than you think in the business world. Everything from email to incident response plans may fall on your shoulders and you need to be able to write clearly and communicate effectively in this medium.
- The easiest way to improve your writing is to write a lot. Keep a journal and make it a daily practice. This can help you reduce stress, improve your mental state and keep track of important events in your life.
- Writing tends to fall into four phases, research, writing, editing and publishing. Don't expect your first cut of anything to be great and don't skip the editing phase. Your readers will thank you.
- Small sentences are effective. Keep your sentences short. They are easier to remember and digest that way.
- There are some basic grammar rules that can help you improve your writing, but the number one rule to remember is that if it sounds awkward read out loud, there's probably an issue with one or more of your sentences. Figure out which one and fix it.
- Learn to use PowerPoint presentations to your advantage by carefully planning your message, refining your content and making sure you are not reading slides verbatim. Strive for fewer but better slides. If you feel you need supporting data, move it to an appendix.
- You need to make peace with email. Like it or not, this will probably be a big part of your job. Don't make it the only part though. Keep sentences short and come up with a workflow that lets you spend more of the day on deep work.
- Read as much as you can to expose yourself to other styles, voices and material that may help improve your own writing.

# ONLINE RESOURCES TO IMPROVE YOUR GRAMMAR

## Grammar Girl

Grammar Girl (https://www.quickanddirtytips.com/grammar-girl) has been around a long time. The tone here is fun and the tips are easy to understand and useful. This is also a podcast in iTunes, Spotify and other platforms. You can use this site as a reference for specific questions or just subscribe to the podcast and learn something new every day.

## Daily Writing Tips

Daily Writing Tips (https://www.dailywritingtips.com) is an easy reference site to simply type in your search term to get a list of posts related to your question. Stuck on the differences among to, too and also? This is the site to help you sort it all out.

## The Purdue Online Writing Lab (OWL)

The Purdue Online Writing Lab (OWL) (https://owl.purdue.edu/owl/purdue_owl.html) is a free service of the Purdue University College of Liberal Arts. It covers everything from grammar to writing to tips on writing job search materials like cover letters.

# RECOMMENDED READING

Bernoff, Josh. *Writing without Bullshit: Boost Your Career by Saying What You Mean.* HarperBusiness, 2016.

"Best Time to Send an Email." *Yesware*, 25 February 2020, www.yesware.com/best-time-to-send.

Curtis, Todd, et al. "Productivity App: Text Expander." *TextExpander*, textexpander.com/.

"David Allen's GTD® Methodology." *Getting Things Done®*, 22 June 2019, gettingthingsdone.com/.

Girl, Grammar, and Books by Grammar Girl. "Grammar Girl." *Quick and Dirty Tips*, www.quickanddirtytips.com/grammar-girl.

"How to Stop Wasting 2.5 Hours On Email Every Day." *Forbes, Forbes Magazine*, 13 July 2017, www.forbes.com/sites/annabelacton/2017/07/13/innovators-challenge-how-to-stop-wasting-time-on-emails/.

King, Stephen. *On Writing: A Memoir of the Craft.* Scribner, 2010.

Newport, Cal. *Deep Work: Rules for Focused Success in a Distracted World.* Grand Central Pub, 2018.

Osman, Hassan. *Don't Reply All: 18 Email Tactics That Help You Write Better Emails and Improve Communication with Your Team.* CreateSpace Independent Publishing Platform, 2015.

Rubin, Danny. *Wait, How Do I Write This Email?: Game-Changing Templates for Networking and the Job Search.* Danny Rubin, 2016.

Strunk, William. *The Elements of Style.* BoD—Books on Demand, 2020.

"Write Your Best with Grammarly." *Grammarly*, www.grammarly.com/.

Zinsser, William. *On Writing Well.* Harper Paperbacks, 2013.

# Say What? Verbal Communication Skills

# 6

*Your voice is the fastest way to make others know, like,
and trust you. It's the greatest tool of influence you will
ever own.*
~ Roger Love, voice coach

Our voice reveals a lot about us. You can usually tell a person's emotional state and maybe even a bit about their personality just by hearing them speak. Introverts generally speak softer and a bit quieter than extroverts. Extroverts tend to be louder. Of course, there are the obvious things as well, such as being able to tell if someone is a male or female, young or old.

Verbal communication is simply the use of spoken words and language to share information and ideas with other people. We spend a good amount of our lives using verbal communication. Referring to Chapter 1, there are instances where verbal communication will be intrapersonal, interpersonal, group communication, public communication or mass communication. In short, we talk to ourselves, 1-1 or with small to very large groups. The language you use needs to get progressively more formal depending on how many people you are trying to connect with and what message you are trying to convey. Obviously, self-talk is a lot less formal than what we might say on the BBC news.

How we speak and what we say has a great deal to do with how people will perceive and interact with us. This is why it's so important to consider how strong your verbal communication is and where you can make improvements. In this chapter, we will examine how to project confidence and increase your gravitas, how to improve your speaking voice and how to make your phone conversations, storytelling, meetings and presentations more effective. We will also cover some public speaking scenarios and how you can get more comfortable giving public presentations.

DOI: 10.1201/9781003100294-7

# PROJECTING CONFIDENCE AND AUTHORITY WHEN YOU HAVE NEITHER

You're going to have to deal with a lot of different people in a security leadership role. Most of them are not going to report directly to you. To have credibility, you'll need to project an air of confidence and authority but avoid sounding overly arrogant. Fortunately, there are many ways to make subtle improvements with your verbal communication. Note that making certain changes like eliminating filler words can take time to get consistently right. When we are not present and aware when we speak, we are likely to revert back to old habits that have been with us for a lifetime. Also note that I'm not suggesting you change the way you speak in every setting. The more formal the setting, like a board meeting or on stage at a conference, the more you are going to want to consider not only what you say, but how you are saying it.

## Eliminate Filler Words

Filler words ("umm," "like …," etc.) tend to be used while you're thinking of what to say next. These filler words also make you seem unsure of yourself. Try to eliminate these words completely, even if you only replace them with silence while you get your thoughts together. This takes some practice, and I haven't mastered this myself because it also requires being very present and calm while you speak. Again, it's hard to be always "on" with good communications. Awareness is your first step, so don't expect to be perfect all the time. What should you use in place of filler words? Try silence. Don't be afraid of silence, there's not enough of it in the world. Once you've gathered your thoughts, carry on with the conversation.

## Being Loud Is Not the Same as Being Confident

Don't confuse speaking loudly with speaking confidently. While speaking loudly enough to be heard is important, your volume matters less than your pitch and cadence (the rhythm of your voice). To project confidence, be sure to keep a low pitch and a smooth cadence when speaking. Maintaining a low, steady tone of voice is a great way to sound more confident. Just make sure you vary the pitch enough to avoid speaking in a monotone.

A low dominant tone that makes you sound confident comes about naturally when you release the tension in your voice. The muscles in your shoulders, neck, jaw and throat all affect your voice tone so you want to relax those muscles when you're speaking. To get those muscles relaxed here's a simple exercise you can try:

1) Start by taking in a slow, deep breath.
2) As you exhale make a noise that's half yawn/half sigh and soften any tension you feel in your jaw, throat, neck and shoulders. There's an added benefit that you'll feel more relaxed and less tense overall.

This exercise will help you become aware of any existing tension in your voice so that you'll be able to relax when speaking. It will also help bring your awareness to the present moment so you can be at your most effective when you talk.

## The Importance of Breathing

Taking slow, steady, deep breaths as you speak ensures you will maintain a slow, steady, deep voice. Some speakers, especially nervous ones, almost run out of air and need to catch their breath because they are taking fast, shallow breaths. Shallow breathing is common in high-stress situations, such as presenting to a large group. Shallow breathing disrupts the smoothness of your voice and, ironically, can also make you more stressed than if you were taking slow, deep breaths. The key to getting this right is to make sure that you're not just breathing into your chest but instead are breathing deep into your belly and diaphragm. You should be able to feel your belly and lower rib cage expand out as you inhale and fall as you exhale. People often breathe in the reverse of this optimal manner, though we were all born breathing this way and somehow learn not to do it later in life. Together with good posture, this breathing technique will also aid in projecting your voice if you have a large room or large audience. Your voice is like an instrument, getting the right air flowing in and out will greatly enhance how you project your voice and how it sounds.

## Avoid Uptalk

When people are uncertain of what they're saying, the pitch of their voice automatically goes up at the end of a sentence. This is often referred to as upspeak or uptalk. You can hear examples of this all the time when people ask questions. The problem is that if you uptalk when you are making a statement, you project uncertainty. It makes you hard to understand, because listeners have to process if you were asking a question or just making a statement. If you want to add an extra touch of authority to your speech do the opposite and end your sentences by lowering your pitch slightly. Even if you're asking a question, this slight downward inflection is going to make you sound more confident in what you're asking.

## Staying Present

Many problems like speaking too fast or breathing incorrectly come from losing our connection to the present moment. People speak, but they are not fully aware and present. They may be wrapped up in thinking about what to say next or what everyone thinks about them. This takes away from the moment and makes it easy to lose your train of thought. Try to just relax and stay present. Focus your attention, slow down a little and enjoy the moment.

## Baseline Your Voice

For any of these techniques to become part of your natural way of speaking it's important to practice them. An easy way to start the process is to record your voice speaking in a normal, casual way. Listening to a recording of your voice will make you aware of some of the mistakes you're making and areas you can improve upon. Do this on a regular basis and it will be much easier to cut out bad speaking habits and replace them with good ones. You'll have the added benefit of seeing how much you've improved since you started.

# COMMANDING AUTHORITY AND GRAVITAS WITHOUT SOUNDING LIKE A DORK

The more you work with the most senior people at your firm, the more you will want to pay attention to your perceived level of gravitas. Gravitas was one of the ancient Roman virtues that denoted "seriousness." Having gravitas at work means you are taken seriously, and you are trusted and respected. Gravitas will help you persuade business leaders and help you be taken seriously by upper management. The good news is that gravitas is something you can develop. However, it is important to remain authentic to yourself and your personality. If you stray too far from your real personality, you will seem artificial and inauthentic.

Starting with your mindset, realize that being a security leader requires you to make connections on a personal level to a variety of different stakeholders, including senior management. These people are likely to have their own executive presence, which includes their own level of gravitas. You also want to be perceived as someone who has authority and commands respect. This means that you should think about your personal gravitas and if it's sufficient in important settings like meetings with senior business leaders and the board. I wouldn't worry a lot about gravitas outside of important settings, since we all have enough other things to worry about in life. But there are times where you need to be serious and to be taken seriously, and that's where gravitas helps.

Of course, gravitas can encompass more than just your voice. Your appearance and the way you carry yourself can all influence how others perceive your gravitas. Your emotional intelligence and your ability to cope with stressful or demanding situations also influence your perceived gravitas. Remember, you want to be a calm and rational voice in the organization. This should come out in your voice, your demeanor and your appearance.

Some methods for being more conscious about your gravitas are discussed below.

## Dress the Part

If you want to be perceived as an executive, you're going to need to look like one. This can vary greatly from if you are working in a bank or a startup technology company.

Try to take the lead from other senior executives and especially your boss. Don't go too far off in the wrong direction. If you work in a tech firm, it's going to come off as weird if you're the only guy in a suit. In fact, you might even be seen as not fitting the culture if you're too far outside of what's accepted as the norm. The rule of thumb is, don't be the sloppiest guy on the management team and try to strike a balance. This stated, if you dress a little nicer, you will feel more professional, and this will come out in your communications and overall confidence. There's a psychology behind this, but the clothes you wear and the way you groom yourself will actually change the way other people hear what you say. Think of speaking to an investment advisor wearing a t-shirt and sweatpants versus one wearing an expensive suit. It may not always be a conscious decision, but you're more likely to listen to the guy in the suit.

## Mind Your Pace

Executives don't look flustered or rush into meetings late and they typically don't speak too quickly or too slowly. Keep an overall pace that exudes some level of authority in both your speaking voice and in the way you carry yourself. You'll want to have a slow, measured pace and a calm air of authority. Think of yourself as cool and calm under pressure and very much in command of yourself, your team and the situation.

## Say It Clearly

Could you imagine if the CEO of a company used a lot of fancy MBA terms that no one on his management team could understand? What if this person gave orders to their direct reports that they didn't understand? And what if they also spoke to shareholders using acronyms or terms they didn't understand? How well do you think this business would run? How effective would this CEO be, despite having an expensive education and potentially high IQ?

Great leaders aren't going to try to confuse or impress people with their technical jargon. They are not going to talk over everyone's head or in a way that isn't inclusive of everyone in the room. They are not going to try and make others feel stupid or ignorant. This goes back to knowing your audience and making sure that your message is tailored to *their* preferences, not yours.

No jargon, no acronyms, no tech talk.

## Watch Your Tone

Some research has found that a deeper tone suggests a higher level of gravitas. Be careful here, as you don't want to take on an artificial tone or pitch. You're also not going to be able to maintain an artificial tone in all situations. This was famously discovered with Elizabeth Holmes, CEO of Theranos. She adopted a very deep voice that seemed to disappear whenever she was drinking, excited or under pressure. This is not the image you're going for, so speak at a measured pace and pitch and just try to be natural.

# IMPROVE YOUR SPEAKING VOICE

*When you sing a song the melody is important. Why is it that most people are singing boring songs. They stay on one note, they have no interesting dynamics, and they bore the listener. I say, pretend you are singing while you are speaking. Move it around, shake it up, swoop, dive, soar. Let your voice be as interesting as you are.*
~ Roger Love

Do you own a smartphone or tape recorder? Good. Now take a recording of your voice speaking normally. Now for the bad news. Yes, that really is how you sound. Many people are shocked to hear their voice on a recording. When you speak and hear your own voice, bones and tissues tend to change the lower-frequency vibrations, making the voice in your head sound fuller and deeper to you than it really is. Which one is more accurate? Sorry, it's the recording.

People generally don't like hearing their own voice because it sounds so different than what they think they actually sound like. The bad news is that there's probably not a lot you can do to change your overall voice without significant practice and effort, but there are some simple things you can do to help you leverage the voice you already have that don't take as much effort.

Again, the best improvement you can make to your voice is to make sure that you're breathing right. This means breathing through your diaphragm. To breathe correctly through your diaphragm, simply inhale and let your belly rise, and exhale and let your belly fall. Not only does this help your speaking voice, but it helps relax us physically. Diaphragmic breathing is the number one improvement you can make to the way you talk and how your voice sounds and projects. This only takes a little practice and being mindful.

If you are concerned about making big improvements to your voice, you could always get a vocal coach. This isn't as expensive as it may sound, and many have reported good results in as little as a session or two. But if you're just looking for some additional guidance and some easy things to try for easier improvement, consider the following:

- **Posture**: Believe it or not, posture can have a big effect on your tone. Keep your back straight and your shoulders back, down and relaxed. Your head should be retracted a little. A useful trick is to imagine a helium balloon is attached to your head, pulling it upward.
- **Pitch**: Do you have enough variety to keep your voice interesting but try not to be too high or too low in your range? Generally, you're best aiming for something towards the middle of your range and with just enough variety to avoid monotone speaking.
- **Pace**: The speed of your voice impacts how intelligent, nervous or excited you seem to your listeners. If you talk too quickly, they won't be able to mentally process everything you're saying. If you speak too slowly, you may seem that you're not that smart or may be perceived as boring. There's no perfect speed, so you'll have to experiment and see what works best for you.

- **Warm up**: Just like an athlete warms up before a game, you will be at your best if you warm up your voice. Hum, sing the Do-Re-Mi song, or just start putting your voice through some paces by singing scales. You can also just read some things aloud if singing isn't your thing. You can also try some tongue twisters like Peter Piper picked a peck of pickled peppers. These exercises get your vocal cords warmed up and ready to go.
- **Stay hydrated**: believe it or not, drinking enough water ensures that your vocal cords are also hydrated. The key word is "enough," since your vocal cords are actually located down your windpipe. You can't pour water down your windpipe without some big problems, so there is no way to keep your vocal cords hydrated other than making sure you're drinking enough water that hydration gets there through the bloodstream. Also, avoid things that can dry out or damage your voice, such as too much alcohol, smoking and too much hot coffee or tea. The tea with lemon and honey theory is busted; this can actually make your voice worse.

# PHONE CONVERSATIONS

Office phones are becoming a more uncommon means of communication in the office. In fact, a few companies have completely done away with employee voicemail in an attempt to encourage other communication mediums and save costs on expensive corporate voicemail systems. As more meetings are happening on Zoom and Microsoft Teams, you may find that your time spent on an actual phone call is pretty less, although it is still possible to use tools like Teams as a voice call over the Internet.

Studies have found that relying on the voice is just as effective as, if not more so than, relying on a video Zoom or Teams call. This was found to be especially true when you're on a large Zoom call with many attendees. Think about it, if you have 30 people on a Zoom call in the multiple person view, this can be overwhelming to our senses. We have to listen and watch and process what 30 people are doing at the same time. Conference calls allow you to focus more on what is being said than how everyone's office or home office looks and all the other visual stimulation that comes from these calls.

Regardless of how much you use a phone, here are some tips to make better use of this medium. Good phone skills matter, especially communicating clearly and knowing when to speak and when it's time to listen. Here are some general tips to keep your use of the phone more effective.

## Does It Make Sense to Use the Phone?

Sometimes short messages and informal questions are best suited to email, instant messaging or texting. However, there are situations where this can prolong the conversation or lead to confusion and mixed signals. For more complex topics or things that take longer to explain, you might be better served by picking up the phone. Phone calls have a real-time feedback loop. Even if you can't see the person, you can hear the tone of their voice and ask questions. I know that when I see a lot of long emails on the same subject coming in, I like to just call someone and talk through the problem rather than trying to read up and understand a lengthy email chain. Better yet, I usually ask that person to summarize the email chain for me. It's very satisfying to have one phone call and then subsequently delete ten unread messages in your inbox because the problem has been resolved by a simple conversation.

## Stay Focused

If you're on an important phone call (and if it's not, please feel free to hang up), resist the urge to multitask. I've seen people trying to carry on a phone conversation and read email at the same time. I don't care how good you are at multitasking; it is next to impossible if you are going to get anything out of either task. Try reading a book and listening to a different audiobook at the same time. Sorry, expert multitaskers, this doesn't work. If you read email while you're on the phone, you are going to do both poorly and probably get called out on the phone call for not paying attention, since that's where it will be the most obvious.

One exception to the multitasking rule might be if you choose to take phone calls in the car. My advice in this situation is at least don't initiate calls from the road. Even with a hands-free set, it can be dangerous trying to pay attention and both drive and talk. I get it, sometimes you're stuck in traffic anyway and it's not that dangerous when you're mostly at a full stop or inching forward at 2 miles an hour. Use your judgment and let it go to voicemail when it's not the right time to pick up the phone. You can always call them back later. If it's an important call, the caller deserves your full attention anyway.

## Have a Clear Purpose

If you are calling someone else, prepare an agenda or at least a list of points to cover. Keep yourself on track and make sure you hit your objectives. This may sound basic, but you'd be surprised how many people just pick up the phone and make a call without thinking about why, or even worse they end the call and later realize that they forgot to cover their most important point. Take a moment and think beforehand and keep some notes handy.

## Have a Professional Phone Manner

When you can't see the person you're speaking to, your voice somehow needs to convey authority, empathy and trustworthiness. You can achieve this by paying attention to your delivery and what you say. Speak slowly and clearly, especially if you're discussing information that the other person may not understand.

In the absence of nonverbal cues, such as facial expressions, it could be a good idea to add a feedback loop by pausing and asking if the person you're talking to understands you so far. Otherwise, carry on and use simple and straightforward language. Give one idea or piece of information in each sentence and try not to ramble.

## Listen Actively and Empathically

Obviously on a voice call with no video, you need to do a lot of listening to keep track of the conversation. Use your active listening skills to make sure you stay engaged in the conversation. If you didn't catch something, it's perfectly fine to ask the person to repeat it.

Security people deal quite a bit with upset or otherwise flustered users. You might need to use patient, empathic listening, especially if the other person is angry about something. The more the person on the other end of the phone doesn't feel they are being heard, the more difficulty you're going to have with the rest of the conversation. Obviously, if they are reporting an incident this is a different conversation than if they're complaining about your security controls locking out their account for the 100th time this week. Combine active and empathic listening in a situation like this by restating what the problem is and saying how you understand this could be frustrating. This lets the person know that you're paying attention, and that you hear and understand their frustration. Always remember, people want to be heard and understood. Everyone wants to be acknowledged and have their voice heard. Some people will actually vent, then thank you for letting them vent and acknowledge that there's probably not much to be done and thank you again for listening.

Show that you're still engaged with the call, even when your caller is doing most of the talking. This can be as simple as saying "yes" during pauses, but it's better to say something which demonstrates that you've been listening carefully using active listening and asking questions to help clarify any points or open-ended questions to drive the conversation deeper.

## Avoid Interrupting

Again, people like to be heard. If someone has called you and wants to say something, let them finish before saying anything of your own. Interruptions break the other person's flow of thought and can be perceived as rude. Of course, polite interruptions might be warranted every now and then like if you really do have to get to another meeting. Don't just cut them off and run though; apologize and set a time that you can get back to them when you're not as rushed.

Phone calls can actually be the perfect antidote for the "Zoom fatigue" of having too many video conference calls. There's not always a lot to be gained from adding video in most cases, and there might be something to lose. Selfishly, I like to pace around the room when I'm on a call, so I'd take a phone call over a camera-on Zoom meeting any day.

# SPEAKING UP IN BUSINESS MEETINGS

For better or worse, our leadership readiness is often measured by our willingness to speak up in business meetings. How we come off in these settings can have a big impact because they happen so frequently. While presentations and board meetings can be very high-stakes conversations, the day-to-day communication that happens in most business meetings is how most executives will perceive the "real" you.

Speaking up in meetings also means you want to know what to say and when to say it. This is easy using the active listening skills we learned earlier. Pay attention to what everyone else is saying, listen, ask questions and engage in order to add value and make a meaningful conversation.

If speaking up is something you're not that comfortable doing, I have some suggestions for speaking up more often without seeming artificial.

1. **Start with your mindset**. Be confident in yourself and expect to speak up in meetings. You have been hired for your expertise and to bring value. People want and need to hear your insights. Go prepared to add value and plan to actively participate rather than just getting through. This doesn't mean you have to speak up in every meeting. The last thing you want to do is force yourself to say something when you don't have much to contribute on a given topic. The more you're ready to speak and engaged in what's being said though, the more likely you are to have a thought or comment you really do want to share.

2. **Don't overthink it**. You might overthink trying to come up with something profound, impactful or impressive. If this is the case, you might also wait too long and think and plan until the meeting is over and the opportunity is lost. It's important to share your thoughts and ideas without over-editing them or limiting how you express yourself. Don't let the opportunity slip to share what you know with everyone else. This is especially a problem with introverts. We tend to do a lot of internal processing before we open our mouth.

3. **Ask questions**. If you are an introvert like me, you'll likely be reflective, strategic, thoughtful and a good listener. My experience is that it's best for introverts to do a little research in the lead-up to the meeting. Investigate the subject under discussion and plan a few things you might want to say or ask. This is a great way to show that you're attentive, engaged and interested. Sometimes, introverts have great thoughts that go through their heads, but by the time you work up to saying something the moment has passed. This is too bad, because we usually have some really good insights compared to the people who just start talking before they're done thinking. Preparing in advance, no matter if you're introverted or extroverted, will still help you have some idea of how to add value to the conversation. For introverts, it lets them do a lot of mental processing in advance. How many of us have found ourselves bouncing from one meeting to another and barely even knowing what call we're on or what the purpose of the meeting is? It happens to me quite a bit, but if I do have the chance to prepare, I know that I will have

questions and be more engaged than if I have to spend the first few minutes wondering what the meeting I'm in is all about.

4. **Get comfortable speaking up**. Some people are not very comfortable with even informal public speaking at a business meeting. Hopefully, this isn't you, but know that this gets easier with practice. If you have some hesitation before you're about to speak, pause and breathe. Take a deep breath and find your center. Then speak anyway. As a senior security leader, you are expected to bring your voice, your presence and your perspective.

5. **Avoid saying, "I disagree" or similar contrary positions**. People hear terms like "I disagree" and immediately feel confronted. They may even be dismissive and stop listening to you altogether thinking you are arrogant. Why not try something like "I have a slightly different perspective I'd like to share ..." When you do disagree, don't make it personal. You don't want to put down another person's ideas or beliefs. Use active listening and empathy to try to understand the other person's perspectives. Then gently offer your perspective, the logic behind it and then try to come to an agreement. It's easy to get too stimulated trying to be right and then shutting down other perspectives.

When should you hold back from speaking up? I used to have a joke about never saying anything to make the meeting longer than it already is, but realistically there are a few times where you might want to hold back.

1. You're speaking up just to say "something" but nothing of any real merit. Add value or don't speak if it's just not your topic. It happens, you end up in a meeting "just in case" any security issues come up rather than because you really need to be there. Still, if you do have a non-security perspective that might be valuable for others be sure to share it. Don't lock yourself in the security box like that's all that you know; you've been to a lot of places and have seen a lot of things. Share your knowledge and your perspectives on all relevant topics.

2. You're bringing up something that might be best left to a 1-1 meeting and not shared in front of a big audience. Be mindful that certain topics might be best handled outside of a meeting in a less public setting.

# RUNNING EFFECTIVE MEETINGS

*Meetings should be like salt—a spice sprinkled carefully to enhance a dish, too much salt destroys a dish. Too many meetings destroy morale and motivation.*
~ Basecamp founder and CEO, Jason Fried

If you work in a corporate office setting, especially in large companies, you are going to spend a lot of time in meetings. Some of these meetings will seem excruciatingly long and go nowhere. There's not a lot you can do about this other than walk out, but you can definitely choose to not be part of the problem by running your own meetings well and leading by example. A well-run meeting combines all the writing, speaking and listening skills that we've been discussing in this book.

People hold meetings for many reasons. Some are called together to share information. Some are intended to make decisions. When it comes to your meetings, start with your purpose. You need to have a clear objective, which also means you know what a successful outcome looks like. As basic as this sounds, a lot of people don't seem to take this step. They see a problem or have a topic, but they don't take this to the next step and think about why they want to get everyone together. They don't think about what they specifically want to discuss, and they don't visualize the outcome of the meeting and what success looks like. If you can't figure out what your reason for holding a meeting is, then please take it off the calendar. Everyone will thank you for freeing up their time. If you have a standing meeting that doesn't have any active topics to discuss, either canvas everyone for topics or cancel it. The first rule of meetings should always be: make sure you need a meeting.

Once you understand your objective, strongly consider if a meeting is really the best medium to use. When I worked at a large investment bank, we used to calculate an average cost of meetings given a certain number of associates, VPs and managing directors who would be present. If you are going to pull 50 people into a meeting, consider if this is really necessary and what this is going to cost in terms of lost productivity for others. There are a lot of meetings held that should have been emails, phone calls or chat messages. Save people time and productivity for more important tasks.

Lost productivity aside, meetings can be a very powerful communication mechanism. They are held either in person or by video or audio call, such as on a conference call. This means that there's a strong feedback loop and a fast exchange of ideas and information. If you are in person or on a video call with cameras turned on, you can also read body language, hear everyone's tone and read other nonverbal cues. But because there are many people and personalities involved, meetings can also go off track and become free-for-all circuses. They can also not have enough relevance to individuals to keep their interest. This means that you need to prepare for meetings in advance and manage them actively once they are underway.

You want your meetings to be clear, to the point and get things done. When you've determined that a meeting is the right way to go, there are four easy steps to running better meetings. These steps include:

1) Prepare
2) Have a clear objective
3) Share an agenda
4) Manage the meeting actively

Step 1 to running a good meeting is to set yourself up for success by preparing. Why are you scheduling this meeting? Have you considered alternatives like email? Do you need a face-to-face discussion? Do you really need the input from all the people you are going to invite? Are you inviting too many or too few? No one likes meetings that waste their time. If you've ever been in meetings where a key decision maker wasn't invited, this can be just as frustrating as progress stalls and follow-up meetings need to be scheduled. The only decision made is to have another meeting. Equally frustrating is when there are simply too many people invited and opinions and input run amok, or apathy sets in and few are paying attention to anything being said.

The next step to running effective meetings is to clarify your objective. Are you looking for consensus? To generate ideas? Is your objective to communicate status? Make a decision? Having this kind of clarity ahead of time will ensure that your meeting is on point and will also help you steer the conversation if things go too far off track. Define a concrete objective and don't schedule a meeting until you have it clear. In fact, hold off on scheduling that meeting if you don't have time to prepare or if email might be more appropriate, like for simple status updates.

After you decide on a clear objective, set an agenda that is shared in advance. Having an agenda shows people that you are organized and have a point in calling a meeting that will not waste their time. It also allows them to consider if delegation might be appropriate to one of their staff. Most important, having an agenda gives you a framework to make sure that you stay focused on the objectives of the meeting. It's your "cheat sheet" to keep the meeting on track. Agendas should be simple. They can be as simple as a few bullet points. For longer meetings, like full- or half-day sessions, you may want a detailed agenda that timeboxes certain topics into very specific slots. This not only helps you visualize the flow of the meeting but can help you gauge when a meeting is getting too far off track so that you can either recover time or skip some items and focus only on the critical elements.

I like to keep the agenda in the meeting invite itself so that I don't have to stumble to find it when the meeting starts. However, a lot of people just blindly click the accept button on a meeting request, so if it's important for people to understand the agenda in advance, you can follow it up with an email restating what's being covered. This way they will have something in their inbox that they're more likely to read.

The final step is to manage the meeting and follow your process. I recommend a simple process along the lines of starting by noting attendance, summarizing the agenda for everyone, taking each agenda item one by one and then summarizing the key points and action items at the end of the meeting. For important meetings or meetings that have action items, follow up with a written summary and highlight action items and who's responsible for following up by a specific target date.

Be careful with scheduling a meeting. Don't schedule 30 minutes if you know you need an hour and don't expect that people will stay in your meeting after you run out of time. Most executives are in back-to-back meetings and they can't give you even an extra 5 minutes without being late for their next meeting. It goes without saying that you need to start and end the meeting on time. This means you'll also need to rein in tangent conversations and superfluous topics and make sure that everyone sticks to the agenda. It's OK to have a little small talk, especially right at the beginning, since this helps set the tone and break the ice. Just don't let time get away from you.

Be mindful of the clock as your meeting progresses. If people show up 5 minutes late to your 30-minute meeting, you now have only 25 minutes to get everything done. This brings up a point worth considering, which is that you should probably not wait for latecomers. Starting on time, even if you know a few attendees haven't arrived, lets the people who did show up know that their time is valued, and it will hopefully help latecomers realize that showing up late to your meetings means that they're missing out on information.

After the meeting, unless it's something routine like a staff meeting, you should follow up with an email summary highlighting any action items that people have and the due date. A few bullet points recapping what was said or decided is all that's necessary. I've seen some people try to capture almost every sentence, which no one wants to read. You want to capture decisions, problems and solutions with action items. That's all you need.

Meetings represent a unique way of bringing people together and are a powerful communication tool. A little preparation like considering the right attendees, crafting an agenda, keeping your meetings to the point and following up with action items will help the whole process run smoothly.

# VIRTUAL MEETINGS

When I originally planned this book, I was only going to lightly touch on the subject of virtual and online communications. But then things shifted from not only road warriors being virtual but to almost everyone being virtual. Zoom became a household name (and a verb), as people worked almost exclusively from their homes. At the time of this writing, I have no idea what the future holds for office work, but it seems apparent that remote working is here to stay for at least a portion of the population. Some offices have closed physical buildings completely and others, like Shopify and Facebook, have told employees they can be remote forever.

People's ideas about remote working will be forever changed after COVID-19. Many will question if they really need to show up somewhere in person. They will likely work in offices less, or not at all. The dated concept that you don't care about someone when you don't show up in person will go away. Computer-based communication will be the norm. I do think that in-person meetings will happen again, but their frequency might be nowhere near that in the past.

But is a Zoom or Microsoft Teams meeting different from just meeting in person? The short answer is yes. There are a number of variables that come into play with remote video sessions. For starters, your connection to the audience is much, much weaker. People are more likely to be multitasking and only lightly paying attention. Distractions such as children and pets abound. We were already in an age of distraction, so this all seems like almost the last thing we need. In worst-case events, leading a Zoom conference without being able to read the room is like talking into a black hole, especially when no one has their camera on and everyone's on mute. In these meetings, body language and audience feedback is minimal.

If you've worked in large, global companies or companies with many offices, many meetings have always been at least partially virtual. When you're in 100 countries, it's not practical to think that every meeting will be in person. Typically, these organizations are always running video calls, though if you work out of a well-populated office, there could still be many of you in a conference room in person and other attendees on a video link. For others, this is a very new concept. My current organization was almost 100% on site before the pandemic. This can be a big adjustment for some.

So how can we be more effective in this format? What are the best ways to grab and keep attention and make sure that your voice is heard?

## Pay Attention

I'm going to start with the hardest piece of advice here. You need to pay attention. Yes, it is tempting to check email, the stock market or the latest news but science has shown that multitasking simply doesn't work. You will not communicate effectively if you have only been 20% involved in the conversation. Worse yet, your days will pass in meetings that offer very little benefit and just burn calendar time.

You'll want to minimize your distractions as well. Put your phone in Do Not Disturb mode if it's an important meeting. Turn email notifications off. Spending some time tailoring notifications from phones and computers will help you realize that the vast majority of them are nothing but needless distractions in the first place.

Did you know that Zoom is offering a new feature called "attendee attention tracking," and it can only be disabled by the meeting host? If the host is sharing their screen and you click out of that screen for more than 30 seconds to go check email or surf the web, the host can be alerted. While this feature was originally intended for teachers working with students, don't be surprised if this catches on in companies skeptical about the whole work-from-home thing.

The bottom line is to treat virtual meetings as seriously as you would treat an in-person meeting. This means, don't be late. Don't multitask. Contribute to the conversation in a meaningful way and pay attention.

## Be Active and Be Heard

Once you're in tune with what's going on in the meeting and conversation, it's time to actively participate. As a senior security leader, you are going to need to ensure that your voice is heard and that you are adding value. This doesn't mean to break in constantly and talk all over the other participants. It also doesn't mean repeating things that have already been said or don't add any value. Finally, don't speak up with controversial issues that might be better off handled in a 1-1 conversation. Make sure you review the "speaking up in business meetings" section of this book for other thoughts on this subject.

## Take Care of the Elements under Your Control

There are several elements in a virtual meeting that are under your direct control and can help you look more professional. While many of us are working casually at home during COVID-19, you want to be as effective as possible, and you don't want technical issues to take away from your message.

The elements under your control are as follows.

## Start with Your Mindset

In a virtual meeting, it is easy to forget that you are still being watched if your camera is on. Don't be caught multitasking and try to remain present and mindful of the conversation at hand. You don't want to be unprepared if someone asks you a sudden question. Again, think of a virtual meeting as an office meeting and you'll be in the right frame of mind for what you should and shouldn't be doing. Would you bring your lunch to a conference room meeting and eat in front of everyone? Probably not. Act the same as if you were in the office.

## Using Cameras

Every company seems to have its own culture around the use of cameras. Keep in mind that meetings could include customer-facing sessions and want cameras to be on to keep things more personal. Some companies may also have unofficial guidelines or etiquette for online meetings. For example, participants may be asked to turn off their cameras or mute themselves while others are speaking. Interjections may use the chat feature or the raise your hand feature.

Keeping the camera on serves to keep you honest with multitasking but also creates a stronger connection and communication feedback loop with the person or people on the other end of the conversation. Online meetings can feel very distant and impersonal. This is doubly true when you can't put a face to a voice or read body language. I freely admit I usually default to what others are doing on the call before I turn on the camera and you can do the same. It's equally awkward to be the only one with a camera on and active.

Try to position yourself about 2–3 feet away from the camera and try to center yourself in the image. Your camera should be at eye-level so that when you speak, you appear to be talking directly to your audience. You should be the focus of what people see, not your background or other distractions. Especially look straight into the camera when you're speaking. This may seem a little strange because your eyes may naturally drift to looking at the images of other people in the meeting. Just be conscious of where the camera is actually located on your computer. I admit that this one is kind of tough to follow. Position yourself so that the camera picks you up from a head and shoulders angle or waist up if you have the room. If you have two monitors, be conscious about which one has the camera. This is the new virtual way of making "eye contact" and will provide you a much better connection with your colleagues. Some people recommend getting a standalone camera, which will usually be much higher resolution than a webcam.

You might think about buying a premium camera if you are going to be on a lot of remote meetings. The built-in camera, even in a high-end MacBook, is typically of not very good quality. You can now pick up a decent 1080p or even 4K camera on Amazon for around the $100 mark. Embrace the new virtual world and consider good, better and best. Good is your built-in camera. Better is a USB webcam, like a decent Logitech. Best is a professional camera, which may also cost quite a bit. Some of the real high-end webcams can cost $1,000 or more. If that's too rich for your blood, go with something like the 1080p Logitech cameras like the C920, which is fine for most people. You get what you pay for. Pick a good brand at a competitive, but not cheap, price. I bought an allegedly 4k camera that was nowhere near as good as the C920 camera from Logitech and I ended up getting rid of it altogether.

## Use Good Lighting

Try not to be in a room that is too dark or too bright. You want your face to be seen clearly on camera. One trick is to position a lamp aimed at you from behind the camera the way they would do on a movie set. In either case, make sure there is enough lighting that people can see you clearly. You can easily do this by seeing your own video stream on the computer monitor. While I am a fan of natural lighting when working from home, this may not be enough on an overcast day or if you are working in an area without access to much sunlight. In my opinion, LED lighting tends to work the best in these situations. Many webcams are starting to come with LED lighting built right in, so buying one of these eliminates the problem and also keeps your work-from-home setup simple.

You can also consider using a ring light, which many professionals use. Does this seem a little overboard? Maybe … but just don't be the person who is lost in the shadows either. Figure out some good lighting and use what works for you without making you feel like you're being interrogated by the CIA in one extreme or that you're in a dungeon on the other extreme.

## Have Good Sound

The built-in PC or Mac microphone is not always ideal if you will be doing a lot of talking. A standalone or clip-on mic can help with this. I prefer using Bluetooth headsets and since I have a Mac, I just use the AirPods Pro which have the added benefit of noise cancelation when the neighbors decide it's the perfect time to start mowing the grass next door.

When you're not talking, I recommend keeping your mic on mute. Be careful with the mute button though. This is something that needs a little active management. If you leave the mute button off all the time, too much ambient noise can be a distraction for others and also make your image jump to the screen even when you're not talking. If you leave the mute button on all the time, you are obviously not making your voice heard, contributing to the meeting or being an active participant. You also don't want to be told over and over again that you're speaking while on mute. We all do it now and then but try not to be "that person" all the time.

There are a number of ways to reduce background noise. Noise canceling headsets like the AirPod Pro can also filter out ambient microphone noise when you speak. A new application called Krisp (https://krisp.ai) promises to take the noise canceling concept to your phone or PC regardless of if you're using headphones or the built-in speakers. If you have a quiet room, then built-in PC/Mac speakers will probably be enough, just make sure you are close to your microphone when you are speaking so that people can hear you. Needless to say, you want to make sure your Internet connection is fast enough for all this video and audio traffic or that will also come out in your conversation in the form of blips, gaps and freezes in the conversation.

Built-in microphones on our computers aren't very good and may not be in the most convenient location. If you're primarily working from home, you may want to invest in a good headset or even a standalone microphone. All the major live meeting platforms compress audio, so the higher quality the input, the better you'll sound overall. I imagine that even when people go back to physical offices, there will probably still be a lot more remote meetings. You might as well make the most of it and get yourself a good setup.

## Avoid Distracting Room Backgrounds or Virtual Backgrounds

If you are using a virtual background, make sure it is a professional one. Pretending that you're on the Starship Enterprise is not going to add a lot to your credibility. If you are using the actual room with no virtual background (my recommendation for most cases) then make sure there are not a lot of distracting items in view. Again, you are supposed to be the focus, not your background. Having a window with bright sun coming in or any light source that may be brighter than the foreground doesn't translate very well to video and will make you look overexposed.

If you do use a virtual background, consider picking up a cheap "green screen" from Amazon. Virtual backgrounds can often show a lot of odd video anomalies. Anyone who has been on a Zoom call knows exactly what I'm talking about here. People move and either part of their face disappears or creates other video anomalies. Buying a green screen, should dramatically reduce the number of weird situations like this that can arise in this new format. For me personally, I prefer to use the blur background effect or just use the real background with a basic room décor that's not too cluttered. If you really do prefer virtual backgrounds, try the green screen. You'll be surprised how much of an improvement it can make.

## What You're Wearing

I know a lot of us can dress in business casual, or even just plain casual, while we're working from home. While I personally do not get overly dressed up since that works with my company culture, think about your own outfit and how it makes you look. Take your cues from your peers and your boss. We all want to be comfortable while we work from home, but don't be the most shlumpy person on the call. As a senior security executive in your company, you will want to make sure that you look the part. This said, as already stated don't be the only one wearing a suit when even your boss is wearing jeans.

## Slow Your Speech Down

OK, I freely admit that I am horrible at this. In casual conversations we speak quickly at about 170 words per minute or even higher. That's too fast for video calls. Video calls are still unnatural for us. Zoom fatigue is a real thing; there's a combination of a lot of elements that just make the experience … well, weird.

Find the speaking pace that's right for you. You want to be natural, but maybe deliberately be a little slower. You can find out how fast you talk easily by recording yourself speaking naturally for a minute and sending the recording to a text-to-speech app. Your Word processing program can quickly count how many words you spoke in that minute. If you are speaking higher than 170 words per minute, you're probably speaking too fast for your remote audience to comprehend every word. Alternatively, if you're like me, you probably know that you're speaking fast already, and you can just consciously slow it down a bit.

Sometimes, in lieu of slowing down, you can simply pause a bit more often. I personally find that this works better for me since I'm naturally such a fast speaker. Instead of slowing down the way you talk, just come up for air more frequently and give your audience a chance to catch up to what you've said. A well-placed silent pause can serve almost as well as slowing all your speech down.

Also make sure to speak clearly and enunciate your words. Saying words like "goin'" instead of going can be magnified in an online meeting and make what you're saying very hard to follow.

## Zoom Feature

There's a somewhat unknown feature hidden in Zoom that you might find useful. It's called "touch up my appearance" and is found under the Preferences and Video menu. It will give you an instant digital face-lift. Hey, it couldn't hurt, right? While you're there check all the other settings too. Zoom is always adding new privacy and usability features.

## Avoiding Last-Minute Technical Issues

Make sure you have a good wireless (or wired) connection and are away from any major noise sources like construction or your neighbor's lawn mower. In a home setting, you probably want to shut the windows before the meeting starts. If you don't have a fast Internet connection, consider upgrading if you will be in a long-term, work-from-home setting. A wired connection is best, but if you have solid WiFi service and a strong, consistent signal, then wireless is just fine.

When possible, try to log into your meeting a minute or two before start time. This is doubly true if it is a meeting that *you* are hosting. I've found that right before the important meetings it's always time to either re-authenticate or apply some critical system update. Don't be late to your own meetings.

Finally, close other windows and applications. These can create a lag on the system. As an added bonus, this will take away a major source of distraction and you'll pay better attention to the meeting.

## Leverage the Chat Window

One big differentiator with virtual meetings is the ability to use a chat window. This is quite useful for sharing information like a document or web link that might be useful to other participants. I've also found that chat windows give people an easier way to share their thoughts. In a big meeting, some people prefer typing out their thought or idea more than trying to break into the conversation to speak. I've found in some of our biggest, state-wide meetings, people seem much more comfortable raising questions and issues through the chat window rather than by speaking up in person. For this reason, I consider chat a big advantage of video meetings over in-person meetings.

## Hosting a Meeting

If you are actually running the meeting, there are a few other items to consider. The number one consideration, as previously mentioned, is to be on time to your own meetings. The longer you keep people waiting for a video call to start, the more they'll assume the meeting simply isn't happening and they might not log in at all. All the other rules of good meetings still apply to virtual meetings:

1. Have a formal agenda and share it in advance. Even if it's a few bullet points, make sure it is clear why people are attending your meeting and have respect for everyone's time by keeping the meeting on topic.
2. Start and end on time. Basic advice, but again you have to do it. Respect everyone's time by starting and ending on schedule. If you're running the meeting and you know you'll be late, pass the responsibility to a trusted colleague or let everyone know the time will change. If you finish your agenda early, give everyone their time back rather than trying to find additional topics to cover. Everyone will appreciate time back.
3. Keep everyone engaged. Pull others into the conversation by asking open questions or asking what they think about certain points that have been made. It's easy to be distracted in online meetings, so strive to keep everyone active and engaged in the conversation.

Video meetings are not like in-person meetings. They are an entirely new interactive experience, which requires a little forethought. This format is inherently awkward, and you may need to make some changes to adapt. The lack of audience response and difficulty making direct eye contact makes it much harder to read the room and get feedback from the audience.

# GIVING PRESENTATIONS

*Designing a presentation without an audience in mind*
*is like writing a love letter and addressing it "to whom it*
*may concern."*
~ Ken Haemer

Whatever company you're working for, if you have senior management's ear, you are going to need to create and deliver some PowerPoint presentations at some point. A lot of people have mixed or even outright hostile feelings towards PowerPoint, but there are ways to make using this tool less painful. In this section, I will go through some basics of PowerPoint and putting together and delivering a good presentation. The goal of this section is not to discuss the mechanics of how to use PowerPoint. There are plenty of good books and courses on that. The goal of this section is to think about how we leverage tools like PowerPoint to be more effective communicators.

PowerPoint is part of the Microsoft Office suite of tools and is standard in a lot of corporate American offices. There are several other competing tools, like Apple Keynote and Google Slides. We will focus on PowerPoint in this section, but the reality is that the problem isn't with the tool; it's with the user. Switching tools will not fix the problems I see with most presentations. Fixing your approach to the tool will work wonders though.

There are two common scenarios for using PowerPoint slides in corporate settings. One scenario is where the slides are essentially handouts. You speak to the slides, but they are really meant to capture a lot of raw data in a visual format. This type of PowerPoint deck can serve as a sort of more visual Word document and it is possible that it will simply be passed around in email rather than having any formal speaking presentation involved. Status slides like this are generally packed on purpose since the slides are probably not going to be shown as an overhead presentation to an audience.

The other scenario for PowerPoint is where you are giving a talk in front of people and slides are shown on an overhead display and are there to support and drive the conversation, not to convey tons of information. Avoid trying to combine the two in a scenario where you have packed status slides that are shown on an overhead display that no one can read.

I've seen some fantastic presentations where the speaker is backed up only by pictures in the slides that helped illustrate talking points. There were either no words at all or just simple captions on photos that help illustrate the subject. The speech is not encapsulated in the slides and the presentation becomes more about the speaker and the story than the slides. Obviously, if your scenario is presenting to the board of directors, this probably doesn't serve as a good handout and could even be perceived as unprofessional. I will discuss what makes a good board presentation in a later chapter.

Used as a medium to convey visual information, PowerPoint can be a dangerous tool. I've seen some people using six-point fonts to cram as much data as they can on a single page. I've even seen one firm assemble 100-page decks with every single page pretty much in the same, small font size. This was obviously not something that would be displayed to a large audience on an overhead display. The idea was that people would be sitting around a table and get handouts or else read it on their iPad to keep up with the main points. Here's the problem though: it's just too much data no matter how you get it. I am not a fan of busy PowerPoints that throw the reader a ton of data without telling a story or making a point that will be memorable. People want great stories and a great speaker, not computer visuals that make you feel like you're spending the afternoon at the optometrist.

With a metrics-oriented PowerPoint, it's OK to go a little smaller on the font to try and cram a bit more information on the page. But I'm willing to bet there's some basic editing that could make this more easily digestible and a lot less of an eye chart. It's a balance though. You don't want to have a lot of slides either, and if they can be consolidated, they should be consolidated. Just don't be part of the six-point-font club. Spend extra time taking text *off* the page rather than trying to fit everything on the page in a tiny font. I bet if you look carefully, you'll find all kinds of content people could live without.

# PRESENTING TO A LARGE VIRTUAL AUDIENCE

We already covered video meetings earlier in this chapter. If you are presenting in more of a one-way fashion to a large audience that doesn't involve as much interaction, there are a few other considerations to keep in mind. Presentations like this introduce some interesting challenges. You can't read the room. Many people may have cameras off, so you can't see their reactions or body language. Even if they have cameras on, it's just not that easy to see more than a few people at a time. This can make it a bit stressful presenting, as you may wind up feeling that no one is really listening.

So how can we solve this problem? How do we relieve our anxiety that nobody's paying attention? And of course, how can we make sure we are getting our message across? Virtual platforms introduce a few interesting ways to address this. It's not perfect, and it will never replace being with a live audience. But we need to make the best of things, so here are three ways to ensure you're getting through to everyone.

1) **Open with impact**. In these sorts of presentations, the audience will typically remember a good opening and a good closing. Starting out strong, maybe with a personal story or impactful statement, can help your audience realize that this might be an engaging presentation and that they'll want to pay attention. Summarizing the objective of the presentation and what they'll learn by listening is a great way to open strong.

2) **Use the chat function**. Despite online meeting platforms being very simple, we all have some anxiety around the listener's experience. Being able to engage chat for simple questions like what they're hoping to get out of the presentation can not only confirm that everyone can hear you but also help you steer the presentation in ways that the audience would like to see. Don't start with "can you hear me?" type questions, as it's a weak start and reinforces the fact that there's a big distance between everyone. A great opener is for people to add what they are hoping to get out of the presentation, which gives you a chance to do some last-minute tailoring for your audience.

3) **Keep it conversational**. Even if it's hard to engage the audience in a virtual presentation, you can do your best to keep things conversational.

4) **Engage the audience**. When possible, you can make a presentation more interesting by bringing the audience directly into it. Informal polls, a Q&A session or encouraging the use of the chat window can all be ways to get people more interested and engaged. Asking rhetorical questions that don't require an answer can be a reasonable substitute when you can't directly engage the audience. These questions sound a lot like "do you remember when …" or "have you ever had this problem before?" Rhetorical questions keep people mentally engaged since it forces them to think of their answer.

5) **Follow up**. Even if it's not practical to engage the audience in your presentation, you can encourage a feedback loop by emailing a summary to the attendees and asking for any feedback on future topics or improving what was already discussed.

Virtual presentations are inherently awkward. It's easy to forget your audience is even there. Try to focus on how you can serve them and what value you can add and you will alleviate some of your own anxiety and give everyone a better experience.

# MAKING POWERPOINT HURT LESS

Most people create PowerPoint decks by jumbling a bunch of seemingly random slides together and then calling it a presentation. This is a mistake. A good presentation starts with planning. It takes some time and effort, but there are ways to approach this tool that will make you stand out from the crowd and not be the speaker for yet another bad presentation.

Here are a few tips to do better with PowerPoint.

## Audience First

Of course, by now you know that you don't start any elaborate communication without considering the audience first. How else would you know what information you need to present, how you need to present it and other factors like what information they already know and for how much time you'll have to discuss it.

Think about what your audience *needs* to know and how they would best like to receive this information. For example, if you were putting something together for a client meeting to help them understand your security program, you would probably have a small time slot and want to keep the information very high level since the knowledge level about security will be mixed or nonexistent. You might want to consider what the 1–3 points are that you'd like to get across to this audience and then start planning your presentation from there.

## Storyboard It

Honestly, one of the best tips I ever had with PowerPoint was to put the tool away and get your high-level ideas down on 3 × 5 index cards. There's something a little more visceral about doing it the "analog" way. Each slide should have an idea or two in bullet points and all the slides should be arranged to collectively tell a story. Make sure your slides flow from one to the next. This is where storyboarding really helps. You can try different orders for slides without letting the tool get in the way of the process. It also helps you see a little more clearly when there's not a good transition from one slide to the next. Without using a proper transition, you really do have a collection of random material bundled together into what is a presentation in name only. As basic as it is, do you have a beginning, a middle and an end? Many presentations just have one slide after another. They begin and end randomly.

There is, of course, a thumbnail view of slides you can use instead, but in the very early stages of drafting a presentation, you are better off just jotting down some ideas on each slide, rather than getting into design. This lends itself to index cards very nicely. Using the thumbnail view in PowerPoint tends to make you bounce from the big picture view of what you're doing down to the little picture view of what's on each slide and how it looks. Resist the urge to get into the details until the big picture is clear.

## Images and Graphics

When you're working with PowerPoint it's best to factor in that this is not just a written medium but also a visual one. Straight text with no charts or graphs can get tedious for the audience. PowerPoint is not the right medium for lots of text. You'll want to strategically pick some images, charts, clipart or other graphics to break things up a bit. PowerPoint now comes with a feature called Design Ideas that can suggest some layouts to spice things up a bit. Like everything in life, this is a balance. Is this the board of directors' deck? That's probably not the right place for that purple background, white text and Word Art fonts. Also, the more formal your company is, the more they may already have some corporate branded templates that are either required to be used or are strongly suggested. Generally, you want to go with the corporate look and feel and not be overly creative if this is the case.

# A NOTE ON VISUAL COMMUNICATION

There's an old saying that a picture is worth a thousand words. This saying implies that complex ideas can be conveyed by a single image such as a photo or infographic. A picture can sometimes convey a message more effectively than a verbal or written description, which is why you should also consider visual communication skills and where to use them.

Visual communication can represent information in an easy-to-digest manner. Visual communication can complement verbal or written communication forms, such as a PowerPoint slide, and can greatly enhance the point you're trying to make. Think about adding a chart to a report. You can write a lot of raw numbers but showing them in a chart may get your point across far quicker. Visual aids like this can serve as a reminder or reinforcement of information. Video and other animations are also forms of visual communication.

Visual communication does not have to be used as a standalone communication method, although it certainly can be used this way. One of the strongest examples of standalone visual communication in the business world would be an org chart. Can you

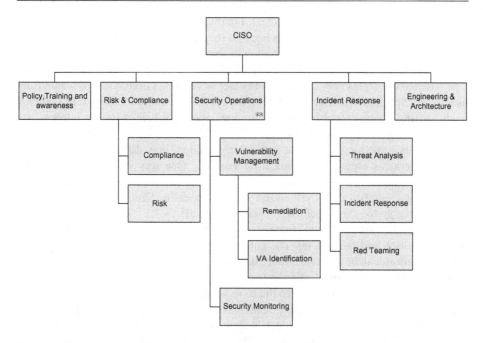

**FIGURE 6.1**  Some communication is best represented visually. Something like an org chart would be very difficult to understand without a visual aid.

imagine trying to communicate an organizational structure without a visual aid like an org chart? For anything but the smallest teams, this would be a lot to ask (see Figure 6.1).

Visual communication is often used to enhance written or verbal communication and can break up the monotony of seeing a lot of text. Using graphics to enhance reports, PowerPoint presentations and other communications with an image, chart or other graphics can really increase the chances of having your point understood quickly and clearly. Some other examples of using visuals to enhance your communication include:

- Data visualization
- Process flow diagrams
- Swim lane diagrams
- Heat maps
- Symbols and icons
- Using color to draw attention to certain points
- Infographics
- Process diagrams and flow charts
- Mind maps

Visual communication provides a lot of benefits by helping you relay messages faster and more succinctly. Visuals are also processed much quicker by the brain than text or speech. They help ensure that a consistent message is heard by your audience by giving less room for interpretation. Visuals spell things out in black-and-white and make points that may seem a little obtuse crystal clear. Finally, retention of information tends to be much stronger when using visual aids because they provide a mechanism to refresh the information without re-reading or listening to something more than once.

To use visual communication effectively, you should consider that its primary function is to make complex subjects simple. In other words, don't overdo it. You want visual aids to be helpful, informative and easy on the eyes. You don't want them to be complex or distracting. Done correctly, visual communication can help maintain interest in a subject and help you retain information.

If this is something you're not particularly good at, how do you go about getting some better ideas around communicating visually? One way is to study how other people use visuals. Try to note when you see someone use their own visual aids effectively. What about their use of visuals did you find appealing? You'll probably find some examples that go too far as well. Why was it too much?

You can also use visuals to create a sort of brand for the security program. Having a common look and feel for reports, policies, your Intranet site and your awareness communications can help people immediately recognize your area, the same way that a corporate logo easily helps identify your company.

Finally, take some time to learn the tools you already have at your disposal. If you manage a big team, you might have someone who's skilled at tools like PowerPoint, LucidChart or Microsoft Visio. Honestly though, these tools are easy to learn if you just dedicate a little time to them. LinkedIn Learning, Udemy and many other online sources have cheap or free classes you can take to learn some of the features that can really make your message more impactful through charts, diagrams or pictures.

The downside of visual communication is that it can be limited in its depth of information. Visual communication is great for making complex material simple and memorable, but if you try to convey too much information in a visual format it will rapidly go from helpful to being a hindrance. You don't want visuals so busy that it becomes an eyesore or over-stimulating for the observer. Remember, the goal is clear communication, not pretty graphics. Use any combination that makes your message clear and memorable, but don't try to squeak too much data into something just for the sake of using visual communication techniques. The technique is never the goal; a clear message is the goal.

# DELIVERING PRESENTATIONS

Once you have your presentation compiled, it's time to deliver the material. By now, you have thought through whom your audience is and what they want to know. You've storyboarded some ideas and put together a set of slides that don't provide too much information and not too much raw text. Now it's time to present the material.

The good news is that if you can have a conversation, you can deliver a presentation. I emphasize the word *conversation* because another issue where people miss with PowerPoint is reading slides to the audience like they can't see what's right there in front of them. Don't read or memorize every line in a presentation word-for-word. This just doesn't have the same impact as a conversation or personalized story. Reading slides verbatim is monotonous and will not engage your audience.

Instead, I recommend treating presentations as more of a conversation. Conversations are engaging and don't insult the audience's intelligence by having you simply read what they can already see. If you haven't done a lot of presentations before, I recommend doing a practice session or two in front of the mirror so you can see how you are doing. This will also make it a bit more of a muscle memory exercise when it's time for the real thing.

Some other thoughts on using PowerPoint presentations include:

1. If you start out with an overhead PowerPoint presentation saying, "I know most of you probably can't see this, but ..." then you have already failed. Why would you display anything that you know most people can't see? This doesn't sound audience friendly, does it? If you know your PowerPoint will be on an overhead display, aim for simplicity, fewer words and just enough material to drive the conversation. In this style of presentation, you should not be reading the slides anyway; you should be speaking to your audience.
2. Don't forget to use BLUF (bottom line up front). Start with your most important point; everything that follows should support it. Don't save the conclusion for the end but do provide a summary at the end that restates your main points.
3. Support one main idea on every slide. Once you've nailed down the objective of the presentation, make sure that every slide that follows moves towards that objective. If it doesn't support the main point, get it out of there. Your slides should have a natural flow and if they don't, make adjustments. Again, storyboarding helps.
4. Tell a story. Make sure you have a beginning, a middle and an end. It's not only OK to use metaphors, but I find that it really helps people understand difficult security concepts by relating them to something that everyone already knows.
5. Show some enthusiasm and vary your tone. No one likes monotone PowerPoint presentations.
6. Don't have tons of slides, though there's no harm in having reference slides in an appendix that you can quickly bring up should they be needed. Aim for fewer but better slides.

I'm sure most or all of you have heard about TED talks. Between TED and TEDx, the independent spinoff, you can find unlimited ideas about what good presentation delivery looks like. The best TED talks also take you through a single story and they don't stray far from a handful of points they are trying to make. They are so good because they are time-bound, simple and typically only making a single, impactful point. Good TED presentations also don't use a lot of slides that require you to watch the screen instead of the speaker. This has the added benefit of allowing people to listen to the talks on a podcast without feeling like they are missing anything.

While you may not think this style of talk works in a corporate setting, it does teach us a lot about great speakers. It teaches us that people want to hear great stories and listen to engaging speakers. The visuals are not the star of the show. Think about that while you agonize over the look and feel of your slides. Instead of working on slides, you might be better off polishing your voice and how you will present your story. Do you think you might just be stuck with a boring subject for an indifferent audience? Consider this as a challenge. Even accounting numbers can be interesting if you present your *insights* rather than your *information*. Spend the extra time to make your story relevant and engaging to the business. This is way more important to the audience than information on a page and some pretty charts. Even if your slides are just a bunch of charts and metrics, what is the "so what" for your audience? What can you tell them that only you would know? What's *not* captured in the slides and what *insights* can you pull from all this raw data?

To be more engaging, think about what the audience needs or wants to know. What do they want to know from you? What questions might the audience have and what would you like them to understand? What's your key message? If the audience only remembers one or two things, what would you want it to be? Again, focus on providing insights, not information. You can include information to support your insights, but don't just throw together a bunch of information and hope the audience has a few takeaway ideas. They won't.

You don't need to be Steve Jobs to be a great presenter. You just need to make connections with the audience, leave them with valuable insights and be somewhat memorable. An effective presentation gives the audience what they need to know. It doesn't have to do any more than that.

# WHAT'S YOUR STORY?

> *The best speakers in the world are the best storytellers.*
> *They have a gift to not only tell a great story, but also*
> *share a lot of the details that many others wouldn't.*
> ~ Larry Hagner

Great presenters and great leaders use storytelling to connect and make a point at a much more personal level than facts alone can achieve. Everyone loves a good story. People tell business stories to communicate and to connect. Business stories are different from regular stories, in that you tell them with a goal instead of for entertainment. When used correctly, this technique can create a personal connection between your audience and add a lot of power to your message. Stories can inspire, influence opinions and drive change.

Business stories tend to fall into a few basic categories. These range from background stories, where you are detailing your experience and who you are and where you've been. These stories help build rapport and better connect with people. Other stories can detail why you came to the company and what you intend to do. These stories help prove you don't have a hidden agenda and can be trusted. Stories can also be teaching stories, for example, they can be about a time that something didn't go as planned and what you learned from the experience and how it might apply to a decision that's on the table right now for discussion.

So rather than walking your business leaders through what they should be worried about in the event of some hypothetical security incident, it might be much more effective to walk them through an actual security incident. One of the best things about corporate storytelling is how it can bring a generic topic to life.

Consider this example. Rather than saying how the business should be concerned about third party security risks, you could walk them through a big, public breach like Target. In the Target breach, a heating, ventilation and air conditioning (HVAC) vendor compromised the Target cardholder network when someone stole credentials. The breach affected as many as 70 million customers and 40 million credit and debit card accounts and resulted in more than $300 million dollars in fines. The CIO and CEO were both forced to resign. This story is a great way to illustrate why a company should assess all vendors that have network connections, even if they seem low risk. What could be lower risk than an HVAC company, after all? Walking through this story would be much more effective than simply saying how all vendors with network connectivity need a security assessment regardless of their perceived risk because I said so.

Storytelling is a great technique to use when facts alone fail to convince someone that you're right. People are always more responsive to personal narratives than to plain facts and statistics. Storytelling is useful for anyone who presents updates and reports, or wants their ideas to influence decision-making.

# Building a Business Story

A business story can be constructed from a simple framework for your supporting facts. I like to think of business storytelling in four elements:

- **Setting**: Your setting is your context for the story. A setting can be a place, a point in time or anything that establishes the context for the message you are trying to convey.
- **Characters**: Your characters are those who are directly involved in a situation. You don't have to be a character in your story; it only needs to be the people affected by the situation. In the Target example, the characters are actually two companies.
- **Conflict**: Once your setting and characters are established, the conflict details the problem that needs to be resolved. Conflict gives your audience a reason to care about the characters.
- **Resolution**: The resolution is how the conflict is ultimately resolved. Just like in real life, not all stories have happy endings. A resolution in this context just means an end to the conflict—positive or negative.

As always with communications, storytelling is an audience-first exercise. You need to have a good understanding of who you're communicating with, what their background is, how many details they will tolerate and what types of narratives would resonate best.

There are some important points to keep in mind with corporate storytelling:

- **Have a point**. Don't tell stories that go nowhere and or take too long to get somewhere. Tell stories to make a well-defined point and are directly related to what was already being discussed. Make sure the story helps enhance and not detract from the conversation that was already in progress.
- **Provide enough background**. You want to provide just enough detail without becoming bogged down in them. Details should be both relevant and necessary to the points you're making.
- **Action**. Typically, some action needs to take place or it's not much of a story. For example, you might detail what happened during a security incident, even if this includes some mistakes that people made.
- **Result**. What happened at the end of your story? What should everyone learn from this? What is the key takeaway or the proverbial moral of the story?
- **Humor**. Use humor strategically, but don't use humor to be too edgy. A well-placed laugh can help you connect better, make you more interesting and help liven the mood.

Finally, be authentic. If your story is drawing from your own career, don't be too afraid to tell details that show poor judgment or mistakes on your part. We all make mistakes, and this can help people relate to you on a personal level. Try to end on a positive note though, even if there were some problems along the way. For example, if things ended poorly what positive lessons can be applied to a new situation?

# BETTER PUBLIC SPEAKING: FROM CONFERENCES PODCASTS AND INTERVIEWS

You might be asked to speak at a conference, podcast or webcast in your career. If so, congratulations! These are great opportunities to share your knowledge and market yourself a little. Public speaking can boost your confidence, improve your communication skills and help you boost your social network. If you work in consulting or ever want to become a consultant, public speaking can be a huge boost to your credibility as well. Public speaking helps you learn how to deal with stressful situations, or at least situations you may perceive as stressful. Once you get on the main stage at a big conference to talk, imagine how much easier your next board presentation is going to feel!

While I've done a lot of public speaking over the years, I'm not naturally drawn to it. But when someone asks me to speak, I will generally go along with it if my schedule allows since I think it's important to give back to the community. I've given mainstage keynote talks, I've been on panels and I've been interviewed one-on-one. If I can do this, anyone can. Trust me, I mean anyone.

A lot of people are afraid of public speaking. In fact, many people are more afraid of public speaking than of death, snakes or heights. I personally think that most people who say they're not "good at public speaking" or are afraid of it, simply don't do it enough. A 1960 study titled "The Visual Cliff," by Eleanor J. Gibson and Richard D. Walk, found that there are only two fears that we have from birth. One is fear of falling and the other is fear of loud sounds. You might note that both of these fears are directly related to survival skills. Everything else is a learned fear. If you're afraid of public speaking, you have somehow learned this fear from experience or from watching other people. It's also possible that the fear is just entirely in your mind because of how you imagine things might unfold. Fundamentally, it's a fear of humiliation that makes most people never get on the stage or in front of the camera in the first place.

Speaking in public does not represent the possibility of actual physical harm. Really. So, it may take some practice, but it's possible to become comfortable with public speaking. In fact, you can become very comfortable at it. The only time I did worry about speaking in public, I was at an ISACA conference for a talk that was supposed to include a partner speaking with me. My partner had been given the wrong time for the event and was nowhere near the Jacob Javits Center in Manhattan. Our talk was supposed to go on in 15 minutes and my co-speaker wasn't there yet. We had a 45-minute block on the main stage in front of a lot of people. We had put a lot of effort into getting the session on the agenda. I had to either go on solo and hope for the best or forfeit the session altogether since they couldn't move other sessions around for us. As far as speaking events going wrong, this would be a worst-case nightmare for a lot of people. But I didn't want to lose our slot and our only chance to speak, so I got up solo and just started, not even sure I could fake my way through or not.

This event was supposed to be a back-and-forth conversation with some questions and answers we would ask each other. So that plan sure had to change quick. I also knew I had nowhere near enough material to fill our 45-minute slot. So, what did I do? I got past the fear and just started, without any real indication that this was going to end well. After about 15 minutes, my co-presenter finally ran up to the stage and joined me. He had somehow defied the odds and the traffic of Manhattan and made it. We not only got through the talk, but we also got a lot of really good feedback on the session. Even when things seem to go wrong, they can be recovered and even your worst case is not that bad.

If I can get through the ISACA meltdown, you can get through efforts that are better planned with ease. Sure, things can go wrong, but it's actually pretty rare and you can minimize your risk by preparing and practicing. A lot of the reason my talk went well is because I got past fear and just started talking, trusting that this was somehow going to work out for me. While it was tempting to cancel our session, I knew that we both would have been disappointed. However, everything that I had done to prepare had to be thrown out the window and I had to improvise using not much more than the subject of our talk.

Today, I am often called upon at the last minute to fill in for a speaker who cancels. I am known as someone who can step in at a moment's notice without much preparation and get the job done. I don't say this to brag. I'm saying this because as a pretty extreme introvert I can promise you that if you get enough practice speaking, there is no reason you couldn't reach the same level of comfort that I have.

Like all the communication skills we talk about in this book, the key to getting better is by practicing, not just by reading about it. You won't get better at driving a car by reading about it either. You might learn some concepts, but you can only implement these concepts with hands-on practice.

If you want to get better at public speaking, I recommend working on these five steps.

## Step 1: Start with Your Mindset

What are you trying to improve? Fear of speaking? Depending too much on your notes? Too many "uhs" and "ums"? Pick some goals to focus on improving. The more you can focus on specific problems, the faster you can find solutions and improve. "Getting better at speaking" is not a tangible goal that can benefit from any specific measures or practices. You'll want to define some tangible objectives and measure progress. This could include:

- Fear of getting in front of an audience in the first place
- Not depending on notes
- Eliminating filler words (uhs, ums)
- Speaking slower (this one is still a problem for me!)
- Speaking faster
- Having more melody in your vocal range, avoiding monotone
- Pacing back and forth and other nonverbal problems

Once you've outlined some objectives, you can look for opportunities to practice. Keep in mind, you don't always want or need to practice in front of a live audience. If you do have the opportunity, you can record what you say and review it later to find areas you might want to improve.

## Step 2: Prepare

No matter if you are on a simple panel or a keynote speaker, you need to prepare at least a little. While it is possible to go into a talk cold without preparation, that is never my preference. You owe it to the audience to prepare. You owe it to yourself to prepare as a professional. You should do a lot more preparing than the actual talk. Are you planning to talk for 30 minutes? You should probably be preparing at least 90 minutes. This is more relevant when you're not as comfortable with the process. Over time, you will find that you can prepare for most things quickly as long as you're already familiar with the subject matter.

Preparation might include preparing a formal keynote and PowerPoint, which takes a lot of work. Or you might be on a simple panel discussion, where you will at least want to understand the topics and questions being proposed in advance and what you might say back for an answer. In any case, the more you prepare, practice (the next step) and visualize your success the more you will likely have a strong presentation.

If you remember from the PowerPoint material earlier in this chapter, you'll want to make sure you are presenting *insights* rather than information. The preparation phase is where you want to get this right. You're an expert on your subject, what insights can you share that would be thoughtful and relevant for your audience?

## Step 3: Practice!

Preparation is different from practice. In the practice phase, you are taking what you have prepared and running through it all as if it were a live session. Again, most people who say they're "not good at public speaking" simply don't do it enough. And they never get any better because fear holds them back from trying. If fear is holding you back, consider starting out on a panel discussion. If you're not familiar with this format, A panel discussion (or simply a "panel") involves a group of people (usually 3–6) talking about a common topic in front of an audience. You've probably seen many of these at conferences and on webcasts.

I love panel discussions because they are low pressure and low preparation. In this format, the group sits down and talks about their views on a subject. If you don't have anything strong to say, you can sort of punt to one of your other panelists. Easy. I do these at a moment's notice and have even filled in with no notice when another panelist wasn't able to make it at conferences. They can be fun, but they're especially fun because I don't have to burn a lot of cycles preparing for them. You simply talk. Obviously, if it's a topic that you have limited familiarity with, you'll probably want to pass.

Fortunately, most panel discussions are not held at the last minute with no preparation. They are usually planned well in advance with the moderator sharing some thoughts on questions and the panelists agreeing or changing questions around a bit. Typically, this can all be done over the phone, email or Zoom. I've found that it helps to write down some notes on points I'd like to make for each question. If you go through this exercise, you probably won't even need your notes. The discipline of writing down some answers on paper is enough for you to get comfortable visualizing that if you were on stage, video or whatever medium; this is what I would say to answer this question. Resist the urge to take extensive notes and bring them with you. In the heat of the moment, bullet points can serve as fast reminders of what you wanted to say, but you will not have time to read through notes.

Keynotes and solo sessions are different. In my first keynote speech, I spent a bunch of time preparing a PowerPoint deck and my basic talk. Then something interesting happened. I did a dry run to practice. I was given an hour to fill and my talk was just about 20 minutes. I was supposed to be on the big stage at a well-attended conference and I didn't have enough material. Good to know! This is why we practice.

Part of the problem is I'm a fast talker (thank you, New York City) and I like to get to the point quickly. But filling 20 out of the 60 minutes wasn't going to work for me, the conference organizers or the attendees. I had work to do. You don't want to find this out at the event while you're on stage. In a keynote format if you aren't talking there's total silence. There are no other panelists to jump in and save the day. In my case, the mistake I made was that I had spent a lot of time on overhead-friendly PowerPoint slides and thought I had more than enough material to fill the time. I didn't.

I'm grateful I figured this out before going live and I learned a valuable lesson about doing a dry run alone even if you feel silly talking to a mirror or pretend audience. Steve Jobs practiced in full presentation mode exactly the way he would as if he were in front of an audience of thousands. He practiced relentlessly weeks ahead of a product launch. That's why it looked so easy when it was showtime. If you've never watched a full presentation when he was Apple's CEO, go find his talks on YouTube. A lot of people think he was a natural, but the reality is he made it *look* natural by practicing a lot. Michael Jordan looked natural when he got on the basketball court because when it was showtime it *was* natural. And it was natural because of all the practice that went in beforehand.

## Step 4: Get Some Feedback

You can be overly harsh or underly harsh on yourself. Neither view is productive, when what you really need is an objective view. Go get some independent feedback or give your talk on a smaller stage in front of a neutral audience. I've never used Toastmasters before (https://www.toastmasters.org) but I hear this is a great way to get practice and feedback without putting your reputation on the line in a formal setting.

You should consider feedback as a "continuous improvement" exercise. Assume you're starting out in need of at least some improvement and take feedback seriously. Make improvements along the way and don't take any of it personally. Every talk you give, you will get better and better if you incorporate this feedback loop and do a little deliberate practice.

## Step 5: Pace Yourself

I mentioned that I talk faster than I should, so this is something that I try to keep improving. When you're speaking, the listener needs to process what you are saying, so speaking too fast doesn't give them time to process what you're saying. This can be hard for some people, and it's hard for me. If you want to make some immediate improvement, incorporate some pauses and stops. Even if you are speaking too fast, this will at least give your audience a chance to catch up to you and it may be easier than trying to change your personality overnight.

# OTHER CONSIDERATIONS

In public speaking situations, not only do you need to understand your audience, but you should also understand any constraints like time/geographic constraints, technological requirements, and any contextual factors which could influence the way the message is received.

These parameters are important and can influence not only what you are going to say, but how you should say it to be most effective. For example, if you are asked to give a quick, informal update on the security program in a town hall, here's how you can break down some parameter considerations before you even open your mouth. You will then tailor your content based on these parameters.

1) **Time**: This isn't a keynote at a conference. You've been asked to give a brief update. Think of a few big highlights showing progress of the program or achievements you're proud of that can be shared in a few minutes.
2) **Audience**: A town hall is generally a very large mixture of backgrounds. You'll want to minimize technical jargon and acronyms and be inclusive in what you say.

3) **Detail**: Make sure you're crediting the whole team, not just yourself. You've been asked for a brief update, provide only enough detail to highlight the programs and their benefits. For example, you wouldn't just say: "We've deployed a modern intrusion detection system." While potentially true, it doesn't really convey any real benefit and sounds like an update made for your team rather than for a diverse audience. How about something more like: "We've deployed a security monitoring tool that will help us detect and respond to sophisticated attacks that might have gone undetected in the past. This will allow us to accelerate the company's Internet expansion strategy and move more aggressively into digital sales."

# YOU'RE STILL NERVOUS? THAT'S GREAT!

*Let fear be a counselor and not a jailer.*
~ Tony Robbins

We already covered how public speaking is scarier than death for some people. So, this means that, even after several speaking engagements, there might still be some nervousness involved before a public speaking event. Congratulations! You're human. Everyone has at least some jitters. You can use this to your advantage rather than letting it derail your presentation. This is all a mindset problem and can be turned around by reframing the situation.

A lot of stage performers use nervousness as a cue to keep them focused and spend more time preparing and rehearsing. In his book *Psyched Up: How the Science of Mental Preparation Can Help You Succeed*, Author Daniel McGinn talks about pre-performance rituals for people who have to perform in high-stakes situations. One interesting point made in the book is that being calm and being nervous are worlds apart from each other on opposite ends of the spectrum. Don't expect to go from nervous to calm, it's going to be too hard. A more realistic approach might be going from nervous to excited. According to the author, this can actually increase your performance. The takeaway here is don't try to calm down because it probably won't work and you'll get either frustrated or more nervous. But sitting there being nervous probably won't help much either, because you'll start to visualize all the things that could go wrong and how you'll look like a fool in front of all those people in the audience. Instead, put your focus into preparation and visualizing how everything will go very well. Some performers and athletes use "warm up rituals" that help them reframe nerves into performance energy. Visualize your success. Listen to a song that pumps you up. Tony Robbins jumps on a mini trampoline before his live shows to help not only get excited, but no doubt to harness his own nerves. Don't get nervous, get excited.

What do you do if you find the anxiety overwhelming and none of the above is working for you? Try some deep breathing. Breathwork is amazing, literally acting as a remote control for your nervous system with the right practice. There are calming breaths and energizing breaths. One pattern that works very well for calming down is "box breathing" made famous by the Navy Seals.

Seals are put into some of the most stressful situations any human being will ever face. Because of this, they've done a lot of research into what works and doesn't work for reducing stress in combat or other dangerous situations where ordinary people might panic. They manage their physiology to better control their psychology. And it works.

Box breathing is as simple as it is effective. Here's how it works:

1) Breathe in for four seconds.
2) Hold your breath for four seconds.
3) Exhale for four seconds.
4) With your lungs empty, hold your breath for four seconds.

4-4-4-4, in, hold, out, hold and repeat. You can also gradually increase the size of the box if it suits you, so 5-5-5-5 and so on until you have a comfortable number. The great thing about breathwork is that your breath is always with you, so you can practice this right before that big board meeting just as you can do it right before going on stage.

Interested in learning more about breathwork? I've included some resources at the end of the chapter and recommend *Breathwrk: Breathing Exercises* on IOS devices, by Breathwork Inc. (https://www.breathwrk.com). They have a number of different calming and energizing techniques to try for a nominal subscription fee.

---

# SUMMARY

---

In this chapter you learned that how you speak can determine how you are perceived. One of the most common forms of communication we have is verbal communication. How we speak and what we say has a great deal to do with how people will perceive and interact with us.

- While you can make subtle changes to your speaking voice, avoid doing anything overly dramatic.
- There's a certain level of gravitas, or seriousness, expected from security leaders. Minding your tone, pace and overall look can help project an image of authority.
- Your voice is like an instrument. It benefits from being warmed up and in tune. Pitch and pace and other factors determine how good you sound.
- Phone calls are sometimes a welcome change of pace from Zoom calls and can be just as effective, especially with bigger audiences.
- Storytelling is a powerful means of conveying basic information through a narrative. Stories can help you build a personal connection faster than just presenting facts and figures. Stories are typically formed by describing a setting, characters, conflict and a resolution.

- Meetings can waste a lot of time and you do not want to be part of this problem. You should always make sure that a meeting is warranted before scheduling one. Preparing, having a clear objective, sharing an agenda and managing the meeting will all help keep your meetings on track.
- Presentations are generally used either as handouts or as visuals on a screen. Both formats dictate how much data should be put on a single slide.
- Speaking up in business meetings is expected of senior leaders. Make sure that when you do speak up that you are adding value.
- Preparation and practice are key when you're giving more formal presentations.
- Don't expect to turn your nerves off if you're speaking in public. Instead, use this to build energy and enthusiasm.

# REFERENCES AND RECOMMENDED READING

Berkun, Scott. *Confessions of a Public Speaker.* O'Reilly, 2011.

Carnegie, Dale, and J. Berg Esenwein. *The Art of Public Speaking: Enrich Your Life the Dale Carnegie Way.* General Press, 2018.

Duarte, Nancy. *Resonate: Present Visual Stories That Transform Audiences.* John Wiley & Sons, 2010.

Gallo, Carmine. *Talk Like TED.* Pan Books Ltd, 2016.

Hawkins, Charlie. *First Aid for Meetings: Quick Fixes and Major Repairs for Running Effective Meetings.* BookPartners, 1997.

Heath, Chip, and Dan Heath. *Made to Stick: Why Some Ideas Take Hold and Others Come Unstuck.* Random House Books, 2009.

Herold, Cameron. *Meetings Suck: Turning One of the Most Loathed Elements of Business into One of the Most Valuable.* Lioncrest Publishing, 2016.

Lent, Richard M. *Leading Great Meetings: How to Structure Yours for Success.* Meeting for Results, 2015.

Reynolds, Gretchen. *Presentation Zen.* New Riders, 2007.

Roger Love. "Online Voice Training & Singing Training." *Roger Love*, 14 Sept. 2020, rogerlove.com/.

# Communication Superpowers

7

This chapter covers a few skills that I consider to be "communication superpowers." These are the skills I wish I had much earlier in my career and the skills I still strive to develop and keep sharp. They include reading body language, saying no, negotiating, the elevator pitch, managing salespeople, remembering names, asking better questions, conflict resolution, dealing with people you can't stand and asking for help the right way.

## A CRASH COURSE ON NONVERBAL COMMUNICATION

*The most important thing in communication is hearing what isn't said.*
~ Peter Drucker

**Why nonverbal communication is a superpower**: Most communication is actually nonverbal. This means that the words you're saying may not convey anywhere near as much information as your nonverbal cues. Learning to read people's cues and learning to be conscious of your own can help you be a more effective communicator.

Most people think of communication as being either verbal or written and using words and language. There have been several studies on nonverbal communication trying to gauge its importance. While there isn't 100% agreement, most of the studies agree that something upwards of 70% of communication is nonverbal and it may be up to 93%. There are many books on nonverbal communication, and this is a big subject. This section is intended as a brief overview to give you a flavor of what nonverbal communication is and how it can have an impact on your other forms of communication.

Mastering nonverbal communication takes understanding two perspectives. One is mastering how to read nonverbal cues in others, and the other is mastering your own nonverbal language. When you use body language that is relaxed and open, you facilitate communication by putting others at ease. This can also work against you if you are perceived as not open and approachable or you have body language that sends the wrong signals. On the other hand, reading these signals in others can help you understand when someone is disengaged, angry or otherwise not receptive to what you are trying to communicate.

DOI: 10.1201/9781003100294-8

The thoughts that are going on in your mind tend to come out in your body language involuntarily. If your mind is full of anxiety, stress or even anger at someone, you can rest assured that it will come out in your nonverbal cues. Sometimes these cues are painfully obvious to people and other times they may just come off as a "bad vibe" that people can't quite pinpoint. Most body language comes out subconsciously. Although body language can't always be controlled (try to keep a poker face when your kids or the family dog runs excitedly to greet you) you can influence and be aware of your own body language and learn to read it in others.

When you're running a security program, you can expect some difficult or tense conversations. Ideally, these situations will be resolved calmly. But sometimes conversations are complicated by feelings of nervousness, stress, defensiveness or even anger. And, though we may try to hide them, these emotions often show through in our body language. Some signs of this include:

- Minimal eye contact
- Arms are folded in a defensive posture
- Tense facial expressions
- Body is turned away from you

When you notice that someone you're talking to is disengaged, you're in a better position to do something about it. Pull the person back into the conversation by asking a direct and open question. If you think that someone is angry, it might be better to come back later and try again.

# How to Read Body Language

There are several considerations when you're looking for body language cues from people. Elements of nonverbal communication include tone, facial expression, fidgeting, hand gestures and posture. These elements can help you both communicate more effectively and understand to read the hidden messages in other people's communications. While you are watching for all these subtle cues, you are also giving off your own cues.

Let's examine a few of these in a little more detail.

## Use Vocal Dynamics

Speaking style, pitch, rate and volume all contribute to understanding a speaker and have nothing to do with what is being said. The same sentence can come off as factual, sarcastic, angry or any number of other emotions depending on *how* you say it. Changes in tone may suggest that a person is getting irritated, impatient or frustrated. Changes in tone can also indicate that what is being said and what was meant to be said may be different. For example, in explaining that a project may be delayed your boss might say "no big deal" but perhaps their tone went from neutral to cold and sharp. If this is the case, it might be a bigger deal than they're letting on to you.

## Facial Expression

One of the best body language indicators is your facial expression. Facial expression is closely tied to our emotions and can provide a lot of information about what we are really thinking. Facial expressions are also mostly involuntary, which means you might not be able to control them. Whether you are able to control your expression or not, you should be aware of what signals you might be sending. When reading someone else, you can put a lot of credence into various facial expressions as registering excitement, disappointment or even lying.

Along with facial expressions, head movement can be one of the easiest nonverbal cues to read. For example, when people are nodding along with what you are saying, it's a good indication that they understand and agree with what you're saying. It's also a good indication that they are actually listening to what you're saying and following along. Confusion is another obvious emotion that is easy to read.

Do you want to know the ultimate body language tip that will make you more approachable to others? Put on a smile. Not a forced smile, which can look creepy. Put on a genuine and heart-warming smile. Science has measured some 19 different smiles, believe it or not. Rather than trying to understand and master all these different types, why not just think of something pleasant and crack a heartfelt smile. There's an old saying in the qigong community that you should "smile from the heart." The advice for this technique is not to intellectualize what it means or try to learn some special technique. That doesn't work with this one; you're supposed to just do it. Look deep into your heart and start to smile from the inside out. You'll find that the rest happens naturally. It also has the added benefit of boosting your mood quite a bit. Give it a shot.

## Fidgeting

Fidgeting can express boredom, disengagement or anxiety. Anxious fidgeting occurs because the body has elevated levels of stress hormones. If you are reading this signal in someone else, perhaps it's time to wrap up the current subject or come back to it at a better time. If someone is this uncomfortable, you shouldn't take it personally unless they're like this every time you meet them. That's one of the challenges with body language; sometimes something seems to be sending a positive or negative signal, but it's more of a personality quirk. Remember to factor this in whenever you are trying to read the tea leaves with body language signals.

The ultimate "I'm not listening to you" body language signal is messing around with your smart phone while people are talking to you. I participated in a communication exercise once where we were secretly instructed to look at everyone but the person talking. You could play with your phone, look away or otherwise be completely disengaged in the conversation. The speaker was simply told to keep talking no matter what the listener did but was not informed about what the listener would be doing. Exaggerating this level of disengagement is quite disconcerting if you're in the speaker role. If someone is doing this to you for real and you need to get through to them, sometimes a great technique is to simply stop talking but don't leave the room. Just stand there and wait for them to finish what they're doing, such as replying to text messages.

It can help underscore that their behavior is rude and will hopefully encourage them to put away the phone for a minute or two and listen.

## Posture

Slouching is the opposite of sitting or standing up straight. Slouching and other bad posture may indicate the listener is anxious, unfriendly, bored or uninterested in the conversation. I think it's also easier to pay attention and be more engaged when you are not slouching, so mind this signal in your own body language. This said, sometimes people are just taking a break or exhausted. Read this signal carefully and the meaning of slouching as one of several possible meanings.

## Eye Contact

Avoiding eye contact signals that a person does not want to engage in a conversation or is somehow uncomfortable around you. This is a favorite tactic of students who don't want to be called on by the teacher. It's a nonverbal cue that asks to please pick someone else, anyone else but me. Avoiding eye contact can also be a form of submissiveness. Subordinates may sometimes avoid eye contact to indicate submission or respect.

When you're minding your own eye contact, you want to be careful not to stare or overdo it. Eye contact, when employed correctly, shows confidence and can create much deeper connections with people. They say the eyes are the mirror of the soul. Making eye contact is a very powerful way to connect. However, when it is overdone, it can be creepy and disconcerting for the person being stared at. Don't force making eye contact, or you're probably leaning towards the latter.

# Your Body Language

Positive reinforcement of body language, combined with the words you're speaking, can add strength to your verbal messages and help avoid mixed signals. While you shouldn't go overboard with any of these techniques, some general tips to keep in mind are:

**Negative body language:**
- Avoid crossed arms as they can sometimes seem defensive, but they can also indicate that you're cold!
- Avoid fidgeting, staring and the other issues already mentioned.
- Avoid having a tense facial expression.
- Avoid keeping your body turned away from the speaker.
- Avoid having poor eye contact.

**Positive body language:**
- Use a firm (but not overly firm) handshake when you're greeting someone.
- Keep an easy-going posture without slouching.
- Maintain an open body position (arms unfolded).
- Maintain an upright posture.
- Keep your facial expression relaxed and open.

- Keep your arms hanging relaxed by the sides.
- Maintain regular eye contact.

It's worth noting that there's another type of nonverbal communication and that's touch. I would personally stay away from this completely in an office setting. While this does vary in different cultures, my perspective in the United States is that you should simply stay away from unsolicited touching with the sole exception of the handshake. In a post-COVID world, maybe even that one is off limits now.

# SAYING NO

**Why saying no is a superpower**: In the book *Anything You Want*, author Derek Sivers states that if you are asked to do something, the answer should either be a "hell yeah!" or else it's a "no." This rule has served me well. That is, when I actually follow it. Derek's philosophy is that if you're "busy" all the time, you are simply out of control of your life. How many commitments come your way that you reflexively say yes to when you probably should have said no? It might be that product demo you're not really interested in seeing. A favor for a friend. Speaking at a conference that doesn't really do much for your schedule. It might even just be meeting some friends when you're not in the zone or are already overcommitted.

Giving yourself the permission to say no is potentially *the* superpower of the distracted age. It is the ultimate productivity hack. It will help you remove the things that don't add enough value to your life and don't move your security program or other goals forward. There is a limited amount of time in the week, and you want to be protective of your calendar and time. Why do we say yes to things that don't excite us? It might be our default reaction if we are an accommodating personality. It might be that we don't want to disappoint the person asking us. It might be because we were taken by surprise and didn't have enough time to think about the answer that came out of our mouth. So how do we say no gracefully without being a jerk? Here are my top three tips.

## Use a Personal Policy

We're security professionals, right? So, set a policy and enforce it. This is one of my favorite techniques because it's so simple and it fits our personalities. You can have a personal "policy" like denying all social requests until your big project is complete. I had a policy to limit social engagements while I was writing this book. And yes, I did break it now and then, but only when the answer was a "hell yes." Just like any policy, there can be exceptions that should be handled with some conscious thought and deliberation. A policy is intended to handle the majority of issues that come your way. It's intended to handle the simple things, so that you can focus on the complex decisions when they arise and avoid decision fatigue. Personal policies don't have to be overly rigid. Having a policy and sticking to it makes rejection for the requestor a lot less personal. A lot of people got it right away when I said I wasn't taking on anything new until my big project was complete while I was writing this book.

Conversely, you can also have a policy that you say "yes" under certain circumstances. You might have a policy that you'll be happy to meet friends on a weeknight, but your weekends are for you and your family. You can set a policy that you will only look at new vendor products when your other four product implementations are complete, or only after new budget frees up to spend. Get clear on your priorities, put a policy in place and stick with it. This is also called a personal boundary, and they're fantastic when you follow them. Having a personal policy eliminates uncertainty when someone asks you to do things that are not in alignment with your values and takes the struggle out of saying no because the answer is already decided in advance.

## Decline Gracefully

Always go with your instinct. If someone asks you to do something that you're just not that into, your answer should probably be no. It's not hell yes, right? But what if you're feeling awkward about it? It might be a friend or someone who did you a favor and you feel you owe them. If something doesn't fit into your personal policy or if it's something you're on the fence with, make sure you buy some time and think about your response. It's OK to simply ask for some time to think about it and tell the requester you'll get back to them. Of course, if you're sure right away that this isn't for you, it's probably best to just say no on the spot. Don't put off difficult conversations, get them out of the way and move on with life.

For the decisions that you've had some time to think about but don't want to go forward with, it's best to let the requestor down gracefully. You can start by complimenting the request and letting the person know you're flattered by it. Then let them know that you've thought carefully about it and it's just not the right opportunity for you right now. You don't need to go into a lot of subsequent details or excuses. In some cases, this just opens you up for a rebuttal to all your reasons for declining. Stand firm, and simply emphasize that you gave it some thought, and this is your answer. Letting the requestor know you've put some thought into it helps them at least understand that you considered the request seriously and hopefully provides them with a gentler letdown.

## What If You Should Have Said No but Didn't?

I don't like backing out of things I've already said yes to, but sometimes it's a necessity. When I started writing this book, I was already committed to several outside projects. I started to realize that I wasn't being honest with my team of collaborators that were expecting me to contribute to their project. The reality was that I always had a conflict on the calendar and my "free" time was consumed with several other priorities. I should have said no, but I didn't. If I had viewed the project in hindsight, I would have said no, but I didn't consider all the more important things I had going on in my life.

Remember that sometimes if you provide too many details on why you're saying no that you open yourself up to rebuttals. However, sometimes it's also completely appropriate, such as when you owe a better explanation than a one-word answer. In my case, I fessed up by explaining that given my new job, the book and all my other prior commitments, I was not going to be able to give this project the attention it deserved. It worked out well, because we were actually able to come to an agreement where I could take on an extremely lightweight advisory role that had a flexible schedule and light workload of only a few minutes a month instead of a few hours a week. This was completely acceptable to me, since it addressed my real concern of time commitment. Of course, had I thought this whole situation through BEFORE saying yes, it would have been much easier! Not having enough open bandwidth is always a good excuse in the security world because it's always true.

## What If It's Your Boss?

Can you say no to your boss? Of course, you can! There just a little bit more of an art to saying no and you don't want to call this favor in very often. The boss is always right, and they can certainly assign you work so you don't want to make saying no a reflexive answer. When it comes to power struggles, you are going to lose this one every time. What you can do in this situation is think about your reasons for wanting to say no. Is it because you're already overworked and can't delegate the task? Is it a task that doesn't have much to do with the security program and will serve as a big distraction? Understanding your reason for wanting to say no is important so that you can define what success in this confrontation looks like. Is success really saying no to a new assignment? Or is it just avoiding an already overcrowded list of things that need to be done? Once you understand your reasoning, you can then reframe your answer appropriately.

Make sure you have a valid reason to say no with your boss. If you can't delegate a task and don't have the time for it, you can negotiate a truce by trying to take something else off your list to make room for the new task. Perhaps you can even delegate one of your older tasks to take on the new one. Is this new task really less important than the others you're already working on? If you don't currently have the skills for the assignment, could you acquire them quickly? Is there anyone you could think of that might already be in a better position for this assignment? This last tactic is especially effective because it doesn't leave your boss with a task that still needs to get done and no idea about who can handle it.

The wrong reasons to say no to an assignment include projects that might seem too challenging, projects that aren't in your job description and personal reasons like upcoming vacations. In the event of a personal conflict like a vacation it's OK to remind your boss that you had something planned, but it's not OK to say no to the assignment because you want your personal life to take precedence. Most bosses will not ask you to cancel vacation plans unless it's really, really critical. In this case, you can try to negotiate the timeline of the project so that you're hopefully not working on your vacation.

If you've hit the point where no is the right answer to give your boss, you'll need to explain your logic and present this carefully. The situations where you would be warranted to push back on an assignment include if there are literally not enough hours to get the work done by the deadline (be prepared to show your math!), there's too big of an impact on a more critical project or when you simply do not have the right skills to do the work and also can't acquire them quickly.

Be careful with this one. You don't want to be perceived as difficult or overly aggressive in your pushback. Help your boss think of alternatives to get the work done more effectively by someone else.

# NEGOTIATING LIKE A BOSS

> *My father said: you must never try to make all the money*
> *that's in a deal. Let the other fellow make some money*
> *too, because if you have a reputation for always making*
> *all the money, you won't have many deals.*
> ~ J. Paul Getty

**Why negotiating is a superpower**: You may think that negotiations are for lawyers or FBI agents. But maybe you should broaden your view of what you consider to be a negotiation. Business is full of negotiations. You will likely need to negotiate your budget or even your salary. You will have to negotiate promotions on behalf of your team. You will need to negotiate remediation timelines owned by groups not under your direct control. Do you still think you don't need negotiation skills?

A negotiation is simply a way that two or more parties settle their differences. It is the way that a compromise or an agreement is reached. In fact, I think a lot of people hate the idea of negotiating because they think everything will result in a winner and a loser (win–lose). In reality, negotiations can be win–win, lose–win, win–lose or even lose–lose. It's rare that everyone wants the same outcome and agrees on it right away, so negotiation skills are important to master.

What's a good framework for handling a negotiation? Here's a simple one to help you think about the negotiation process.

1) Prepare
2) Keep communication crystal clear
3) Double down on active listening
4) Collaborate and use teamwork
5) Focus on problem solving
6) Use decision-making abilities

## Prepare

Preparing for a negotiation means making sure that you have a clear understanding of what your objectives are and at least an idea of what the other side of the negotiation is looking for. You want to know whom you are negotiating with and where they might be willing to compromise. You'll also want to have some idea of what your requirements are and what you're willing to let go. In other words, you want to know your non-negotiables before going into a negotiation.

It helps to write these down on paper and force-rank them. Are you negotiating a new job offer? Maybe you're willing to let go of some aspirational salary requirements in place of some extra vacation time. Get clear on what you "must have" and what you would "like to have." Later in the negotiation process, it might pay to give a little back to get a little back.

## Keep Communications Crystal Clear

Negotiations are high stakes in that there usually is an outcome. This type of communication isn't just about getting your point understood; you are looking for tangible results and an outcome. You might be negotiating what's for dinner or who cleans the dishes, or you may be negotiating a salary or something where the stakes are much higher. In every case, there is usually an outcome that you're looking for, so keeping communications on both *sides* crystal clear is important. There may be multiple mediums used to help with clarity, like a written follow-up that recaps a verbal conversation. You'll need to weigh every word carefully and make sure that you are making your position very clear. Conversely, you will also want to make sure you understand the other side very clearly. If you're not sure about something, ask clarifying questions, get it in writing or do whatever it takes to make sure you completely understand the other position or proposal.

## Active Listening

Completely focus on the negotiation while it is underway. Use all your active listening skills from Part 1 and understand the other side's position before preparing your response. For a negotiation to be successful, you must be able to focus 100% on the other person's position and understand them thoroughly, including their own non-negotiables. If you fail to give this your full attention, you will likely miss important details or points that could give you the insight needed to succeed. This could also lead to frustration on both sides if misunderstandings occur due to lack of paying attention. If someone catches wind of a lack of attention, they may draw conclusions that you don't care about negotiations and walk away entirely.

## Collaborate and Use Teamwork

One-sided negotiation can become confrontational very fast. One-sided negotiation is where one side tries to push their agenda without giving anything back. Collaborate and work together for a much smoother process. You are likely going to have to meet the other party halfway, so you need to work closely with them to understand their "must haves" and make sure that they hear and register your "must have" items. Teamwork and collaboration help create a sense of trust, which helps provide positive energy to get to a successful conclusion. Push too hard for your agenda only, and you are likely to alienate the other side.

## Focus on Problem Solving

Using problem solving is where you can get creative about approaches to solutions. Both sides are probably going to have some issues that need to be worked through. Finding a mutually agreed upon solution might be tough, but by collaborating from the previous step and problem solving together you have a much stronger chance of a win–win outcome.

Working together demonstrates a commitment that both sides want a mutually agreeable outcome. Again, one of your best outcomes here is a win–win situation, where both parties feel like they got their non-negotiable, must-have items.

## Use Decision-Making Abilities

Once you have all of your must-have boxes checked and the other side feels good, you've reached the end of negotiations. This sometimes involves a trade-off. Maybe you had to let something go that you wanted but you got something back in return. This is what negotiation is all about and it is the only path towards win–win negotiations.

Of course, not all negotiations end well. Sometimes the outcome is win–lose or even lose–lose, where both parties walk away dissatisfied and no deal is struck. That's OK, not everything is meant to be, and no framework guarantees your success. If this is your case, simply move on to whatever's next and try to learn from the experience.

---

# MASTERING THE ELEVATOR PITCH

---

**Why the elevator pitch is a superpower**: Another communication superpower is the elevator pitch. The ability to give brief but impactful summary of what you do, or other short but impactful stories, can leave a lasting impact on the most important people you meet.

The idea behind an elevator pitch is that a good pitch should last no more than 20–30 seconds, or about the time it takes to ride in the elevator. Elevator pitches are handy for any number of scenarios, for example, bumping into a senior executive at your company who wants to know more about what you do. You might also be asked about how a critical effort is going. You might even just not want to squander a chance meeting with the CEO, COO or other senior business executive. These chance encounters are great opportunities, but they are over fast and it's easy to lose the moment. Therefore, they need to be sharp and succinct. And planned in advance.

What's my favorite use of the elevator pitch? When someone asks you the age-old "tell me about yourself" in an interview. This question is often mis-interpreted as a cue to go into detail about your experience and all the jobs you've ever had in chronological order and what you accomplished in each one. Twenty minutes later, the interviewer wishes they hadn't asked you the question. The "tell me about yourself" question is really a unique opportunity to use the elevator pitch to be memorable. Think which answer is more effective to this question: "well, I started back in high school ..." or "I'm a business-focused CISO who bridges the gap between the data center and the board room ..."

If the idea of thinking on your feet *and* coming up with something short, impactful and memorable on the spot like this seems difficult, here's an easy framework that can help.

# Prepare

Like most aspects of communication, a little preparation goes a long way. Rather than assuming that you won't need an elevator pitch or that you'll wing it should the opportunity arise, be a deliberate communicator and assume you'll need one at some point. Some possible topics to consider include:

- Tell me about yourself? (the famous interview question)
- What do you do for a living?
- Are we secure?
- What does your group do?
- How are you protecting us?
- What do I need to know about cybersecurity?

If you journal (and as previously recommended, you should), these are great topics for your journal. Strive to keep your answers short and punchy. Remember the urgency of the elevator: "well, this is my floor ..." and then your opportunity is gone. Did you manage to get something memorable across? Writing down example elevator pitches forces you to think through and even visualize your answers to these questions. The minute you can visualize yourself answering these questions with powerful, simple answers in person is when you know you're on the right track.

# Start with Your Audience

Did you just bump into the CEO in the elevator? This person is probably business savvy, but not tech savvy. Did you bump into the CFO who asked you to tell them about yourself? This is probably the time to talk about the value you add to the company. If they're HR, this person is probably not tech savvy. If it's the CIO, they are obviously technical. Whatever the case, you probably want to minimize technical detail since it rarely helps an elevator pitch. Did you ever hear the term "lost in technical details"? Remember, keep it short, simple and memorable.

# Think Bigger Than Yourself

One way to craft a good elevator pitch with some punch to it is to think like you're a startup trying to sell your services. In other words, what is your unique selling proposition (USP)? What are you doing that's special or different from everyone else? If you were asking for someone to invest in your department or your company, why would they do that? If you're at an interview, why should they hire you above all the other candidates? This is never going to be about experience and what you were responsible for; it's always going to be about what you can and will do if they choose you over the alternatives.

There's a great elevator pitch story about three bricklayers. They were each asked what they do. This parable is actually rooted in an authentic story. After the great fire of 1666 that leveled London, the world's most famous architect at the time, Christopher Wren, was commissioned to rebuild St Paul's Cathedral. As the story goes, Christopher Wren asked three bricklayers what they were doing. The first replied: "I'm a bricklayer. I'm working hard laying bricks to feed my family." This isn't a terrible answer; it's certainly admirable to take care of your family and work hard. The second bricklayer responded: "I'm a builder. I'm building a wall." This is also an acceptable answer, maybe with slightly bigger picture thinking of not just putting down one brick after another to make a living. However, the third answer in perfect elevator pitch format was this: "I'm a cathedral builder. I'm building a great cathedral to the Almighty." Wow. Great answer, right? But what makes this a great answer? It's relatively straightforward:

- **The right mindset**: A positive attitude and pride in what you are doing
- **Big picture orientation**: Being able to see the end result and how your work contributes to a purpose
- **Connection to a mission bigger than yourself:** Being connected to a mission and values that are much bigger than yourself

# What's Your Best Outcome?

When you're delivering an elevator pitch, you should have some idea of your outcome before you start talking. Are you looking to make a sale? Are you looking to gain support for a security initiative? Are you looking to get a senior executive ready for your board presentation? Maybe, you're just trying to land a job interview. You should have some idea of what you're after and be able to pivot the conversation based on that desired outcome. In some cases, there may not be a strong outcome, you just want to give a plug for the security team and the great work they're doing. That's OK too. Gaining and maintaining support for the program is definitely part of your job. Whatever the case, you should have some idea of what a great outcome would be so you can pick and deliver the right pitch.

## Align Your Voice to Your Message

The most important aspect of a great elevator pitch is aligning your voice to your message. Consider your tone, pitch and body language at least as much as the words you say. People will remember how you made them feel much more than what you said. A great pitch that falls flat on delivery won't make a strong impact. Strive for both: a great pitch with a great delivery. You don't need to be artificial or overly "salesy." Just be yourself, but make sure you're showing energy and enthusiasm and aligning message, body language and tone.

## Want to Sound Spontaneous? Practice!

How do you bring this all together quickly in the heat of the moment and still sound natural? Practice! Practicing is not the same as preparing. In the preparing phase, you think about *what* you are going to say. In the practice phase, you are focusing on *how* you are going to say it.

Again, when someone asks a question like "tell me about yourself," resist the urge to share your life story. You'll want a few sentences that have some impact and are memorable.

A simple response to a family member asking what you do could look like this:

> I advise businesses on how to protect their most valuable information from cyber-criminals or disgruntled employees while still complying with security and privacy regulations.

In an interview, your answer might look more like this:

> I use exceptional communication skills and a deep knowledge of cybersecurity to partner with C-level executives, business leaders, regulators and auditors to build and sustain best-in-class cybersecurity programs. This helps them focus on their business rather than the technical aspects of cybersecurity.

You obviously don't want to use this example outright; instead, think about what your own skills and benefits are and make your own pitch that plays off your unique strengths. The idea of an elevator pitch is to be fast, relevant to the listener and memorable.

# MANAGING SALESPEOPLE

**Why managing salespeople is a superpower**: If you work in cybersecurity, you are going to have a lot of vendors and salespeople trying to market their products to you. It's just a fact of life. There are something like 2,000-plus vendors out there, and they are all trying to sell their products and get on your calendar. While I expect a lot of consolidation in this industry, the number will stay large for a long time.

I get tons of vendor requests and they can be time consuming and annoying to manage. Here's the stress-free way of dealing with these requests.

- **Be polite but firm**. If you're not interested in a product or solution, simply say so or say that you already have the product space covered. Just be clear and firm and say no thanks.
- **Decline demos if you don't have a genuine interest in the product**. Salespeople need to sell. If you sit through a demo, they will push you for next steps and try to get you closer to making the sale. Sitting through demos for products you aren't ever going to buy isn't a good use of anyone's time.
- **Don't fall for artificial deadlines or other urgency tactics**. Some salespeople will try using a deadline like an impending price increase to push a sale through. Usually, they are just trying to get their own quarterly sales targets hit. It's usually a big red flag if they try to push you for an immediate decision on the spot.
- **Don't ask questions**. If you get part way through and realize you're not interested, don't ask questions. Just look to end the meeting as quickly as possible if you realize it's not going to be a fit. If something isn't going to be a fit for your organization, just say so and look to end the meeting.
- **Be repetitive**. If you need more time or aren't ready to decide, say so. And say it again if the message isn't getting through. In fact, keep saying it until the message does get through.
- **End the conversation**. When they won't take no for an answer, your answer can still be no. Just end the conversation if you find that they are not taking your negative response seriously. This one is thankfully rare to use (at least for me), but your own experience may vary.

My final tip is to cut them a break and try to keep it cool. Don't personalize it. Aggressive salespeople are just doing their job the best they know how. If someone annoys you, that's no reason for you to be anything less than professional in return. Be the bigger person and always be polite. People talk a lot and it's not that big of an industry. Incidentally, I do cover some better methods for approaching sales later in this book.

# REMEMBERING NAMES

*A person's name is to him or her the sweetest and most*
*important sound in any language.*
~ Dale Carnegie

**Why remembering names is a superpower**: Few things will help you connect more directly with an individual than using their name. Forgetting a name puts you at a distinct disadvantage when trying to communicate.

I used to be horrible at remembering names. But do you know why people aren't good at remembering names? Because you tell yourself that you aren't good at remembering names and then you never make the effort to try. While I'm not perfect, I've gotten much better at this skill.

How can you be a strong communicator if you don't remember the names of the people you are communicating with? Remembering someone's name makes them feel valued and helps you create a greater personal connection. In fact, a person's name is the greatest connection to their own identity and individuality. Using someone's name is a sign of courtesy and a way of recognizing and acknowledging them. It is a direct path to them, personally. It pauses the action when you address someone by name and then say what you were going to say. Try it.

I worked with one CIO who had an uncanny ability to remember everyone's name. In fact, he never generically said "the business," but would rattle off individual business leaders by name and function. He also knew names up and down the ranks and would routinely call out good work by name in big events like town halls rather than thanking "the project team." This was unbelievably powerful in a big organization where you can sometimes feel like a number. It's worth spending time in this area. Even if you think you are horrible at remembering names, here are some easy tips and tricks that can help you get better at this skill fast.

## Start with Your Mindset

The most important first step, as always, is to check your mindset. You can remember names. In fact, *anyone* can remember names, and anyone can get better at remembering names. Make it a commitment to improving this skill and realize that the purpose is to create better connections and be a better communicator. If you think you can't or won't remember names because you're bad at it, then expect to stay bad at it and simultaneously lose an important edge in communicating. Once you understand your motivation and acknowledge that improvement is possible, you're halfway there.

## Pay Attention!

The second tip is to concentrate when you are being introduced. You can do it; it's only a few minutes. Put the phone down and pay attention. We live in an age of distraction. If you are daydreaming, using your phone or thinking about what you will say when introductions are finished, you will likely miss out on your best chance at actually hearing and remembering someone's name the first time around. Most of the time, remembering names isn't a memory problem at all. It's a *focus* problem.

Clear your mind, be in the moment and let go of your thoughts about how you know this person or what you should say when they're done speaking.

## Face Associations

Try to find an unusual feature when you are introduced. This could be ears, hairline, forehead, etc. Then create an association between that characteristic and the person's name in your mind. The association may be to link the person with someone else you know with the same name. This is a great way to remember someone, because seeing them will automatically remind you of their name.

## Repetition

When you do learn someone's name, look for excuses to practice it. For instance, "It's nice to meet you, Sharon." You should try to use the person's name a few times throughout the conversation and again when you part ways. The more you repeat the name, the more likely you will remember it.

## Mental Associations

Of all the tips to remember names, this one works the best for me. Think of a connection between the person's name and anything at all that you already know. I mean absolutely anything. If someone's name is Jim and they are especially fit, you might associate them with a gym. Maybe someone named Bill from accounting reminds you of your own outstanding bills. Sure, this sounds silly, but it works. Give it a try and remember that you don't need to share these associations with anyone. They are exclusively for you.

## When All Else Fails ...

What do you do when none of these methods work? Just ask them! The easiest and often best thing you can do when you forget someone's name is just be honest and ask. Most people will be understanding. Hint: the longer it takes you to ask after repeatedly meeting in person, the more awkward this will feel. Don't wait until your seventh interaction to fess up that you don't remember their name.

## Use Their Preferred Name

This one is more of an etiquette tip, but some people prefer nicknames or abbreviated versions of their full names. If someone prefers to be called Michael, don't use Mike. Use the name they gave you when you were first introduced. This is also typically the name they use to sign their email communications. Be conscious of this, I have known a Donald that most certainly didn't want to be referred to as Don and a William who no one in their right mind would ever call Bill or Will. Other people don't care as much, but generally they still have a preferred name. I'll answer to Jeffrey, but most people simply call me Jeff.

# CONFLICT RESOLUTION

*If you have learned how to disagree without being*
*disagreeable, then you have discovered the secret of*
*getting along—whether it be business, family relations,*
*or life itself.*
~ Bernard Meltzer

**Why conflict resolution is a superpower**: Conflict is no stranger to the cybersecurity field. Accepting this fact and putting together a strategy to respond to conflicts will make you more effective when the proverbial crap hits the fan.

There are a lot of conflicts that can arise in the cybersecurity field. The business may need to move forward on a project that has an unacceptable level of security risk. Critical patches may need to be deployed that will require other, already overtaxed areas, to do the work. You may even run into situations where management wants to push you harder with a security effort that is already strained, and the team is working overtime.

So how do we handle the inevitable conflicts that come up as part of our job? The first step is to adopt the right mindset and realize that not all conflict is bad. Conflict is a necessary part of working with teams that have diverse backgrounds. Conflict can even be healthy and lead to better outcomes. If everyone just agreed with each other without considering multiple different perspectives, the best decision might never be made. The key to conflict resolution is learning how to manage a conflict effectively so that it can serve as a catalyst to organizational improvement rather than be a source of aggravation.

As a communication superpower, conflict resolution should be high on your list of skills to develop. This skill will serve you well, make you more effective in your role and, if handled with grace, will make you someone who gets stuff done even when the odds are stacked against them.

There are five basic strategies for dealing with workplace conflicts when they arise. They are avoiding the conflict, accommodating it, competition, collaborating and compromising. Some of these strategies are more effective than others, so let's review what they are and why this is the case.

The first strategy of avoiding conflict is considered a lose–lose strategy. By not dealing directly with an issue or conflict, no progress is made, and the issue is not resolved. Ignoring an issue does not make the issue go away. Of course, not every battle is worth fighting. Sometimes, as they say, you just have to let it go. There are cases where this is the right course of action. You might use this strategy when you simply don't care about the outcome enough to get into a conflict. You might also use this when you know you won't win the conflict, like if the CEO or the board has set a direction at odds with your own. This is not a battle you are going to win, so the best course of action would be not to engage in conflict. It could also be a grey area, where the conflict might not be worth the disruption and you feel neutral about the outcome.

Unlike avoiding conflicts, when you accommodate a conflict one side or the other will essentially give in to the other side. This is a win–lose strategy where someone wins one side and the other side accommodates the winner, basically waving the white flag. This is the "you win some, you lose some" strategy. It's a valid strategy, too, as there may be times when you either owe someone a favor or where teamwork is more important than being right. Of course, the best time to use the accommodating strategy is when you realize that you're wrong and the other side is right!

Competing is another win–lose strategy. You'll want to employ this strategy for important issues. The business wants to take down those pesky firewalls that slow down their Internet access? Yeah, you're going to have to fight this one and, frankly, you need to figure out how to win. While competing is an important conflict resolution technique, you don't want to use it with every conflict, or you'll just be seen as a bully who has to get their way every time. It will work for a while, until it doesn't. You'll find that if you use this strategy as a last resort, people will respect the issues where you are making a bold stance more than if you pushed for your way every time.

Compromise can be seen as win–win, although there are circumstances where it can be seen as lose–lose or even some neutral ground where everyone is sort of, but not really, happy. Still, compromise is an important strategy. It is a meeting-in-the-middle strategy where both sides can move on from an impasse. It is also a good strategy when time is of the essence and a decision needs to be reached quickly. Both sides give a little and, even if they don't see themselves as winning, no one sees themselves as losing.

Collaborating is a win–win strategy that aims to find a solution that is beneficial to both sides. This is the most advanced conflict resolution technique, since both sides come out feeling like they either won or that they are OK that both sides won. It is a conflict resolution that makes you wonder if there was ever a conflict in the first place.

Conflict resolution is a clear communication superpower that is worthy of some deep study. There's no one-size-fits-all approach in any of these strategies. The situation must be carefully weighed, and the benefits of a particular strategy should be considered. Is your best strategy to avoid the conflict? Sometimes. Put your foot down and win at all cost? Sometimes. Ideally, we would always seek a collaborative win–win solution, but this isn't always possible.

Another important tip with conflict resolution is to consider multiple sides of the story by asking open-ended questions. For example, if someone has accused a person on your team of being rude or curt, instead of rushing in with a reprimand, try letting them know that you heard about a problem. Then simply ask the open-ended question: "what happened?" You might get a whole different story than the one you heard. I've learned that by not making assumptions about who's right and wrong can be an important mediator. There are two sides to every story, and you'll want to hear both before making any conclusions. Always get the information before acting. When in doubt, use a lighter approach. Remember, communication can't be taken back. Once something is said, written, emailed, it's done.

# ASK BETTER QUESTIONS

*If I had an hour to solve a problem and my life depended
on it, I would use the first 55 minutes determining the
proper question to ask.*
~ Albert Einstein

**Why asking better questions is a superpower**: The right question, asked in the right way and at the right time, is a great tool to help drive an organization forward and pull out the best in people. Questions are great. Questions can help you establish facts, clarify information, reframe issues and can help you engage people in conversations who might not have been otherwise engaged. The *right* questions, however, are a communication superpower and can help you take quantum leaps forward. These types of questions can help provoke new ways of thinking, generate curiosity and open new creative possibilities.

Determining the best questions is a skill reserved for the truly great leaders. It's easy to forget that Apple was once on the verge of bankruptcy before it became a trillion-dollar company. Steve Jobs is credited for turning the company around in a process that started with a simple but profound question: "Who is Apple and where do we fit in this world?"

This question eventually led to the "think different" campaign and in Apple trimming 70% of its product line to focus on the 30% that better aligned with their values. When asked how he answered the "who is Apple?" question, Steve Jobs' reply was:

> What we're about is not making boxes for people to get their jobs done—although we do that well. But Apple is about something more than that. Apple's core value is that we believe people with passion can change the world for the better. Those people who are crazy enough to think they can change the world are the ones who actually do.

Finding your own great questions to ask can take some time and deep reflection. You can ask great questions of yourself, your team or your company. Asking amazing great questions is a skill like any other skill, it takes practice. Here are a few thoughts on how to up your game on asking better questions:

- **Use open-ended questions**: Using yes/no questions will get you yes/no answers. Ask questions that stimulate the conversation and can't be answered with a simple yes or no.
- **Go deeper**: You can use techniques like the five whys (Figure 7.1) to go deeper on a subject. Using this method, you keep asking "why" to get to the root cause of an issue. For example, you may have a problem that a customer is angry at your firm. Why? Their delivery was late. Why was their delivery late? The order took much longer than expected. Why did it take longer than expected? The warehouse incorrectly listed the product as available.

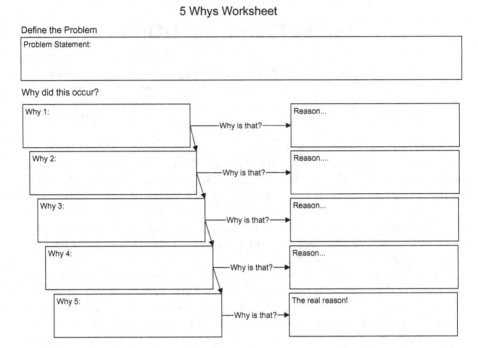

**FIGURE 7.1**   The five whys iterative cycle can help you quickly get to the root cause of a problem. The answer to each "why?" question helps form the basis of the next question.

Why was the inventory system incorrect? Because it's a manual process and someone forgot to make the update it. In this case, you can go a few whys in and come up with the conclusion that the inventory system could use some automation.

- **Ask more questions**: They say that the best way to ask a lot of great questions is to simply ask a lot of questions. Asking a lot of questions can get you used to the process and will invariably lead to a few great questions.
- **Leverage the silence**: People are sometimes uncomfortable with silence. Ask a question and wait for the response. Then wait some more. Many times, the person you are questioning has more information they're willing to share and waiting brings it out from them. Interrogators use silence very effectively, since people feel a need to fill the holes in the conversation and will bring out the critical bit of information.
- **Never interrupt**: If someone is responding, let them finish all the way. Interrupting will disrupt their chain of thought and you may not be able to get it back.

Asking better questions can help you get better answers. Think about ways that you can leverage the collective brainpower of your organization by asking your own better questions.

# DEALING WITH DIFFICULT PEOPLE

**Why dealing with difficult people is a superpower**: Let's face it, not everyone was meant to get along. Sometimes you have to work very closely with someone in the organization that either you can't stand or can't stand you. Personality problems like this don't need to get in the way of progress; there are ways to improve your communication when the personalities don't get along.

Let's face it, not only is dealing with difficult people part of the job, but sometimes you *are* the difficult person. When someone is overly aggressive or defensive, good communication isn't possible. You can take one of four approaches when coping with difficult people: do nothing, walk away, change your attitude or change your approach in dealing with them. The good news is that if you master certain communication skills, you can overcome confrontational situations and turn them into good communications.

Difficult people come in all shapes and sizes. They can be the know-it-all, the character assassin, the unpredictable time-bomb or the whiners. They can be overly aggressive, constantly complaining, spreading rumors, constantly criticizing or someone who interrupts your every word when you speak. This is hardly a comprehensive list. Any organization is going to have a few people that either you rub them the wrong way or vice versa. This cast of characters can ruin the atmosphere in any workplace and make your already difficult job even more difficult. There's an approach that won't take away all the pain but will at least make your odds of a successful communication a lot higher.

## Mindset

Always starting with your mindset, ask yourself why this person is difficult and also *how* they are difficult. Do they just like to take over projects and steamroller over everyone? Are they obnoxious know-it-alls who challenge everything you say in an effort to be the smartest person in the room? This audience-first approach will help you craft your communication in a way that gets you in, gets you out and increases your odds of getting through to them. It might also help to prepare and visualize how you'll handle any interaction. What are some brief subject-changers or conversation-enders that you can use to remove yourself from a bad conversation? Having these ready in advance not only helps you prepare if something does go wrong but also helps you have the confidence that things might actually go OK.

Don't just consider their mindset, check your own as well. If you're going into a conversation looking for trouble, you're going to find it. Are you calm? Are you willing to accept that the person you will interact with is also another human being deserving of respect and professionalism? Make sure you are in the right zone before engaging. In fact, consider showing some compassion. Instead of letting angry or defensive feelings about someone overcome you, try sending them thoughts of compassion. Maybe they're difficult people because they've had a difficult life.

# Approach

There are only two things you can control when dealing with difficult people. Your thoughts and your actions. You'll want to keep both in check when dealing with difficult people. But the first step is considering if this is typical behavior for someone or might they be having a bad day. If it is unusual behavior, maybe avoiding confrontation is the best way to go.

Even obnoxious people tend to have a range of behavior. If they cross that range with you, confrontation might be warranted under a few circumstances. If their behavior is getting in the way of you doing your job or if this person is eroding your credibility, then taking some action could be justified. There are a few approaches depending on what kind of problem person you're confronting.

- **The bully**: Don't fight, and don't retreat. Usually, the bully is looking for you to either fight or walk away. When you do neither and remain calm, it throws them off their game. Instead, ask them if they're trying to intimidate you or make fun of you and wait for the response. The key hear is not to stick up for yourself and not back down.
- **The character assassin**: These people are usually jealous, devious, envious or all three. The best defense against any character assassination is to maintain your integrity. Fighting back tends to backfire. If you hear that someone has been saying unflattering things about you, respond that you're disappointed to hear that and carry on with things. Don't give it another moment and don't treat it like it matters.
- **Other bad behavior**: Some bad behavior is just that, bad behavior. This might be people who are constantly interrupting, making inappropriate jokes or complaining too much. Calmly confront the person in private, like in your office, where no one else can hear. There's a chance that this person truly doesn't know they're being rude. Give them the benefit of the doubt by addressing the problem privately.
- **Angry**: Telling an angry person to "calm down" is almost always the wrong approach. Anger can cloud your rational thought and judgment and telling someone who's upset to calm down can have the opposite effect of what's intended. A better way to deal with this situation is to remain calm yourself, use empathy and say something more along the lines of "I understand you're upset, but let's try to work through this together …"

Dealing with difficult people is more of an art than a science. There is no one method or recommendation that will work for every person in every situation. Don't feel like you need to tolerate bad behavior but be judicious about how and when to confront someone. If things are really out of hand and targeted at you specifically, you might be much better off engaging HR or your manager in the situation to make sure that you're following company policy against harassment in the workplace. There's definitely a line that can be crossed where more formality should be brought to the situation.

# ASKING FOR HELP: THE RIGHT WAY

**Why asking for help is a superpower**: Don't be a lone wolf leader; you're going to need help and support in a lot of areas. Asking for help the right way makes it more likely that you'll get the support you need.

A lot of people are shy about asking for help. They might be overwhelmed, overworked and overstressed, but the idea is that life is hard, and you tough it out and keep going. This is unfortunate since people, for the most part, like to help others. If you ask for assistance, you will generally get it, especially if you ask for it in the right way. Here's a brief framework that will help you ask for help in a way that no one should feel uncomfortable.

The first step in this framework is to consider whose help you need. Are you deploying security software and need support from the infrastructure team? Writing a speech for a conference and need a peer review? Composing a complicated email and looking for an opinion on clarity? Figure out who in each situation is in the best position to offer assistance. Is it someone on your team? A peer? A friend? Your spouse? Don't just consider the most likely person to offer help, consider the person whose help would offer the most value.

People are usually pretty willing to help. However, some common ways of asking for help are ultimately counterproductive, because they make people less likely to want to give help. You want people to feel that they would be helping because they want to, not because they have to help. People want to feel that they're in control of their decisions. Once you have someone in mind, ask yourself *why* this person would want to help you. Is there a way that you could make this a win–win offer? When you are ready to ask for help, consider some tips to make the request successful.

- **Be clear**: When you're asking for help, you need to be very clear and specific about what you're asking. Try to be as specific as possible so they know exactly what they need to do. Be willing to negotiate, too. Let them decide how much support they can offer and try to find a mutually beneficial solution.
- **Don't be apologetic**: Being apologetic about asking for help casts the conversation in a negative light. Show some confidence in the request instead.
- **Consider the channel**: Asking for help is best done with a strong feedback loop. In person is your first choice, but if it's not possible, video call or phone call would be the next two best methods.
- **Make it personal**: Make your request more personal by explaining why the person's skills or expertise make them uniquely suited to assist.
- **Don't call in favors**. You might be tempted to use this strategy in order to sort of force the other person into doing your request. The problem with this is it feels like being forced into helping if you're the other person. While you might have a good percentage of people who feel obligated to help you in some way, who wants to feel obligated? Try to keep it more positive.
- **Don't call and ask for help with something small or insignificant**. The problem with this approach is that by minimizing the request, you are also

minimizing the help from the other person. If this is so unimportant, why are you wasting their time with the request? The other issue is you might have really miscalculated exactly how small the favor is and it could turn out to be bigger than anticipated.

- **Don't say how fun it will be to help**. While it might actually be fun to help on certain tasks, emphasizing this fact will actually take a lot of fun out of it. Also, by telling someone how they're going to feel about something seems a bit presumptuous. People don't like to be told how they're going to feel.

---

# SUMMARY

---

Communication superpowers are the ability to communicate against strong headwinds. These skills will help you even in the most difficult of communication scenarios.

- Nonverbal communication is critical to understand. Your ability to "read the audience" will help you become a more effective communicator. In addition, your own nonverbal communication can help facilitate the conversation with a little practice.
- Most communication is nonverbal. You can pick up on nonverbal signals that communicate everything from boredom to outright hostility.
- Everything from your tone to your posture can indicate different signals. You want to be aware of not only other people's body language but your own as well.
- Saying no to things that don't excite you or help you move the security program forward can sometimes be awkward. Learning to say no to friends or even your boss is possible if you handle the communication right.
- Negotiation happens more often than you might think. The negotiation process of preparing, keeping communication clear, active listening, teamwork, problem solving and decision-making can all help you achieve outcomes that are win–win. However, not every negotiation has a happy ending and sometimes the result is lose–lose.
- The elevator pitch can help you have quick but impactful interactions with senior business leaders. Using the right mindset and a big picture mindset, you can craft small but effective communications that will leave a lasting memory.
- Managing salespeople can sometimes be an annoyance. Establishing some ground rules and keeping it from getting personal can help manage the flow of sales calls.
- Using someone's name is the ultimate form of connection with a conversation. If you have trouble remembering names, this is a skill that can be improved.
- You are going to deal with conflict in your career. There are five ways to deal with conflict: avoiding the conflict, accommodating it, competition,

collaborating and compromising. Each of these methods may have its place, but collaborating and compromising tend to lead to the best outcomes.

- Dealing with difficult people is a fact of life. Doing a little groundwork ahead of time can help deal with these interactions and still get your point across.
- There's a right way and a wrong way to ask for help. Setting this communication up using the right methods will help you get better results.

# REFERENCES AND RECOMMENDED READING

*Dealing with Difficult People.* Harvard Business School Press, 2005. https://www.amazon.com/Dealing-Difficult-People-Results-Driven-Manager/dp/1591396344

Ferriss, Tim. *Tribe of Mentors: Short Life Advice from the Best in the World.* Houghton Mifflin Harcourt Publishing Company, 2018.

Foley, Sharon. *Getting to Yes.* T.F.H. Pub., 2007.

Klaff, Oren. *Pitch Anything: An Innovative Method for Presenting, Persuading and Winning the Deal.* McGraw-Hill, 2011.

Merrill, Arthur A. *Remembering Names: Improvement Is Easy.* Analysis Press, 1985.

Murphy, Jim. *Managing Conflict at Work.* American Media Pub., 1997.

Sivers, Derek. *Anything You Want: 40 Lessons for a New Kind of Entrepreneur.* Portfolio/Penguin, 2015.

# PART 2

# Communication in the Real World

You made it to Part 2, congratulations! You now have a good foundation for communication and how it works and you're ready to apply these skills to the real-world situations you are likely to face as a security leader.

This section of the book walks you through several situations you are likely to face during your security career. No matter where you are in your career, you should find these chapters helpful to make you think about how to approach everything from writing policies and standards to handling incidents to presenting to the board of directors. I've even included some of the more uncommon scenarios like taking part in interviews and podcasts. Here's where you really put your communication skills to task in order to be clear and more effective in your communications.

DOI: 10.1201/9781003100294-9

# Policies, Standards, Guidelines and Procedures

**8**

Policies, standards, guidelines and procedures are the cornerstone of an effective security program. These are special purpose documents that form the foundation of the security program. They are the "rule books" and governing documents of the program to get everyone on the same page. While this is not intended to be a particularly deep dive on the subject, I am providing some of the best advice and tips that I can for crafting these somewhat tricky documents. Obviously, this section focuses primarily on writing skills and applying them to the creation of policy.

For the purposes of this chapter, I will refer to policies and standards, but most of this also applies to guidelines and procedures as well. I will be specific when I mean one of the four document types. This said, policies, standards, procedures and guidelines are not the same thing.

A policy or standard conveys management's view or intention on rules or actions that are expected of everyone in your organization. Guidelines and procedures are a little different. Let's review the differences among each document type and why all these documents are necessary in the first place (Figure 8.1):

- **Policy**: Policies are formalized, high-level statements that outline roles, responsibilities and expectations. They are deliberately written at a very high level and are not intended to need frequent updating. Policies are enforceable and mandatory. There are many policies that might be found in an organization, so the subject goes well beyond cybersecurity. A few non-cybersecurity examples might include a dress-code policy or anti-money laundering policy. Most regulators see policy as management's directive for how they expect employees to act in the workplace.
- **Standard**: The next level down are standards. If policies act as a statement of intent, then standards function to implement that intention. Policies reflect an organization's goals, objectives and culture and are intended for broad audiences. Standards offer the nitty-gritty "how to" achieve the directives captured in policies. Standards are generally mandatory actions or rules that give specific direction. Standards tend to focus on quality and derive their

DOI: 10.1201/9781003100294-10

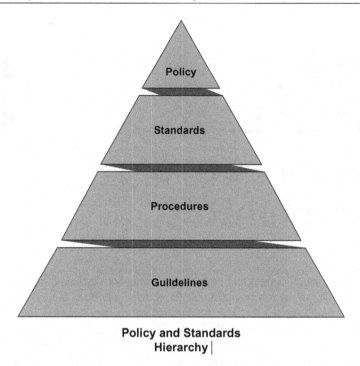

**Policy and Standards
Hierarchy** |

**FIGURE 8.1**   Policies, standards, procedures and guidelines all serve different purposes. Policies are high-level and simple documents.

authority from a policy. Standards also convey what is and isn't an acceptable level of quality, which ensures consistency across the organization. Standards are enforceable and mandatory. I like to think of standards as "implementation standards" in that they are a specific implementation of a policy. A policy might require that all Internet access goes through a secure connection, whereas a standard defines what that secure connection should consist of, for example, a specific Intrusion Prevention System (IPS), firewall and active security monitoring. Configuration standards might even go into very specific product settings, like what settings should be on or off in a Windows server. ISO 27001 is an example of an international standard.

- **Guideline**: Guidelines are recommendations and are typically not mandatory to follow. Guidelines are designed to streamline certain processes according to best practices. They provide specific suggestions and provide enough flexibility to be changed depending on circumstances. A guideline gives general recommendations on how to perform a task and specific advice on how to proceed in different situations. They are not intended to be enforceable or auditable. Note that some auditors in some institutions will still audit guidelines anyway, or at least raise questions on why they're not being followed. An example of a guideline could be a Work from Home Security Guideline. This document would offer suggestions on how to safely configure a home office, but it would be up to an employee to follow the guideline or not. No one is going to force you to configure your personal network a certain way and certainly audit won't be coming to your house to verify the configuration.

- **Procedure**: Procedures are detailed step-by-step instructions. They are intended to dictate a certain set of steps to a business or technical process. Think of procedures as a "cookbook" to provide people with a repeatable and predictable process. Procedures should be detailed enough and yet not too difficult that only a small group (or a single person) will understand them. Procedures are generally mandatory and auditable. You might have a procedure defined for how to reset a forgotten password so that it can be done in a secure, consistent manner by anyone on the service desk.

Despite their seemingly mundane nature, policies and standards form the cornerstone of the information security program. In a large organization, they are instrumental for enforcing global consistency, driving change and launching enterprise security programs.

In addition, many regulations require a documented security program supported by policies and standards. Companies need to be able to prove consistency and articulate their information protection needs. Policies help capture these needs and enforce management's commitment to protecting information assets. Developing, documenting and implementing policy in a large organization can be tricky. If you don't take the big picture into account, such as long-term implications and cost impact, then you might be doing more harm than good by documenting requirements that no one is following. In addition, if user training and awareness, continuous monitoring and robust communication plans are not part of your plan, then your implementation efforts may not be successful.

The policy creation process should be as collaborative as possible. Policies written in a vacuum will die in a vacuum. To create an effective policy, you'll need input from across the organization, active participation in the refinement process, stakeholder buy-in and support before implementation can actually take place. Organizations should define policy approval structures that give adequate representation to key stakeholders, including senior leadership from IT, legal, human resources, business and security. The good news is that once everyone is aligned, driving the security program based on the policy can provide air cover for the security team, implementation consistency and clear expectations.

If your organization is regulated, your policy might need a strongly worded approach that articulates mandatory regulatory requirements. Resist the urge to use boilerplate policy as anything more than a starting point though. Without fully understanding security controls and their potential impact on your organization, you might be saving time up front only to cause a lot of confusion down the road. When it comes to policy, you need to be more practical than aspirational.

Policies are formal statements produced and supported by senior management. They can be organization-wide, issue-specific or system-specific. Policies are meant to be written and then changed infrequently. They are deliberately high-level. Policies also need a flexible exception process with some level of oversight and governance. Even the best policies can't anticipate every possible implementation scenario. A documented exception process should be established in order to review these issues. Too many exceptions might mean that the policy needs to be modified or that enforcement may need to be strengthened.

Policies are more than just words on paper. They are a foundational element of the security program that literally puts everyone on the same page. One of the documents an auditor may request is your written security policy. Large or small, all organizations need a written security policy. It doesn't have to be complex or long-winded, but it does need to be a written policy that has management approval. Why? A security policy documents the organization's stance on issues such as classification of data, data access, appropriate use of technologies such as email and the Internet, etc. Without a written policy, there can easily be conflict within the organization and disagreement over what data should be protected or what services should be allowed.

As a security professional, you will probably be on the hook for writing policies, standards and guidelines at some point. It's worth a moment to consider what each of these documents is, and how to make writing them a relatively painless process.

Policies are formalized, high-level statements that outline roles, responsibilities and expectations for specific programs. Policies may communicate company values and objectives. They are deliberately high-level documents and are not intended to change often. An example might be the Information Security Policy that would typically contain statements like:

> It is the policy of ABC Corporation that information in all its forms—written, spoken, recorded electronically or printed—be protected from accidental or intentional unauthorized modification, destruction or disclosure throughout its life cycle.

Policies essentially establish an intent to do something at a very high level. For example, a policy might capture requirements around password management, personal device use or remote access. These documents will capture minimum requirements across the organizations. In general, policy needs to be written to the lowest common denominator in your company. However, you can and should empower business units to be more restrictive if their needs dictate a stronger policy. You never want to set a policy or standard that reduces security for anyone, but you do need to set the minimum standard. Business units should not be empowered to be less restrictive without seeking a policy exception that has been reviewed and approved by the CISO and senior management.

# DOCUMENT YOUR SECURITY PROGRAM

Auditors look for processes that will assure them that appropriate controls are in place to ensure the integrity, accuracy or privacy of the data being examined. The processes required may vary slightly, but the ones I see auditors require on a consistent basis are discussed below.

One of the primary requirements is a change management policy that details how objects (such as programs and files) get moved into production. This includes the process programmers go through to gain access to modify data on a production system. It's also a good move to have a documented charter for the security function. A charter is a document that is developed by collective groups that clarifies the direction of the security function while establishing its boundaries. I've included a sample charter in the appendix that you can modify for your own needs. Disaster recovery documents show how the organization would recover from various types of disasters and are considered necessary. And of course, you will need a high-level security policy. I will offer several other suggestions later in this chapter for your consideration, but these are all good starting points. Create these documents if you don't have them or revisit and revise them if you do.

Critical processes of all kinds need to be documented. Auditors will look for evidence that these processes are not only documented but are being followed. An auditor may literally watch people perform their jobs to see if they are following the exact steps documented in the process. Therefore, it is vital that the process documentation is up-to-date and matches the actions actually performed. Think about this, if you don't have a documented process it is possible that an auditor will still watch every step and then grade you on any problems they can think of versus any missed steps in a documented process. It's better to take the time and write the documents in advance. But remember, you aren't writing these documents for auditors. Policies and standards serve to get everyone on the same page, not just auditors.

Even as someone who writes quite a bit, I find writing policy and standards a somewhat laborious process. In a sense, it *should* be a slow process. Every word counts in a policy or standard. This is where you are expressing your intent, and for a standard you are expressing very specific things that are expected to happen. Words should be chosen carefully, and you will likely be audited to the content of these documents at some point.

When writing policies and standards, I find that templates are very effective and help you get past that initial blank page very quickly. After all, there are certain elements to policies and standards that should be the same across many documents. Things like scope, intended audience and exceptions should be similar or identical in every document. Spending some time creating these templates and getting these templates right can help avoid starting with a blank page.

To recap, policies should provide information security direction for your organization, include information on how you will meet business, contractual, legal or regulatory requirements and contain a commitment to continually improve.

# THE POLICY AND STANDARDS WRITING PROCESS

You'll find that it's most effective to create templates when you're writing more than one policy document. Templates will help you drive a common look and feel and make sure that you are not missing any elements. It will also help with repetitive or common text that should be found in all policy documents, like exception handling. In general, you'll also want to use the same fonts, numbering schemes, etc., to drive for a common look.

I'd recommend creating a template considering the following additions:

- **Overview**: Explain the purpose and subject of the document.
- **Purpose**: In this section, you should explain why the document is being written and what its intended purpose is. This is typically high-level language that talks about the general topic and protecting company information as part of a broader information protection strategy.
- **Scope and audience**: Remember, we should write for a specific audience. People who read one of these documents, especially business users, will lose interest in documents that have lots of technical detail that doesn't apply to them. When a policy is geared towards all end users, don't add elements that only pertain to IT or Backoffice staff in the same document, if possible. Target your policies with some precision. Is the policy document for all users? Only IT users? Sever administrators? You need to say who is expected to read and follow the policy. Not every document is intended for all employees. An Acceptable Use Policy would be for everyone, but a UNIX Server Administrator Standard would obviously only be applicable for employees that work with UNIX.
- **Background information**: Start your document with reasons why it is created, for instance, to prevent unauthorized access or the misuse of the company's data. If you're writing a standard, this is a good place to point to the policy that it's connected to. In some companies, there is only one security policy and all security standards derive from that policy. Others will have many policies, like password use, information protection and acceptable use.
- **Terminology**. Unless all the terms you're using in a policy document are common and well-known, you should include a definitions section for reference. In this section, list all the terms and abbreviations with their explanation, using simple language. Note that the content of this section will differ quite a bit based on what terms are used in the document. Make sure to have a reference to any words that might be vague or open to interpretation.
- **Contacts**: You should also add information about who owns the policy and policy content. You want to define an authority for any questions on interpretation. Most policy documents should get their authority from the CISO, CIO, Chief Risk Officer or other senior officers in the firm. It's OK to put the role instead of a person's name, as people do change roles. But the functional owners should not change. Additionally, highlight where employees can locate other forms, policies or documentation that the policy references.
- **Exceptions**: You'll want to have an exception handling process defined before publishing any policies or standards. If a business can't comply with the requirements, what happens next? Some companies use elaborate Governance, Risk and Compliance tools like MetricStream or RSA Archer. Others simply document non-compliance and any action plans to get compliant. At a minimum though, you need to consider who approves an instance of non-compliance. It could be the CISO or other very senior executive. Some firms even make the most senior business leader sign off on the risk so that

they understand that it is ultimately their data that may be at risk for failing to implement a security control.

- **Non-compliance**: A section should also emphasize the intention to robustly enforce the policy. This typically reads something like "An employee found to have violated this policy may be subject to disciplinary action, up to and including termination of employment."
- **Review cycles**: Every policy or standard should define a review cycle. Nothing loses credibility faster than a dusty copy of an OS/2 policy floating around that hasn't been updated in five years for a product that (hopefully) isn't even in use anymore.
- **Revision history**: Similar to defining how often a document should be reviewed, you want to include how often it actually has been reviewed and what date. Version control is critical. You need to know that everyone is working from the current version. Make it clear what version the document is and when it was last updated. A sample versioning control table is as follows:

| VERSIONS AND REVIEW | | |
|---|---|---|
| VERSION | COMMENTS | DATES |
| V 0.5.0 | Exposure Draft 1 | 12/5/2018 |
| V 0.6.0 | Exposure Draft 2 | 01/21/2019 |
| V 0.7.0 | Exposure Draft 3 | 02/06/2019 |
| V 0.7.1 | Legal and RM input | 02/29/2019 |
| V 0.8.0 | Exposure Draft 4 | 04/09/2019 |
| V 0.9.0 | Final Draft for Approval | 07/02/2019 |
| V 0.9.1 | Final Draft v2 for Approval | 07/04/2019 |
| V 1 | Final Release Version | 08/20/2019 |
| V2 | Updated for annual review | 08/20/2020 |

# WRITING A POLICY OR STANDARD

When you are writing any policy document for your organization, there are several different considerations. Keep everything reader friendly. A policy or standard shouldn't read like a contract. You'll want to avoid legal language and make sure that policy statements are clear and easy to understand. Consider the audience and whether the policy is written as clearly and simply as possible from the reader's perspective. Long-winded, legalistic or confusing sentences run the risk of not being understood or simply not read. If you have a communications team, it might help if they could review a draft before it is published.

Focus a policy on a single topic, such as password use or remote access. Don't put multiple subjects in a single policy document.

I like using a general framework for putting policy together. For example, consider the following workflow.

The first step is requirements analysis. In this phase, you basically want to pick the subject and gather the requirements for a policy document. If you are in an organization that doesn't have a lot of existing policies, some subjects you will want to consider include:

- An overarching, high-level security policy
- Acceptable use policy
- Data classification
- Physical security
- Asset management
- Cloud usage
- Data retention
- Event logging and monitoring
- Wireless access
- Mobile and BYOD
- Privileged account management
- Password use
- Physical security
- Remote access
- Secure software development
- System patching
- Third party vendor management

Also consider why you need a policy in the first place. Are you trying to solve a specific problem? Are there requirements you need to make clearer with users? Are there legal and regulatory considerations? What about internal audit? This is your chance to pick a meaningful risk area to address and identify specifically what needs improvement or clarification in the organization.

In this phase of writing policy, you want to be inclusive and comprehensive. Speak to your legal counsel and privacy officer. Speak to stakeholders who will have to implement your policy or standard. This needs to be a partnership, don't just create security mandates from an ivory tower and expect everyone to fall in line.

## Compliance

Unless your policy is extremely generic or obvious, there will be instances where the company cannot be in full compliance. This is a balance. If no one is complying with a policy or standard, either people are not taking it seriously or potentially the document is too restrictive. You want to aim for a strong majority of the organization being in compliance and then a handful of situations that need special attention.

How are you planning on measuring compliance with your policy or standard? If the answer is that you either have no idea or compliance cannot easily be measured, you probably don't have a great policy in the first place, and this will come back and haunt you during audits. Vague language will result in vague results and interpretation errors.

There will, of course, be times where the business is unable to comply, maybe due to a legacy system like a mainframe. You should plan for these situations to be managed by an exception or exemption process. This process should also be defined in the policy (see the section on templates earlier in this chapter).

Exceptions should be noted, have a corresponding action plan to address it and be managed actively. Generally, this means that you'll want a path to compliance with regular milestone checkpoints. An exemption, like what you might encounter with a legacy system, means that it will likely never be fixed until the system is retired and there is no expected path to compliance. The risk is simply noted somewhere and reviewed periodically. Having a catalog of exceptions and exemptions can make audits go a lot smoother and demonstrates that you are running a controlled environment.

# Security Controls

The main body of a policy or standard should outline what security controls need to be in place and how they should be implemented. What, specifically, are you asking people to do? For example, if a certain action such as gaining access to a system requires approval, *who* is that approver? You don't need to have a named person (this will make updating policy difficult) but you should have a named function (like the CIO) or at least a department (like the security administration team). Here are a few more suggestions for when you're writing a policy or standard.

**Avoid nebulous terms**: Policies and standards are intended to clarify management directives, not make them more ambiguous. If you're going to use terms like "intellectual property" make sure you attempt to define exactly what this means or at least provide some examples of what it means. You might also want to suggest what department to engage should there be any questions or cases that aren't clear. In the example of intellectual property, you might want to point people to the legal department if there is doubt that something falls into the definition or not.

**Minimize cross-referencing**: Although referencing other policies in a policy may be necessary, you should generally avoid or at least minimize cross-referencing other documents. You want to keep policies relatively standalone so that people don't have to grab four documents before they understand the big picture. That said, it's OK to cite related policies and standards where it makes sense. For example, a data classification policy and a data protection standard are related. You don't want to reproduce content from one document inside the other since it will be hard to manage if it ever changes. In this case, cross-referencing would make better sense.

**Review translations**: If you are having policies and standards translated for a global audience, try to have a trusted point of contact in your company who can review it in the native language for clarity and accuracy. While translation services are very good, you want to ensure that the translation is written just as clearly and concisely as the native language version. You also want to make sure that it is suitable for your company and that it accurately conveys the intent of the original document.

## Lawyers Are Your Friends

Having trouble wordsmithing your policies and standards? Engage the legal team. Their role is to understand the written word in some detail and, even better, the implications of a policy statement and how it could be interpreted. Many Chief Privacy Officers are also lawyers. They make ideal people to help with your review, since they can look at it from both the legal and privacy perspective. Again, don't let the final document read like a contract, but lawyers are extremely good at helping where you think words might lead to misinterpretation issues.

## Length and Policy Bloat

No one wants to read long policy documents. Strive to keep them short and effective. Get to the point quickly and add just enough detail to get the message across. Keep policies and standards simple and avoid jargon and too much technical detail. These documents must be clear, concise and to the point. Use bullet points wherever possible. Again, make it audience friendly and easy to digest. Strive to keep policies in the 3–5 pages range. Standards and procedures are sometimes longer due to the nature of the detail that needs to be included, which may even include screenshots.

Write security policies that make it clear to end users, managers and auditors what is required. There is no need to add a lot of detail on *why* a security control is necessary. Just spell out what is required in the most succinct manner possible.

## Measuring and Enforcing Compliance

How are you planning to measure compliance with your policies and standards? This is a question that needs to be answered or you might have an unenforceable policy, which will lead to audit issues and credibility problems where the document is simply ignored. On the flip side, if you are writing policies that apply to a large company, there might be businesses that need or want to be more restrictive than what would be a good baseline for the rest of the company. Always allow business units to be more restrictive based on their requirements, the local law or other legal and regulatory requirements. But resist making the most restrictive case the baseline for the whole company. You want to set a solid but reasonable baseline for everyone and then leave it discretionary for businesses that want or need to be more restrictive.

## Risk-Prioritization

Take a risk-based approach with your policies and standards. This means that addressing risk should dictate the order that you draft these documents. If the company is struggling with guidance on cloud security issues, make sure you're not busy focusing on less important subjects. The sooner you can get in front of big topics like the cloud, the better. Don't spend your time drafting lesser subjects just because they're easier.

Some other compelling needs for new policies might include new technologies, new laws or regulations or other operational and compliance needs that are not appropriately covered by existing policies or guidance.

Try not to be too aspirational in these documents. I've seen far too many policies where the author has detailed some ideal security environment that would take years of implementation work and the corresponding funding to achieve. You are setting yourself up for audit issues and for people to not take policies and standards seriously. Writing overly aspirational documents will also give the impression that the security group is too "ivory tower" to be taken seriously.

# Spelling and Grammar

Make sure you avoid poorly worded policies and standards littered with spelling and grammar errors. Do your best, and if this is not your strong point, leverage the built-in spelling and grammar checkers in most word processing programs or find someone who can give the document a careful proofread. Don't let this slide. Your grammar issues and lack of clarity will become someone else's excuse for lack of compliance.

# Frequency

Publishing too many policies at once can overwhelm a company and not leave enough time to ensure compliance. If you draft a policy that requires a major change or a long implementation period, then business units may not be able to fund the change until the next budget cycle. That means that a large change required by policy could take a year or more to implement. Policies that require changes to the way people work or other operational processes may need a long runway for implementation. Pace yourself and don't just drop 32 policies on everyone and expect to achieve full compliance in a few weeks.

# Revisions

Most policy documents should be formally reviewed at least annually. That doesn't mean that you must make changes, but you do need to read them and make sure they are still relevant. Policies should be reflective of current business needs and the risks they address.

Factors for deciding to revise a policy or not could include:

- Legal or regulatory changes
- Changes or enhancements to industry standards or best practices
- Integration or consolidation with other existing policies
- Terminology or other content changes

Simple edits for style, formatting, grammar issues or minor clarifications can be handled out of cycle as the need arises.

## "Should" Statements and Other Weak Language

The word "should" has no business being in a policy or standard. Save words like these for the guideline documents. You certainly don't want statements like "Users *should* abide by all laws and regulations …" Really? That's optional? Use "must" and "will" in favor of weak language like "should." Remember, you are setting requirements that must be implemented and that will be audited. Don't make these requirements ambiguous.

## Clarity

Security policies need to be very clear in what they are requiring. Ensure that policies and standards offer specific controls that are required. Don't bury controls in paragraphs of text; always make sure that you can easily scan a document and figure out what's required.

It's OK to add exemptions to certain things in the policy itself. For example, mainframes rarely meet all the complexity requirements of modern passwords. It's OK to call that out explicitly rather than expecting an exception or exemption to be drafted.

## Publishing and Storage

You should have a central repository for all policies and standards that everyone in the company can find. If your company already has a central policy repository, it's best to add security documents here rather than creating something new. If not, you will need to publish your policies and standards on either a SharePoint or an intranet page where everyone can easily find them. Publishing centrally will hopefully discourage people from downloading local copies that may or may not be the most current version.

## Get Executive Sponsorship

The more clout you can put behind a policy, the better. Get the CIO, CFO or CEO to actively support policies. In a bank, auditors take policies and standards to be management's directive on how processes should be run. Make sure your senior executives are on board, as they might well be held accountable for the content.

## Manage Stakeholders

Make sure that you seek out stakeholder input and feedback and consider it carefully. If you are expecting a lot of people to make changes to the way they do things for the sake of security, you need to make sure they are on board with those changes.

## Central Repository

It's best to have a central location for policies and standards. A Microsoft SharePoint server, a governance, risk and compliance (GRC) tool like Archer or simply a website are all good places to store policies and standards. You want to avoid multiple, older versions of documents floating around. It's best to have a single repository that has all the latest versions in one place.

## Balance Risk and Feasibility

Policy requirements need to be practical and sustainable. Don't write aspirational policies and standards that provide a perfect picture of security that doesn't exist and will never be funded at your company. It's tempting for some to document a security nirvana and then either hope or push the business into trying to get to that strategic vision. Other times, a perfect picture is painted and the business may not even realize the policy or standard exists. Until audit shows up to check on implementation. We used to call this breaking IN to jail.

Whatever you document in a policy or standard needs to be implemented or have a path to compliance. If something is far out of compliance at your company, it's OK to provide an implementation period. For example, you might want certain systems to implement multi-factor authentication (MFA) but this is not currently in place. If your company is supportive of getting this in place (meaning they're both supportive of the idea *and* willing to fund it), it's OK to document the requirement in policy and then provide, say, a one-year implementation timeline for compliance to policy. This way, the formal requirement exists but no one is out of compliance until the implementation timeline passes.

## A Word on Cloning

Resist the temptation to download policies and standards off the Internet and adopt them wholesale. Even documents like those provided at the Center for Internet Security (CIS) or the National Institute of Standards and Technology (NIST) should not be adopted directly or rebranded directly as your own. Don't get me wrong, policies and standards from these sources and sources like the SANS Institute (http://www.sans .org/) can serve as excellent starting points for a policy or standard.

The problem is that they probably don't 100% fit into how your organization operates without some rework. They also likely use different language and terminology that may not work for your company. Policy and standards creation takes some thought, collaboration and deliberation across multiple areas. Don't try to short circuit this process in the name of just being done putting a policy in place.

# THE POLICY AND STANDARDS LIFECYCLE

Policies are living documents, meaning that they need to be updated and maintained in order to remain relevant to your company. The lifecycle of a policy can be broken down

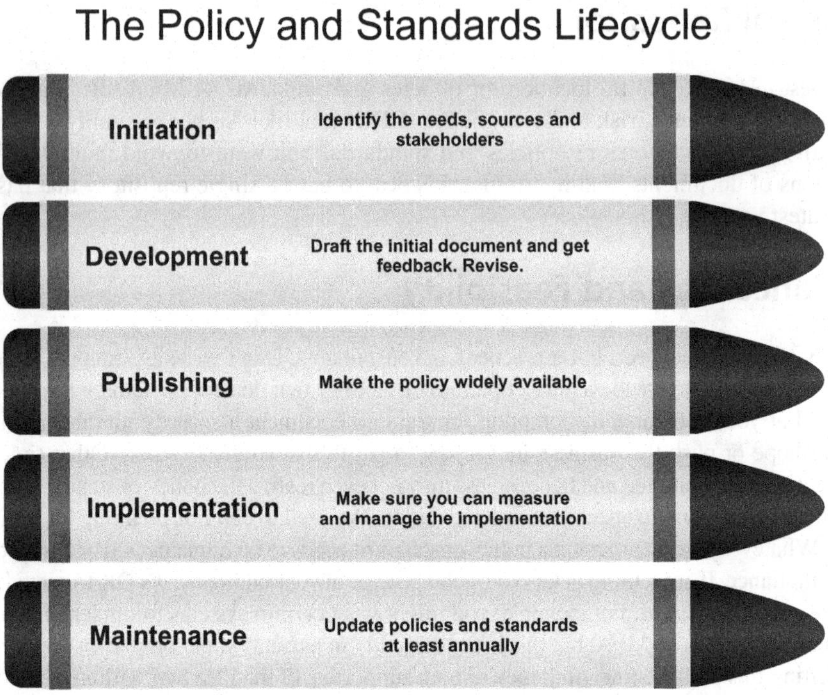

# The Policy and Standards Lifecycle

| Initiation | Identify the needs, sources and stakeholders |
| Development | Draft the initial document and get feedback. Revise. |
| Publishing | Make the policy widely available |
| Implementation | Make sure you can measure and manage the implementation |
| Maintenance | Update policies and standards at least annually |

**FIGURE 8.2** Policies and standards are living documents that have their own lifecycle. Don't skip the implementation and maintenance steps, which are critical.

into five phases: initiation, development, publishing, implementation and maintenance (Figure 8.2).

Each of the phases has key steps and deliverables:

**Initiation**

1. Identify a compelling need for new or updated policy/guidance. Drivers may include new regulatory requirements, technology developments, operational needs and identification of current issues or control gaps.
2. Determine document type (policy, standard, guideline, procedure).
3. Identify sponsorship, stakeholders, subject matter experts and working group members.
4. Develop high-level implementation impact analysis. Are the controls you're thinking about documenting in place already but need to be made more consistent? Or is this something that no one is doing?
5. Obtain agreement to proceed with draft policy (or guideline, standard) from the working group.
6. Prioritize and schedule the work. One person should take the lead as the main author. This doesn't always have to be a security person either. For example, if you are writing a cloud policy, it might make sense to have one of your cloud architects take the lead. They don't need to be a

great writer. At this stage, you want to capture the right requirements, not worry about wordsmithing and grammar.

**Development**

1. Draft the initial policy document.
2. Distribute it to an SME group for review and input.
3. Incorporate their initial feedback (use revision tracking).
4. Distribute it to a larger group of stakeholders for review and input.
5. Send the final draft out for general feedback.
6. Review and, where appropriate, incorporate feedback.
7. Present the final document to the appropriate governance entity for approval. This could be your boss, the policy committee or some other responsible entity.
8. Determine the effective date of your policy. If there are big gaps in compliance, making a policy effective immediately will only create a bunch of non-compliance issues.

**Publish**

1. Publish on your Policy Portal, SharePoint or website.

**Implementation**

1. Post and announce the document to your employees or in-scope personnel.
2. Conduct awareness activities. This could be in-person meetings or online sessions to help everyone understand their role.
3. Initiate projects to close control gaps. These should be funded projects with responsible executives and project managers that track progress.
4. Determine an ongoing review cycle to make sure the document remains relevant (the default review cycle is usually annual).
5. Monitor compliance and the effectiveness of the policy, making sure to document where control gaps exist and what the plan is to address them. You will want to track both exceptions and exemptions.

**Maintenance**

1. Update and revise the policy based on changes in laws, regulations or best practices which would require new or updated guidance.
2. Retire policies that are no longer necessary. As part of the maintenance and review process, policies, standards and/or guidelines may be identified as out-of-date or no longer needed. They should be retired via the same process by which they were approved. The policy owner should propose that a policy be decommissioned when it is no longer needed or is more effectively combined with another policy.

Of these five phases, I've seen way too many security programs stop at step 3, "Publish." This "one and done" approach is almost a guarantee that your documents will quickly go out-of-date and lose credibility. Implementation is another issue. Make sure there is an implementation plan that includes a method to measure and assess compliance. Too many policy documents are published and are never verified. But eventually either an auditor or an external regulator will pick up on this. Policies that have not been implemented are time bombs waiting for auditors to discover. Not following your own policies is inexcusable. Most auditors will see this as a full governance breakdown and view it as worse than if there was no policy at all on a given subject.

# COMMUNICATION CHALLENGES IN GLOBAL ORGANIZATIONS

Even if English is "the language of business," in a global organization you are likely to encounter issues communicating across countries and cultures. There are countries where people need many words to express themselves, like India. Other countries limit themselves to the minimum number of words possible, like the Dutch. There are subtle nuances everywhere that may have an impact on the effectiveness of your communications.

There are several scenarios where you should send communications to every single employee in a company. This includes security training and awareness material, incident notification and policies and standards.

Make sure that when you are creating training and awareness materials that you are taking the time, and expense, to have them translated. When creating these various materials, avoid using any complex language or terms which are specific to your own culture, as this can be confusing for your partners and might lead to unanticipated reactions.

If your company operates in over 100 countries, I am not suggesting that 100 languages are required, but you might consider some of the big ones: Spanish, French, German, Japanese, Portuguese and Simplified Chinese should all be high on the list for consideration. Translations services are surprisingly reasonable and will help you ensure that all policies are crystal clear to your employees. See how your company handles global communications and take their lead on this subject.

# SUMMARY

Policies and standards are an important part of the cybersecurity program and the overall governance model for security.

- There are four main types of documents you will be working with: policies, standards, procedures and guidelines. Each has a purpose and place in your program, but policies and standards are the two most important documents to produce.
- Use templates to provide a common look and feel for your documents. This will also help ensure that important elements aren't left out, like exception handling.

- Policies need to have implementation plans and potentially awareness campaigns. The target audience for your policy or standard needs to know that the document exists. It's best to use a central repository so that the latest version of your document is always available in a single place.
- The policy lifecycle consists of initiation, development, publishing, implementation and maintenance. Policies and standards are living documents. They need to be updated periodically and retired when they are no longer useful.
- Resist the temptation to clone other people's documents. They do not provide enough context for your company's requirements and might do more harm than good.
- Translate your documents if you are in a global organization.

# FOR ADDITIONAL HELP ON THIS SUBJECT, INCLUDING POLICY AND STANDARDS SAMPLES, I RECOMMEND THE FOLLOWING:

- SANS.org (http://www.sans.org)
- IANS Research (http://www.iansresearch.com)

# REFERENCES AND RECOMMENDED READING

Barman, Scott. *Writing Information Security Policies.* New Riders, 2004.

Fitzgerald, Todd. *CISO Compass: Navigating Cybersecurity Leadership Challenges with Insights from Pioneers.* CRC Press, 2019.

Gentile, Michael, et al. *The CISO Handbook a Practical Guide to Securing Your Company.* Auerbach Publications, 2006.

Landoll, Douglas J. *Information Security Policies, Procedures, And Standards: A Practitioner's Reference.* CRC Press, 2020.

# Training and Awareness

# 9

*Failure to give attention to the area of security training puts an enterprise at great risk because security of agency resources is as much a human issue as it is a technology issue.*
~ NIST SP 800-50

*The user's going to pick dancing pigs over security every time.*
~ Bruce Schneier

One of the bigger communication challenges with cybersecurity is the need to reach every single user in the organization so that they understand "just enough" about security to help protect company information. This problem amounts to taking a very complex subject and translating key points in a way that everyone can understand and playing to the lowest common denominator. Right from the start, you know this isn't going to be an easy communication task.

It is well beyond the ability for a single chapter to cover everything you need to know about training and awareness programs. I have included some resources at the end of the chapter if you would like to go deeper on the subject. The goal of this chapter is instead to understand the communication challenges and offer some suggestions on approaching communication in an awareness program, including some key elements you will likely want to include.

Security awareness efforts are designed to change behavior or reinforce good security practices. Awareness is defined in NIST Special Publication 800-16 as follows: "Awareness is not training. The purpose of awareness presentations is simply to focus attention on security. Awareness presentations are intended to allow individuals to recognize IT security concerns and respond accordingly." This is an important distinction. Most security programs will need to employ both training and awareness methods, but understand that they are not the same thing and they have different objectives. (See Figure 9.1)

There are many reasons that you will need to raise awareness of security issues. The list below is not intended to be comprehensive but to try to cover the most common reasons and offer some advice for handling communication in these scenarios. The most common reasons for increasing cybersecurity awareness include:

- Preventing security breaches and attacks
- Implementing security policies, standards and guidelines

DOI: 10.1201/9781003100294-11

# Training and Awareness
# Success Elements

**Continuous
Awareness**

Take advantage of
learning
opportunities

**Approachability**

Make sure people feel
safe reporting issues

**Positive
Reinforcement**

Reinforce good
behavior rather than
punishing bad behavior

**FIGURE 9.1** The three elements of training and awareness success include continuous awareness, approachability and positive reinforcement. A successful training and awareness program uses positive reinforcement rather than trying to punish "bad" behavior.

- Meeting annual compliance requirements
- Raising awareness of phishing emails
- Communicating the right security behavior (e.g., reporting incidents)
- Influencing company culture around security topics
- Using it as a competitive distinction with customers
- Reducing the risk of individuals undermining security controls
- Supporting efforts like National Cyber Security Awareness Month (NCSAM), which falls in October

End users are the weakest link in any security program, so you need to spend time raising their awareness of the most common attacker threats and tactics. Email, which is still the primary vector for most attacks, deserves some special attention. This attention is often in the form of self-phishing exercises but can also take other forms such as computer-based training (CBT) courses.

A training and awareness program will always be a work in progress. It's never something that is finished, and it will never reach 100% where everyone understands everything they need to know about security. This would be true even if employees didn't come and go, since new threats and tactics would still turn up all the time.

# THE RIGHT MINDSET

As always, I like to start with setting the proper mindset. Of course, you will never bring every single employee up to your level of knowledge. That said, it is also not a waste of time to try and help employees better understand their security responsibilities. There is a common belief that users can never be fully trained and therefore minimal effort should go into training and awareness programs. Some even question if these programs work at all. I think this attitude is based on the technologist's preference to deploy software solutions rather than dealing with difficult people problems. But if you don't try to raise awareness and train people, then the attackers will do it the hard way. And when they do, it will cost more money, take more time and reflect poorly on everyone involved, including you. The good news is that the facts actually do support that training and awareness programs are effective. Research shows that these efforts show improvement over time, and things like self-phishing have shown improvement by as much as 87% over longer periods. If you consider that individual employees have been the source of some very high-profile security incidents, trying to protect the human element is never a wasted effort.

Organizations should view security awareness as an ongoing program. It is not just a technology spend. It's probably a good thing that many regulations and compliance requirements mandate that you implement a security education program of some kind. I'd worry that many people might skip it otherwise, and this would be a huge mistake.

# SECURITY IS EVERYONE'S BUSINESS

An awareness program begins with support from senior management. There are not many disciplines at a company that has the need to mass communicate to every single employee the way that cybersecurity requires. Thus, it's important to make sure that everyone is on board with how you're approaching this and what types of messages you are communicating. Having a senior advocate and gaining buy-in for the program are both critical to the program's success. Ultimately, the behavior and priorities of senior management will heavily influence the level of employee awareness and policy compliance, so training and the commitment to security should always start with support from senior management.

Other groups to engage in the security awareness program include Legal, Human Resources (HR), Risk and Corporate Communications. You may also consider having representation from IT, especially factoring in any impact on the help desk. Bringing these groups into the process early will help foster a good working relationship. I've seen programs where this relationship can be strained, so bringing them in early will hopefully help avoid problems later. One way to get all these groups together is to establish an awareness steering committee. This group can be collectively responsible for making sure that communications sent to all employees support company values and mesh with corporate communication policies.

The foundation for a good awareness program is making sure that end user responsibilities are codified in an acceptable use policy (AUP). This policy should ideally come from HR, Legal or Compliance and cover cybersecurity issues. Every employee must do their part in order to achieve appropriate levels of security. Information can be found everywhere in a company and nearly every employee utilizes information in some way to do their job. An acceptable use policy helps outline what responsibilities users have and what behavior is expected from them.

An AUP may go well beyond simply covering cybersecurity issues, which is why it's best coming from HR or Legal. A good AUP outlines responsibilities and consequences for breaking rules, such as warnings, suspension of access or termination. An organization can protect itself from some liability with an AUP that clearly communicates employee behavior expectations. If an employee's AUP violation results in a lawsuit, liability may be shifted from the organization to the employee. For this reason, an AUP is typically signed and acknowledged at the point of hire as a binding agreement for the terms of employment.

There may be a host of other policies and standards that help govern user behavior, but it's best to try and put the most important elements in the AUP since it is signed individually and retained by HR. Other supporting policies that impact most or all employees might include a password policy and remote access. These documents would support higher-level requirements in the AUP like protecting company information but would not require the same level of diligence and individual sign-off as the AUP.

Once you have the AUP and supporting policies and standards in place, you should make sure that everyone is aware of their security requirements and help enforce this throughout the company. Employees at all levels must understand that they play a large part in protecting the company's information assets. This is what will be enforced through the awareness program. The goal is to teach employees that they are a key piece to the security puzzle.

The idea is that through training and ongoing reinforcement, everyone will begin to "think security" as a matter of daily practice. This can ultimately help build a security culture. Good awareness programs help build that culture, but this should also strengthen compliance with your policies.

An information security awareness program brings cybersecurity to a personal level. Everyone is responsible for the security of the information they use, and this will be enforced via your program. The purpose of an awareness program is to teach the audience how to incorporate the rules and procedures into their daily operations. Your program should really answer the question for every individual employee: "what can I personally do to contribute to the protection of the Company's sensitive data?"

All employees (employees, consultants, contractors, temporaries, etc.) should receive some level of information security awareness. The key to being effective is to make security a part of everyone's day without being obnoxious or repetitive. There's an opportunity here. Make sure that you choose training that provides the least amount of information the user needs to know in a very clear format. You need to be realistic about this; if you can train users to be suspicious of email, pick good passwords and one or two other things you are doing an amazing job. Don't think you're going to bring the whole company up to your own personal level of knowledge.

# FIVE STEPS TO SECURITY TRAINING AND AWARENESS

When I consider implementing and running a security awareness program, I like to think of it in a five-step lifecycle. These steps include analyzing, planning, deploying measuring and optimizing (see Figure 9.2).

Ideally, you will dedicate a resource for running awareness activities. This person should be someone who has good communication and soft skills. If your organization can't dedicate someone, you should look for a security awareness platform that can at least automate some of the elements of security awareness training like assigning training and tracking compliance.

## Analyzing

Analyzing is one of the most important steps and should take input from several areas of your company. You want to involve HR, Legal, Privacy and Compliance at a minimum to make sure that your awareness messages are appropriate, factor in requirements like translations and meet compliance requirements like Payment Card Industry (PCI), Sarbanes Oxley (SOX) or GDPR. Some firms choose to combine security

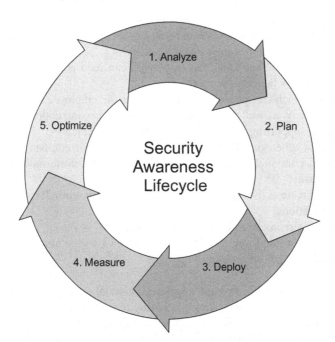

**FIGURE 9.2**   The five-step security awareness lifecycle can serve as a good framework for organizing your security awareness program.

with Acceptable Use training or other privacy training. In this phase you are gathering requirements and making sure that materials produced or obtained meet your requirements.

The analyzing phase is really a requirement gathering phase. These requirements will be the input that drives what success looks like. To be comprehensive in this phase, you should consider:

1) What compliance-driven training is required of users? Is there some overlap with what is required with security and other areas? Be careful here, you don't want to muddy the security material so much that it becomes a compliance exercise, but if there are opportunities to cover, say some elements of privacy with the security training, this might get a lot of support from HR.

2) What is the best mechanism to deliver training? Does your company already have a learning management system (LMS)? Many companies have a central tool, usually managed by HR, that helps ensure that training or awareness material is completed in a reportable manner. An LMS can assign a course and follow up if the user fails to take action.

3) Groups to engage include the technology help desk, technology management, Legal, Risk management, Corporate Communications, HR, Marketing.

You also want to think about what your program will consist of and the frequency. A few common items include:

- Required training, either in person or computer-based training
- Newsletters or email mass communications sent on a regular cadence
- Webcasts or live-in person training sessions
- Pocket guides or handouts that employees can read at their own pace and refer to them whenever the need arises
- Surveys, which can provide benchmarks for the employee attitude towards cybersecurity, as well as their understanding of organizational policies and their role supporting information protection requirements
- Posters or other visual reminders that reinforce your company culture and serve as a continual reminder of your organization's commitment to information security
- Email signatures that give employees regular, subtle reminders of their security obligations
- Security alerts messages for specific concerns
- Presenting at employee town halls, which can be very effective as well as emphasizing the importance of the topic

Companies throw a lot of training and awareness material at people. Don't be surprised if there is some pushback from adding to existing requirements. Therefore, it's best to at least consider some "consolidation" opportunities like working with privacy, compliance or other similar disciplines when you're looking at annual training.

During the analyzing phase, you should also consider what topics are relevant to your organization and what themes you want to cover with awareness activities.

The idea here is to narrow your focus to the important topics. There are many topics you can and should cover, but you can't do them all at once and some topics will result in more risk reduction than others. If you are tracking security incidents closely (and you should be) this can be an invaluable feedback loop for which areas might need more focus from your training and awareness program. For example, if people are repeatedly losing mobile devices, you may want to increase awareness on the topic.

Common themes for these programs to cover include:

- Mobile devices
- Remote access
- Phishing awareness
- Reporting security incidents
- Encryption
- Data destruction
- Wi-Fi security and passwords (including public Wi-Fi)
- Physical device security
- Social engineering
- Email security
- Malware
- Removable media
- Password security
- Social networking risks
- Bring your own device policy (BYOD)
- Cloud security policy
- Handling sensitive information
- Data classification

# Planning

In a field as big as cybersecurity, you are not going to be able to teach everyone everything. I've found that it's best to pick some themes based on risk to your organization. Of course, there are certain subjects that will be common to all organizations. These include email phishing, social engineering and password protection, to name a few.

In the planning stage, you are either drafting your own awareness material or obtaining it from a third party. You should frequently refer back to your requirements gathered in the analyzing phase to make sure there is no disconnect between what your company needs and what it obtains or creates. You will also want to plan *how* material will be disseminated.

You can outsource content creation and there are a lot of great vendors in this space. If you choose to do the work in house, I recommend building a queue of material so that you can get into a regular cadence without scrambling at the last minute to find topics that need to be covered. Having a pipeline of material will also give you some flexibility in that if a topic suddenly becomes more important, you may not need to put something together from scratch.

# Deploying

In the deploying phase, you are actively deploying the awareness material you chose during the planning phase. Sometimes the security group is doing this; sometimes it is the HR group. Deploying in a large company may be in a tiered model, so that everyone does not receive material at the same time. Most often though, you would want to deploy consistently in one shot.

# Measuring the Program

Your security awareness program shouldn't only measure click rates on phishing emails. While this is an important metric, there are several others that can help you measure the overall effectiveness of an awareness program. A few other metrics you may want to capture include:

- **Phishing metrics**: You do, of course, want to include phishing metrics. But again, not just the click rates. You can also measure who reported the phish and other behavior change metrics. We will discuss phishing campaigns later in this chapter.
- **Annual training metrics**: You should strive for 100% compliance with annual training. This can sometimes be tracked automatically if you are using a learning management system. Don't let senior executives off the hook from these efforts; they have access to the most sensitive data in the company.
- **Security incidents reported**: While not immediately intuitive, sometimes when you increase security awareness material, the number of reported security incidents also goes up. The reason is that as people get more "attuned" to security incidents and report more activity. Some of these are not true incidents, but the behavior should still be rewarded.
- **Surveys**: Surveys can provide a good "pulse check" for the attitude towards cybersecurity. Surveys can be compared year over year, especially if employees start as pretty ignorant on basic information security issues.

# Optimizing

Once a training and awareness program is set up, optimizing it is the last step. To optimize, you want a feedback loop so you can be sure that training is appropriate for your users and that the program is effective. With CBT courses, this might be a quiz at the end. For the phishing program, you would monitor the click rates and look for improvement over time. Surveys can also help gage the effectiveness and appropriateness of the materials you're using.

Sending surveys for user input is very effective at helping to optimize the program and they have the added benefit that you can solicit topics that people would like to see covered. You'll want to canvas a lot of people for their opinion, so probably the best way to go is by sending out a survey asking what people like best about your efforts, what they think could use improvement and other topics they would like to see covered. You might also consider asking for something that they learned that they didn't know before your training and awareness efforts. Since this information isn't particularly sensitive, you can just use a tool like Survey Monkey (https://www.surveymonkey.com) to help collect and report on answers. In smaller organizations, email might suffice. Don't expect everyone to respond. A lot of people don't bother to complete survey requests. Keeping it short and focused on important questions will help your response rate, but also having easy-to-answer yes/no questions or dropdown menus will help. Making it easy for the audience and fast to complete the survey and your response rate will be much better.

Some of the more innovative delivery approaches in the awareness space can involve the use of gamification. The idea of gamification is that organizations can make meaningful changes to security behaviors by turning it into a bit of a game. If your organizational culture supports gamification techniques, you can try methods including displaying a leaderboard and awarding digital badges or certificates based on games or completion of certain exercises. One company I worked at gave out a few random trophies of a fisherman if the user did the right behavior of reporting a phish versus simply not clicking on it.

# SELF-PHISHING CAMPAIGNS

If you perform some level of self-phishing against your employees, and I strongly suggest you do, this may present a unique communication scenario. In a self-phishing simulation, either you or a vendor mass-email employees and contractors with a simulated phishing message. When the user clicks on the link, it takes them to a warning page letting them know they fell for the phish. Usually, metrics are tracked and, in some cases, people who click more than once get special attention in the form of additional training.

Even some CISOs get the wrong idea about phishing exercises. Phishing is ultimately an awareness exercise. You are not going to be able to train every single person into not clicking every single message every single time. It's just not going to happen. What you are trying to do is increase awareness that not every email is what it seems and that some could be malicious. Ultimately, you would like to reduce the number of people who click, but it's not realistic to expect the number to go to zero. While some companies have made dramatic improvements, I'm not aware of any that would claim a zero percent click rate.

Choosing the content of a phishing message is a unique opportunity for showing empathy. The company GoDaddy sent out a phishing test offering a Christmas bonus of $650 and asking employees to fill out a form with their personal details during the holiday season. As you can imagine, the people who fell for this phish were doubly upset and the company wound up issuing an apology. You'll want to consider the recipient before sending these emails. While it is true that an attacker could have used the exact same tactic, the attacker will remain anonymous, and you will not. Choose your messages and themes carefully.

Some companies want their HR and communications groups involved in planning phishing campaigns and in selecting the message that will be sent. While this adds a layer of complexity to execution, go along with whatever your boss and senior management would like. It never hurts to have another set of eyes on these messages to avoid any sensitivity issues.

Consider what parameters everyone is comfortable using. What topics you should avoid and what branding you should avoid. One company I worked for wasn't comfortable sending phishing messages that appeared to come directly from the company. Again, even though the attackers could use the same tactic, you need to be aware of your boundaries and respect them. Perhaps in the longer term, you can work to convince them that your ideas for the program are worth pursuing, but never go directly against everyone's wishes without support from the top of the organization.

People who click on a phish more than once or twice generally do warrant some extra attention. I once worked with a CEO that wanted to take away email completely from repeat offenders who had clicked on our simulated phishing emails. The problem was that there were some very senior people who fell in this category and a large number of people overall. I wasn't convinced this was the best course of action from a business perspective, especially since the security group was going to be the face for this aggressive action, not the CEO. Eventually, we talked him out of taking this drastic action and agreed to a more regular cadence for the tests as well as some additional training for repeat clickers. I also know of a large private equity company that sent repeat offenders a voice mail from the CEO, registering their disappointment in the user. While the message was canned, it was quite effective in getting people's attention. Employees even came to the security group asking for one-on-one help to make sure they didn't fall for it again in the future.

One of the problems with these techniques is that no one enjoys being punished and you are also technically punishing someone who would be considered the victim in a real event. I don't believe that many people intentionally click on a phishing link. So, if their intentions weren't malicious, why punish them? But what should you do instead? I recommend some targeted and more in-depth training for them. This is a more positive approach and allows employees to go from being the problem to being part of the solution.

# TRAINING PROGRAMS

Training should include some form of annual, "security 101" training for all employees. This training is usually handled online in large companies in order to scale to all employees. The goal of annual training is to set a baseline knowledge and enforce security topics.

In addition to annual training, there are many situations that warrant targeted training for employees at a deeper level than simple security awareness is intended to cover.

If awareness is the state of simply knowing or being aware of something through observation, training is the more intensive process of teaching how something works or how to deal with a situation. Training programs aim to educate and instruct.

Using the communication principle of considering your audience, training efforts are most effective when they are tailored to individual business or IT needs. Training can and should be customized to these needs. Making sure you have the audience right will also ensure you are providing the right level of detail. Are you presenting basic security information to advanced IT employees? They're likely to tune out from boredom or not take you seriously. Are you providing technical detail to the C-Suite? They'll also tune you out, but it would be because they have no idea what you're trying to tell them.

Some considerations for formal training programs to be tailored to the audience include:

- **New hires and contractors**: All new employees and contractors should be required to complete an awareness-training course within a defined period, such as 30 days.
- **System administrators and privileged users**: These users have a level of access well beyond the standard users and they should be instructed on password management practices, secure software development lifecycle (SDLC), protecting their accounts and secure configuration training on their specific platforms like Windows, UNIX and mainframe.
- **Users who handle sensitive data**: There are many compliance regulations that may require or suggest specific training for users that handle in-scope data like PCI, HIPAA, SOX and FTI.
- **General employees**: You may wish to train general employees on security policy highlights, data classification, AUP, what is an incident and how to report it and regulatory requirements. Some level of training is usually required annually for all employees by most companies.
- **C-level executives**: C-level executives and their support staff (admin team, assistants) represent a unique risk. Sometimes executive level access is delegated to support staff, so training both roles can be important. You'll generally want to do this kind of training in person to ensure communications are clear and it is given "white glove" attention. The training should cover specific behaviors and concerns for their job requirements, which might include travel, mobile devices, etc.
- **Contractors and temporary staff**: Non-fulltime staff sometimes need to be handled separately. Most likely require custom training, via onboarding methods. Legal can provide guidance on limitations and make sure that requirements are included in contractual language.
- **Developer training**: Developer training is tricky, but if you have employees that are creating custom code for your organization, you should instruct them in the secure SDLC and potentially provide them with secure development training. This is a balance since developers are generally very technical people. If you provide training that is too basic, they may tune out from the material. If you go too technical, some of the junior developers could be left

behind. Strike a balance and try to make sure you are covering the most common development language in your business, such as C#, Java and Python and are teaching at a level that is best for the majority.

There are many general topics that can be covered more in depth using customized training. Some other potential topics for all users include policies, reporting incidents, phishing, social engineering, mobile device security, ransomware, passwords and network security. As you can see, there's a lot more to training and awareness than a simple annual course and test phishing simulation every now and then.

# SUMMARY

Training and awareness programs represent a unique communication challenge in that your job is to communicate difficult topics to almost everyone in the company. This means you will need to break complex subjects down into easily digestible components and be realistic about how much material you can disseminate at one time.

- Training and awareness have subtle differences, but you need both to support your security program.
- Trying to communicate a complex issue like security in a way that every employee will understand is a tough communication challenge.
- There are many reasons for having training and awareness programs, including reducing risk and establishing a culture of security.
- Training and awareness programs start with support from senior management and having a solid acceptable use policy in place.
- A good five-step process to implement a program includes: analyzing, planning, deploying, measuring and optimizing.
- Self-phishing is an important awareness exercise that should aim to educate, not punish, your employees.
- Training programs differ from awareness programs in that the objective is to teach and instruct, instead of just raise awareness.
- Specific training should be developed to meet specific needs in your organization, such as secure coding, system administrators and C-level executives.

# REFERENCES AND RECOMMENDED READING

DeZafra, Dorothea, et al. "Information Technology Security Training Requirements: A Role- and Performance-Based Model." *CSRC*, 1 Apr. 1998, csrc.nist.gov/publications/detail/sp /800-16/final.

Herold, Rebecca. *Managing an Information Security and Privacy Awareness and Training Program.* Auerbach Publications, 2011.

*Information Security Training,* www.sans.org/security-awareness-training.

Purser, Steve. *A Practical Guide to Managing Information Security.* Artech House, 2004.

*Security Awareness: Best Practices to Secure Your Enterprise.* Information Systems Audit and Control Foundation, 2005, https://www.amazon.com/Security-Awareness-Practices-Secure-Enterprise/dp/1933284064.

# Driving Change through Metrics

# 10

*When you can measure what you are speaking about,
and express it in numbers, you know something about
it; but when you cannot measure it, when you cannot
express it in numbers, your knowledge is a meagre
and unsatisfactory kind; it may be the beginning of
knowledge, but you have scarcely, in your thoughts,
advanced to the stage of science.*
~ William Thomson, Lord Kelvin, 1883

## WHY METRICS ARE A POWERFUL COMMUNICATION TOOL

Why include a chapter on metrics in a communication book? Metrics are one of the most powerful communication tools at your disposal. They can demonstrate progress or risk in a simple chart or number that everyone can understand. Metrics can help drive change and shift corporate cultures—when used properly. Metrics can also be a liability when used improperly. Of course, a big problem is that measuring security is difficult. But that doesn't mean it's impossible. This chapter discusses some basic principles of metrics, including what makes a metric good or bad and it also provides a few tools and techniques to improve your own metrics program.

Metrics can be used to measure if an organization's security program is accomplishing its goals. They can tell you what is and isn't working so that improvements can be made to get things back on track. Metrics can also just be a bunch of numbers and data that don't tell a story, convey meaning or serve any purpose whatsoever. Metrics can be "skewed" in various ways that can be misleading. An old saying goes "torture the numbers long enough and they will confess to anything." Welcome to a very fuzzy world of metrics, made even fuzzier because at first glance they seem very precise.

The subject of metrics is huge and is also very specific to your organization. If you're looking for a set of exact metrics you should use to drive your security program, I'm afraid you'll be disappointed with this chapter. But I think you'd be even more disappointed if I gave you some completely out of context security metrics that you adopted without any consideration that ultimately made you look silly. There's some

DOI: 10.1201/9781003100294-12

work you'll need to do to find the right metrics for your organization. The good news is that this chapter should give you a good crash course on metrics as well as several starting points to start thinking about how to develop your own metrics program in a way that will help measure progress and drive change instead of just conveying a bunch of numbers and statistics.

The communication principle of understanding your audience is key when you're developing metrics. Think about what your business leaders want or need to know before you simply start measuring everything that can be measured and throwing tons of numbers at them without any context. Generally, business leaders have simple questions like:

1) How secure are we?
2) What are the most relevant risks to our business?
3) What are we doing about those risks?
4) Are we better off than we were last year?
5) How do we compare to our peers?
6) Does the cybersecurity team have the right resources to address these risks?
7) Where are we deficient in compliance with security requirements?
8) How has our cybersecurity risk increased or decreased?
9) How do we manage our risks?

You'll notice that these questions are not sophisticated, but they are very typical of what business leaders ask. Business leaders are going to be a lot less interested in technical details, and therefore technical measurements. Many of these questions have been pulled directly from real-world situations where senior managers (CEOs, CFOs and the board) have been challenging CISOs for answers.

# START WITH YOUR AUDIENCE

Security metric reports are often presented to senior or C-level management, including the board of directors. These are generally busy people in the organization with a lot of responsibility and minimal time and focus to understand technical security issues. They are typically most interested that their systems are working properly and that their most important data is safe.

You might also have specific businesses, technical staff and others who have a stake in metrics. Unfortunately for you, this means you have to do some heavy lifting to make everything tailored to multiple audiences. For example, in a large company you might have several CIOs responsible for individual business units. It is not going to be helpful to go to these leaders with metrics across the entire firm. They are going to want to see specifically what they are responsible for so they can act on what is under their direct control. Include everything in one report and they may assume that none of the problems belong to them. Remember, always make it easy on your audience. Don't make them search for relevant information by digging through pages of data.

On the other hand, if you are presenting the overall risk to the company to the board of directors, you are going to need aggregate data and not present all the individual details that you might have shared at the CIO level. Most boards do not want to see exactly which server in some minor business unit is mitching whatever patch. The board operates at a big-picture, strategic level and these details, while maybe still important to you, are not going to matter to them.

If your audience is technical, they may have a very different set of questions that would require a very different set of metrics. If the metrics you're compiling are for CIOs and heads of various IT divisions, then patching statistics and open vulnerabilities might be completely appropriate. Just be careful, as this gets difficult in some large companies where ownership of a problem may span multiple groups. For example, an application security scan might reveal some underlying problems with the operating system as well as a hosted application. In a large company, the application owner would be responsible for fixing the application issues, but another group might own patching or reconfiguring the server operating system. While these may sound like subtle distinctions, people can get very defensive when you present metrics that may not seem very flattering that are also not under their direct control. This is especially true when it makes them look like they are not on top of their area or that issues have stayed open for a long time (Figure 10.1).

And finally, if your audience is other security professionals or your team, you might want a very different set of metrics and measurements to track. These metrics might

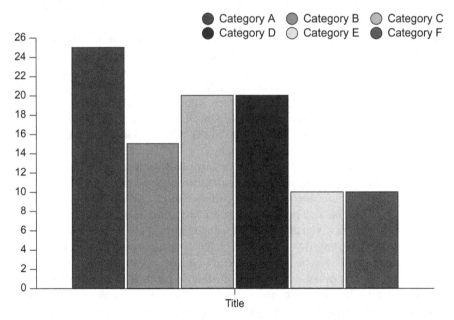

**Vulnerabilities by Business**

**FIGURE 10.1** Visual aids like charts and graphs can help business leaders better understand data. Strive for accuracy and make sure you're considering what your audience needs or wants to see.

show things like what type of attacks you're seeing, how the Security Event and Incident Management (SEIM) is performing or other measurements that only the security group will want to see.

The need for multiple audiences means you're going to need to do some data massaging. If you can find someone to help you, grab a Tableau or Microsoft Power BI expert to help you set up something sustainable. Otherwise, you're going to find yourself running a lot of reports and burning a lot of time on pulling data and putting everything in a readable format week-after-week or month-after-month. Considering your audience is the first, and most important, step of making sure that your metrics program is going to be effective. Next, we can choose some metrics that serve a purpose and serve the audience.

# METRICS THAT MATTER: THE DATA YOU HAVE, VERSUS THE DATA YOU NEED

> *Security is now so essential a concern that we can no*
> *longer use adjectives and adverbs but must instead use*
> *numbers.*
> ~ Dan Geer, 2008

Many security leaders put off developing a metrics program because they don't know where to start or how to assess what metrics they should be measuring. Your security program needs to have a metrics program. If you don't already have one, you will need to establish one eventually and it's best to get this started early so you can show a baseline and some progress. Metrics can also help you identify when risk is going up or when projects are going off track. Metrics are much more than a glorified status report. They can help drive behaviors and change in the organization. This makes them an extremely useful tool when you get them right.

There are two basic types of metrics: qualitative and quantitative. Qualitative metrics are not generally expressed by numbers. Qualitative metrics look a lot like "red, amber, green" or "high, medium, low" or "good, bad and ugly." Quantitative metrics, on the other hand, use a number or a percentage and are more precise in nature. If you are only using qualitative metrics, you are not really using metrics at all. They are highly subjective and not repeatable unless you add some quantitative measures to ensure consistency, such as turning >90% patched into a green rating for server patching (Figure 10.2). Using both a numeric rating and a simple qualitative measure is a great combination, because it's repeatable but also conveys a simple qualitative message of being on track or not. The benefit of qualitative is that everyone gets red, amber, green without much thought processing.

Information security risks aren't always easily quantified. This makes it somewhat different from other types of business risk, like credit or market risk. Credit and market risk can not only easily be put into numbers, but you can also run what-if scenarios like Monte Carlo simulations that can predict the impact of a wide variety of outcomes. The

# Security Performance

**FIGURE 10.2**  Using a combination of qualitative and quantitative metrics can make sure that your data is easily understood without being too subjective.

formality around other risk disciplines makes the typical red, amber, green approach of security issues seem quite primitive in comparison. Some work has been done trying to apply tools like Monte Carlo to cybersecurity, but so far, I'm not impressed. I'm hoping that someone gets this right someday, but I won't hold my breath waiting.

When a metrics program is poorly run, people tend to report the data that they *have* versus the data that they *need*. They measure whatever can be easily measured without considering if the measurement is meaningful or relevant. These metrics fail to tell a story or drive a behavior or even resonate at all with senior management. There are usually a lot of them too. These metrics might seem impressive at first since they present so much data. That is, until you think about what the data means. Some of the most effective metrics actually tend to be quite simple. In some cases, the most effective metrics don't track technical items at all; they track *behaviors* and tell a story.

Let's look at an example of a bad metric and consider the number of threats *blocked* by your security controls. While a lot of people love using this metric, it's bad for several reasons. To start, this metric creates a false sense of security that lots of attacks are being blocked by security controls. The higher the number, the more effective your defenses, right? If so many attacks are getting blocked, everything must be fine, correct? This is not the message you want to convey. This metric doesn't tell a story, nor does it drive a behavior. At least, not a story anyone cares about. What is the message? There are a lot of attackers in the world? Well yeah. But if the number goes higher or lower, what does that indicate? And if it does drive any behavior, it's complacency. Look how many things are being blocked, so we must be secure.

Metrics like this are often called "attack metrics." These numbers are typically the easiest to obtain (making them very tempting to leverage) but they are the hardest to use effectively. While the month-over-month numbers might be academically interesting, the rest is just useless noise. All that these numbers indicate is that the firewall/routers/IPS are all doing their job. Are they blocking sophisticated attacks? Port scans? I'd argue that no one cares, since they didn't get through anyway. If this metric is good for anything, it's to show how crazy it is on the big, bad Internet. Don't bother. Both your senior executives and your technical people already know this.

Other poor choices include statistics such as some raw "number of unpatched vulnerabilities" that provide little information about business risk and show a giant volume of critical findings. These are the "shock and awe" of the metrics world. The problem is that reporting security metrics like this in isolation is not meaningful to a business audience. I've seen some business leaders respond to numbers like this by either wondering why all vulnerabilities can't be patched across all systems at all times or else responding apathetically, asking how many vulnerabilities an acceptable number would be. Neither of these responses will be especially productive in reducing risk in your organization.

While vulnerabilities are a valid topic to track, they don't make much sense unless you add in business risk context. Try to craft a simple story that tells the audience what they need to do or what they need to worry about, rather than just a raw number. Ask the question, "so what?" Here's a metric, so what? Do you really have a good answer? This type of metric also fails to take a business-focused risk-based approach and treats all vulnerabilities equally. Sure, it breaks problems into criticality. But a critical vulnerability on the server that hosts the internal newsletter is not the same as the one that's sitting on your Internet-facing e-banking platform or claims processing system. Critical vulnerabilities on business-critical systems might be a whole different conversation that everyone understands.

Metrics can have unintended consequences if they don't tell a carefully targeted story or drive the right behavior. They can become the equivalent of a "check engine" light that's always on but the car continues to run fine. It's red, but so what? It's always red.

When you choose your metrics, you'll want to run them through a few simple tests and checks to make sure that they are strong and defensible. Then we will discuss how to better leverage these metrics to better tell a story or drive a behavior.

Starting with the basics, metrics should not be:

- Overly complex or difficult to understand
- Overwhelming
- Only qualitative
- Difficult to gather and refresh
- Constantly showing the same result
- Used for reward or punishment

Let's take each of these point by point, starting with complexity. Metrics that are difficult to understand or are overly complex will at best confuse your message and at worse create the wrong impression entirely. Metrics that fall into this category can generally go up or down in numbers and no one really cares, since they are not indicative of any real business risk. In some cases, overly complex metrics don't even make sense to measure in the first place.

Almost as bad as too much complexity is providing an overwhelming number of "who cares?" metrics such as the classic "10,000 vulnerabilities across 2,000 servers" example. While this may be factually accurate, a business executive will have little idea if this is average or bad. If they do know that this is bad, they may have some very uncomfortable questions for you. If your audience was technical, this is still just a mountain of problems and the perception will be that the problem is being dumped on IT operations to fix. This brings up a good point of never using metrics to call out other departments for "not doing their job." It's bad form and will not be a sustainable method for gaining cooperation from all the groups you will need to help you with a good vulnerability management program.

On the other end of the complexity spectrum, qualitative data, while easy to compile, is messy, subjective and imprecise. I've seen metrics programs that consist exclusively of qualitative data. I won't say that all qualitative data is bad, but if you are only ranking everything red, amber, green, be prepared to defend why something is amber instead of red and other pointless arguments like that. Qualitative metrics also don't really help anyone *understand* the risk in a precise fashion. Again, using qualitative and quantitative together can be quite effective. Turning a metric like awareness compliance "green" after it hits 90% is clear and repeatable. Ninety percent is a repeatable measure that an auditor can repeat and "green" is a good way of indicating something has hit an acceptable level. A number even as high as 90% of critical Internet-facing vulnerabilities being patched might not be enough to be green since any one of these open issues could result in a security incident. In this case, you might want 98% or even 100%. The number alone isn't enough, but the number and the color are a great combination for people to easily understand if they are on target or not.

Metrics shouldn't be hard to gather and update. Whatever metrics you choose to track as part of your program, you do not want to have to spend weeks pulling the data from multiple sources only to start all over again when you finish so you can have them ready for next month's report. I remember one metrics deck at a large financial that combined data that was from multiple different months. Vulnerability data might be from March, but incident data went into April. This led to some very odd situations with people looking at old data and making incorrect conclusions. It also made it seem awful silly when the overall metrics report was titled: *March Security Metrics* but included information current in April. Why did it end up this way? That was the easiest way to gather the data and filtering out April was too much work.

We already discussed not including metrics that are always going to show the same result, regardless of if that result is good or bad. This is especially easy to do when using qualitative metrics. I've seen a few CISOs try to use metrics to try to shame people into working on security problems by turning issues red or they would throw in a few that would always be green just so the dashboard didn't always look red and yellow. There's a tendency, since security people are generally a pessimistic bunch, to have all red and amber ratings on just about everything. This takes us full circle back to the check engine light that's always on gets ignored. You want meaningful metrics that show either increases/decreases in risk or else progress and trending with security issues. The absolute value doesn't matter anywhere near as much as the trending.

Finally, there's a saying that "the behavior you reward is the behavior you get." Using metrics to either reward or punish someone (or some department) tends to be a bad strategy because it drives the wrong behaviors. Think of the classic example of an IT help desk measuring how fast tickets are being closed. They may be rewarded for closing tickets quickly and they may get punished for tickets that remain open too long. Do you see the problem with this approach? People focus on closing tickets as fast as they can, regardless of if the problem is resolved or not. Using this metric, you'd be measuring something that seems to make sense but ultimately drives behavior in an unproductive fashion since it will increase customer dissatisfaction with the help desk as their issues remain unsolved.

# MEASURING THE UNMEASURABLE

*In God we trust, all others must bring data.*
~ W. Edwards Deming

I remember once being asked how many gas stations there were in America during an interview. This was before Google and even before the Internet was in widespread use. This was an interesting question. I doubt anyone has the answer at the tip of their fingers without the aid of a computer. While I was working on trying to figure out an answer to this question, the interviewer asked me to stop. Concerned that I hadn't calculated an answer yet, he said to me that there are only three answers to this question. One was to blurt out a number not based on any knowledge, like 4 million. Next, you could give up quickly and say that you have no idea and no idea how to approach the problem. The last solution, and the one I picked, was to start doing my best to calculate how many there are by mentally going through how there are 50 states, some are more populated than others, etc. The interviewer had stopped me because he could see the wheels turning and said that this was the type of person they strove to hire. By the way, run this same question through Google today and you'll get 168,000 and some background on why the number is declining year over year. We've come a long way.

While it may not seem like it at first, almost everything can be measured in some fashion. In Doug Hubbard's book *How to Measure Anything*, the author identifies a reason why some things appear difficult to measure and evaluate: "Business managers need to realize that some things seem intangible only because they just haven't defined what they are talking about. Figure out what you mean, and you are halfway to measuring it." The theory of the book is that absolutely everything is measurable. Still skeptical? Why not pick up a copy of the follow-up book, *How to Measure Anything in Cybersecurity Risk*. There's a whole book on the subject.

If you want to make better decisions, you are going to need data to back you up and support your choices. There are seven principles that can help you figure out how to measure the things that may seem too difficult to measure.

1) If it matters, it can be measured.
2) You have more data than you think.
3) You need less data than you think.
4) We measure to reduce uncertainty.
5) What you measure the most may matter the least.
6) What you measure the least may matter the most.
7) You don't have to beat the bear.

Let's start with the principle that if it matters, it can be measured. We have figured out ways to measure anything from cyber reputation risk to employee satisfaction to the chances and impact of a famine. So, if something matters, it can probably be observed or detected in some way. If it can be detected, it can probably be given a range of possibilities. Once given a range of possibilities, measurement is possible. There are very few things with an infinite number of possibilities, though there certainly may be things with a large number of them.

A lot of people fall victim to the belief that you'll need more data to make educated decisions. If you don't have enough data, don't measure anything and instead pursue more data. They end up measuring everything that can be measured. This is also a mistake. Whatever you are measuring has probably been measured before, and historical data can fill in a lot of the gaps.

The next principle is that you probably need less data than you think you need. In fact, even small amounts of data can be very useful. Remember, if you know nothing about a subject, then almost anything will tell you something about it. In the gas station example, there are some easy "known" elements like the number of people in the US population and the average number of cars per household. You don't need to measure how many gas stations if you already have some basic information that might help give you an estimate.

This brings us to the next point: measure to reduce uncertainty. Notice I didn't say measure to "eliminate" uncertainty; we measure to reduce uncertainty. This is important, as a lot of people can get stuck on precise measurement or trying to eliminate all uncertainty. While this may seem like a great goal, it is probably unnecessary and will prevent you from getting useful measurements faster. Measurements don't need to be precise to be useful or to help eliminate uncertainty. Again, you can probably get pretty close to measuring the number of gas stations just by taking average numbers of how many cars a family owns and the average number of times most people need to refuel in a week.

What about the things that we already measure? The theory goes that what we measure the most may matter the least. What this really means is that sometimes we get hung up pulling metrics and data that don't really tell us that much or don't tell us enough compared to the effort required to pull them. This applies to many real-world scenarios like trying to confirm that every single laptop in a large enterprise is encrypted. This sounds like a good metric, but that last 5% is tough because there will always be a percentage of machines that are turned off and put in a drawer. These machines may be off network for extended periods or simply unreachable for whatever reason. One of the companies I worked at insisted on 100% coverage for encrypted laptops and that last 2% or so took months of effort and at the end of the day probably did not have a material impact on risk. After all, how much risk did a laptop that was powered off and sitting in a drawer really represent? As you might expect, it turns out that while we are busy measuring things that may not matter, we are missing the things that do matter. In fact, sometimes we're measuring the wrong thing altogether. You might be measuring how many people attended an awareness training session instead of how many internal theft incidents are taking place.

One of my favorite sayings is that you don't have to outrun the bear. The story goes that you and your friends are camping, and suddenly a bear attacks and starts chasing you, you don't have to outrun the bear. You only need to outrun the slowest camper. Measurement is a process, not an outcome. It doesn't have to be perfect; it just needs to be better than what you used in the past. In other words, don't let perfect be the enemy of the good.

These seven principles will help you better understand how measurement can be used as leverage metrics to add value, rather than just tediously gathering data and trying to make sense of it all later.

To go much deeper on this subject, I recommend picking up a copy of *How to Measure Anything in Cybersecurity Risk*. Here, authors Douglas W. Hubbard and Richard Seiersen take these concepts a step further. They propose that rather than using the standard high/medium/low risk metrics that are rampant in cybersecurity, cybersecurity can use the same quantitative language of risk analysis used in other disciplines.

# WHAT MAKES A METRIC EFFECTIVE?

*It is better to be approximately right than to be*
*precisely wrong.*
~ Warren Buffett

Good metrics provide clarity and motivate people. They tell stories. They can help move the security program faster. Effective metrics can also create alignment among different individuals, functions and process teams within the organization. So now that we have an idea about what a poor metric looks like, what do good metrics look like? Good metrics are accurate, actionable and relevant. Let's go through a few rules of good metrics.

The first rule of a good metric is accuracy. What's the point of a metric that's not accurate? This sounds simple, but in reality, a lot of the data you see may have hidden complexity to it. Security scanners misclassify things all the time. Share your data with system owners before sharing it widely with senior management. It's hard to come back from data that presents an overly negative picture but turns out not to be factually correct. All it takes is one department head to call out how your security scans have shown AIX (IBM's version of UNIX) vulnerabilities when that operating system isn't even run in the organization. Oops, that's actually a Linux server. Suddenly, all your data becomes suspect and you lose credibility.

The only antidote to this problem is to spend some time on data validation. Make sure that the people responsible for the items you are collecting metrics for agree with the results and have seen the information before it is presented to senior management. There should be no surprises. Your job is to present an accurate assessment of what the environment looks like from a security perspective, not to get other managers in hot water. The "track 'em and smack 'em" approach doesn't get the right kind of support for your program and encourages people to find every fault they can with your data. This way they can passive aggressively send you back to the drawing board with a curt "come back to me with clean data and I'll be happy to help you."

The next rule of good metrics is to keep them actionable. Metrics that are not actionable (meaning you don't know how movement in the metric will change your behavior) are just numbers. When putting a metric together, think about what you would do if the metric seemed way out of line, such as being too high or too low. What actions can be taken to get the metric to a more reasonable measure? What steps could be taken to improve? If there are no real steps or actions to take, you probably have a poor metric.

The next rule of metrics is to keep them limited to a few metrics that are truly important and relevant. While some say the number of metrics you track should be fewer than five, I'm not so sure there are any hard and fast rules with the right number. What you don't want is an overwhelming number of metrics that measure everything just for the sake of measuring everything or presenting data just because you happen to have it. Measure what's important and if it really is important; no one will care how many you are tracking.

Other aspects of good metrics include:

- **They have a purpose**: Metrics should support business goals and regulatory requirements, connecting metrics to the business can help with stakeholder buy-in as well as ensuring resources are efficiently used.
- **They show trending and progress**: Worthwhile metrics must demonstrate that a goal is being met or that a trend is improving or getting worse. You may need to work with the data to be able to demonstrate trending in a way that management understands. Progress in metrics can show that efforts are moving the security program forward. Avoid only showing month-to-month progress; you should really show the historical timeline and greater progress over time.
- **They are simple**: You shouldn't have to spend a lot of time explaining the significance of a metric or providing a lot of background. If you do, you are probably measuring the wrong things or measuring things no one cares about. You don't need to measure everything. You need to measure everything that matters. It's tempting to pull together all the data you can find and produce overly complex metrics decks. The goal of these decks seems to be to demonstrate how busy the security team is and have everyone look at all the problems they have to address. The problem with this approach is that this is about the only message that gets across. Look how busy we are! Obviously too busy to produce a meaningful presentation that can explain risk posture to senior management.
- **They are in context**: Don't just take the results of pen-testing or a security tool and call it a metric. Metrics must have meaning. Answer questions like "why are we collecting this?", "what story does it tell?", and "how does this compare to our peers in the industry?". If you can't answer these basic questions, consider if you really have the right metrics before pushing them up the management chain.
- **They are quantitative**: Again, metrics need to be quantitative so they can be compared across time and organizations. Repeatable metrics avoid being too subjective about the state of the program.

- **They are easy to collect**: The best metric is useless if you can't easily collect and analyze the data. It shouldn't take you a long time to prepare and report your metrics. You should not have to work with multiple departments and then manually compile data. Ideally, you should have an always up-to-date dashboard that anyone in your organization can look at. There are some great tools like Microsoft Power BI and Tableau that can help build a security dashboard using live data.
- **They are timely**: Data should be as current as possible. I've seen metrics presented that were more than two months old. Most of the issues had already been resolved. Don't present old, stale data.
- **Consider using a maturity scale**: Consider metrics that align to maturity scales. This is sometimes the most effective, not because it measures absolute progress but because it shows trending. This is relatively easy to implement using the CMM (Capability Maturity Model) 1-5 scale. Maturity can help you show how the investments and initiatives you are running are impacting the maturity of the overall program.
- **Improve your metrics by giving them business context**: For example, rather than having a metric that shows the number of unpatched servers, break this up by how many are relevant to Sarbanes-Oxley (SOX), Payment Card Industry Data Security Standard (PCI DSS), or Health Insurance Portability and Accountability Act (HIPAA). If you can map them to revenue-generating business processes, that might be even more effective.

# The Key Indicators of Security

Some metrics are a little more dynamic in nature and are explicitly designed to be actionable or show an increase/decrease in risk. There are three types of these metrics that can help alert management to issues. These are Key Risk Indicators (KRI), Key Control Indicators (KCI) and Key Performance Indicators (KPI).

- A **Key Risk Indicator** measures the outcomes you are concerned about and the factors that may influence them. Are security incidents going up? Expressing this number as a KRI helps define an acceptable threshold and a threshold where management may want to act like increasing training and awareness around security issues or implementing additional security controls.
- A **Key Control Indicator** measures the effectiveness of security controls you deploy to mitigate and manage risk. A KCI should provide information on the extent to which a given control is meeting its intended objectives. KCIs are more focused than KRIs, in that they are specifically related to the controls that mitigate a risk. A KCI will indicate that the control itself may be faulty, which will in turn increase the likelihood that a KRI will also change in the near future.
- A **Key Performance Indicator** measures the output of activities we deliver in order to identify our relative performance in their delivery. Mean Time to

Identify (MTTI) and Mean Time to Contain (MTTC) incidents are KPIs that have been around for quite some time. KPIs answer the question, "How are we doing against our goals?"

# THINKING IN PROBABILITIES TO MEASURE RISK

*If we do everything right, if we do it with absolute*
*certainty, there's still a 30 percent chance we're going to*
*get it wrong.*
~ Joseph R. Biden, Vice President
of the United States, 2009

Part of the story that should go along with metrics is the probability of a risk actually manifesting. For example, you might have 300 unencrypted laptops in circulation. Will all 300 of these machines result in lost data? The answer, of course, is it depends. It's possible that it *could* happen. Yes, it is theoretically possible that all 300 will result in a data loss exposure. It's also possible that zero will. So, what is the real likelihood of something happening and how much should we worry?

When designing metrics and talking about risk, it can help quite a bit when you learn to think in terms of probability. Again, risk is something that might happen, or it might not happen. But giving some indication of probability is far more effective than trying to be overly optimistic or pessimistic about the outcome of an exposure. Don't be that person who thinks that no risk is the only acceptable level of risk. You will be alone.

To illustrate this, it helps to assign some numbers to the likelihood of a risk occurring. Since this isn't an exact science, I recommend using a range for showing likelihood. I've used the following to help present some terms to get people on the same page.

- **Certain**: a 100% chance or near 100% chance that something will happen
- **Almost certain**: around a 90% chance, give or take
- **Probable**: 75% chances, give or take
- **Even odds**: 50% chances, right about even or a coin toss
- **Unlikely**: 25% chances, give or take
- **Extremely unlikely**: 5% chances, give or take

Without showing a percentage, if you say that something is "probable" some people might interpret this as almost certain to happen. Others could think you simply mean greater than half of the time. It helps to not leave these terms open to interpretation. Assigning them a percentage of likelihood avoids this problem while still giving people an English word to use rather than a numeric range. You can adjust these percentages to fit your needs or even trim the scale down to three, perhaps to simply use unlikely (less than 25%), likely (around 50%) and probable (75% and greater).

In the outstanding book *Superforecasting: The Art and Science of Prediction*, authors Philip E. Tetlock and Dan Gardner detail the confusion that happened during the Cuban Missile Crisis of 1962 when President Kennedy was briefed that the Cuban invasion of the Bay of Pigs had a "fair chance" of success. This, and other vague language like it, contributed to a disastrous campaign. Later, the general who had written these words, but was not the one to brief them directly to the president, said that his use of this phrase was intended to mean that there was a 3 to 1 chance *against* success, or only "fair" odds of success. The interpretation by the executive team was that this meant a high chance of success and the rest is history. This is why even when you use qualitative language; it's not a bad idea to put a quantitative measurement behind it.

# PUTTING IT ALL TOGETHER

We've covered a lot so far in this chapter. But how do you put all this information together in your own program? Do your metrics have a purpose or a goal? They should. In fact, all metrics should have a purpose and a goal. Remember, it should never be numbers and measurements for measurements' sake. Focus your metrics program on meeting one or more of the following goals:

1. Tell a story.
2. Drive a behavior.
3. Provide a risk-based view of the current security posture.
4. Educate or inform.

Metrics programs that produce pages and pages of data become overwhelming and virtually meaningless to the audience. Don't just publish the data you have; find the data that you *need* that supports one of the objectives above. Here's the challenge. This takes some work. This takes thinking and analysis. But you still need to do it. You need to take the time to find the right things to measure in the right way that accomplishes one of the above objectives. One of the tools I like to use to quickly think through what metric to use is called goal, question, metric (GQM).

# GOAL, QUESTION, METRIC: A POWERFUL TOOL FOR CREATING MEANINGFUL METRICS

When you're thinking about the point of any tracking metric, what is your objective? To show improvement? To drive a behavior? Tying your metrics to goals ensures that you are collecting metrics for a specific purpose rather than just watching the numbers

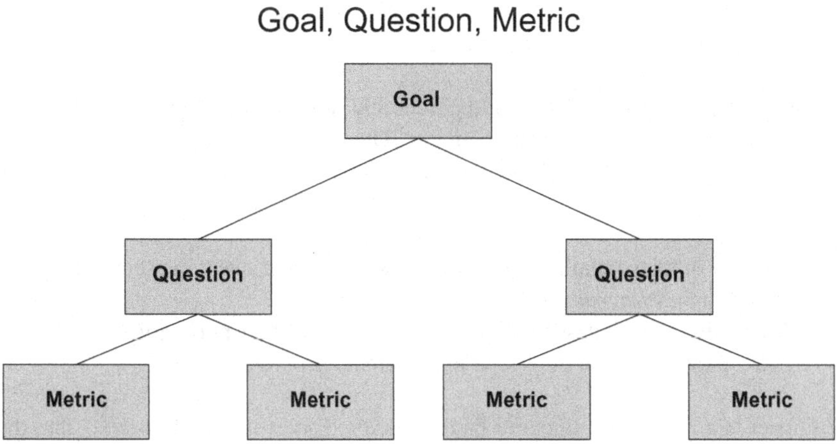

## Goal, Question, Metric

**FIGURE 10.3**  The goal, question, metric technique can help you make sure that you are tracking the right things and that your metrics serve a purpose.

change. A powerful method of coming up with metrics that matter is the "goal, question, metric" method (Figure 10.3).

The GQM approach focuses on identifying goals, developing sets of natural language questions that would allow the intended audience to judge performance based on the answers and subsequently to identify a set of metrics that can drive the answers to the questions.

The GQM approach avoids the all-too-common approach of measuring everything that's easy to measure assuming some of it might be useful to someone. It replaces this with a system of only measuring those things that help answer a question or concern in the first place.

As an example, think of a business situation such as the risk of a data breach that could result in bad publicity, regulatory fines or lost revenue. So, the goal would be to reduce the likelihood and damage from a security breach involving lost data.

What questions come to mind here? A few might be:

- How many breaches do we typically have?
- Where is our most sensitive data stored?
- Who has access to it?
- What fines might result from a data breach?
- How many records could be exposed?
- Is our risk of a data breach increasing or decreasing?
- What are we doing to prevent or limit data breaches?

Next, you could develop some Key Risk Indicators to measure relevant numbers like:

- The number of sensitive data records
- The number security breaches involving sensitive data
- The number of sensitive data records exposed to the Internet

- The number of roles with access to privileged data
- The regulatory fine per data record

Tracking changes in these KRIs would potentially indicate a shift in your risk exposure going up or down. This is just an example. There is no one size fits all for security measurement. Let's run through another example. It helps to phrase risks in this basic format:

- **Risk**: It defines what *impact* will result from what *event* caused by a *threat* exploiting a *vulnerability.*
- **Example**: We may suffer regulatory fines and bad publicity if we lose a backup tape through poor tape management processes.

The impact is a regulatory fine and bad publicity. The event is a lost tape. The threat is potentially losing a tape and the vulnerability is poor tape management processes. Using this example, let's think through how we might want to go about coming up with a metric to reduce our risk of losing backup tapes. We now have a goal for the metric and can start asking some basic questions:

- How many backup tapes have sensitive data?
- How many backup tapes are encrypted?
- How do we protect backup tapes?
- How many people have access to backup tapes?
- How would we know that a tape was missing or lost?

Next, we can develop some KRIs that might help. KRIs should all be measurable and quantifiable.

- The number of backup tapes with sensitive data
- The number of unencrypted tapes
- The number of backup tapes that leave the building
- The number of personnel that have access to backup tapes

You can see that creating metrics is a big subject and will take you some time to mature the program. But once you have them right, they can be the ultimate tool to help you communicate risk in a measurable, quantifiable way that no one can challenge.

# SUMMARY

I know there's a lot to this subject, so I'll end this chapter by repeating that you don't need to outrun the bear, you just want to be better off than you were before. Let your metrics program be a continuous improvement exercise and don't expect perfection right out of the gate.

- Metrics are a valuable tool that can help drive change and shift corporate cultures—when used properly. Make sure that your metrics program isn't just throwing around a lot of data that doesn't have a purpose.
- Your audience matters—business leaders want to answer simple questions like how they compare to their peers; IT leaders will want more technical details; and security people might want details that none of these groups would want. You will need to tailor metrics to your individual audiences.
- Metrics should not be using only qualitative measures like red, amber and green. Use quantitative metrics wherever possible. An even more powerful approach combines the two.
- Try to put your metrics in the business context. You'll want to answer why management would care about the information before you present it to them.
- Metrics should be simple and not hard to gather. Absolute values of metrics don't matter as much as trending. Make sure your metrics are driving the right behavior.
- Anything can be measured. If it is hard to measure, it can be estimated. Metrics and metric data do not need to be perfect, but they should make you better off than wherever you started.
- Metrics should be accurate and have a purpose. They should be easy to collect and timely. The best metrics tell a story, drive a behavior, educate or inform.
- Key Risk Indicators, Key Control Indicators and Key Performance Indicators can be put in place to help dynamically measure when the security program is on track or not.
- Learning to think in probabilities when presenting risks will help management better understand what their real business risk is and how they might want to address it.
- Goal, question, metric is a great framework to start with the problem being addressed before you start designing the metric to support the problem.

# REFERENCES AND RECOMMENDED READING

Bernstein, Peter L. *Against the Gods: The Remarkable Story of Risk*. Wiley, 1998.

Chickowski, Ericka. "7 Tips for Choosing Security Metrics That Matter." *Dark Reading*, 19 Oct. 2020, www.darkreading.com/attacks-breaches/7-tips-for-choosing-security-metrics-that-matter/d/d-id/1339211.

Fasulo, Phoebe. "20 Cybersecurity Metrics & KPIs to Track." *SecurityScorecard*, 8 July 2019, securityscorecard.com/blog/9-cybersecurity-metrics-kpis-to-track.

*The Goal Question Metric Approach*. www.cs.umd.edu/~mvz/handouts/gqm.pdf.

"GQM." *Wikipedia*, Wikimedia Foundation, 16 Sept. 2020, en.wikipedia.org/wiki/GQM.

Hubbard, Douglas W. *How to Measure Anything: Finding the Value of Intangibles in Business*. Wiley, 2014.

Hubbard, Douglas W., and Richard Seiersen. *How to Measure Anything in Cybersecurity Risk*. John Wiley & Sons, 2016.

Huff, Darrell, and Irving Geis. *How to Lie with Statistics.* W.W. Norton & Co., 2006.

Jaquith, Andrew. *Securitymetrics.org*, www.securitymetrics.org/.

Solingen van, Rini, and Egon Berghout. *The Goal/Question/Metric Method: A Practical Guide for Quality Improvement of Software Development.* McGraw-Hill, 1999.

"Why Measure Information Security?" *PRAGMATIC Security Metrics*, 2013, pp. 13–28, doi:10.1201/b14047-3.

Wong, Caroline. *Security Metrics: A Beginner's Guide.* McGraw-Hill, 2013.

# The High Stakes of Incident Response Communication

# 11

*One of the tests of leadership is the ability to recognize a*
*problem before it becomes an emergency.*
~ Arnold Glascow

Security breaches are a matter of when, not if, they are going to happen. Breaches cannot be prevented. You can only mitigate the chance of a successful breach and have plans in place for when they do happen. This means that a good security program will include efforts to protect against breaches, quickly identify when they happen and quickly react to mitigate them.

One of the most important communication scenarios faced by security professionals is during and immediately after a security incident. Security incidents have become a fact of life and while prevention is a noble goal, it is less likely than ever that this will be a successful strategy. In fact, government agencies, banks and even security firms have been breached despite ample people, resources and the financing to implement world-class security programs.

What does this have to do with communication? Poor communication in a significant incident will end your job and potentially damage your career. Communication is never more important than during an incident and you will likely have to use every communication medium you can imagine: paper, phone, text and everything in between. Let's just hope you aren't also on TV for the 5'oclock news. Incidents are a very high-stake game but, like many things we've covered in this book, preparation and practice can go a long way to helping you be ready for anything.

There are entire books written on all aspects of incident management. This chapter is intended only to give you a flavor of some of the communication challenges that can be encountered with incident management. This includes creating an incident response plan and looking at communication before, during and after an incident.

DOI: 10.1201/9781003100294-13

# PREPARING FOR AN INCIDENT

One of the first things I like to do when coming into a new organization is make sure that we are ready to handle an incident. Since incidents can happen literally anytime, there's no rule that says one couldn't happen on your first day or, as happened to me once, right before your first day on the job. The more you prepare for an incident, the more you will be ready when one happens. While there's a huge number of things that can be done to prepare for an incident, two of the most important ones are creating an incident response plan that has a communication component and conducting table-top exercises.

## Writing an Incident Response Plan

> *In preparing for war, I have found plans to be useless but planning to be indispensable.*
> ~ Dwight D. Eisenhower

> *Everybody has a plan until they get punched in the mouth.*
> ~ Mike Tyson

One of the most important documents you may write in a security career is the Computer Security Incident Response Plan (CSIRP), also sometimes simply known as an Incident Response Plan. When you've thought through and defined what an incident is, and what severity and priorities will be assigned, you'll need to formalize these concepts in your CSIRP.

The structure of the CSIRP from one industry to the next should be roughly the same regardless of the nature of your organization. The approach to incident response must be organized, methodical and coordinated. It should contain processes and explicit procedures that are clear and easy to understand. The CSIRP document should also be formalized within your organization with explicit input and sign-off from stakeholders and senior management.

Unless you have unique needs (you probably don't), I'd recommend modeling your plan from the basic structure of the National Institute of Standards and Technology (NIST) framework, which defines the following phases:

1. Preparation
2. Detection and analysis
3. Containment, eradication and recovery
4. Post-incident activity

More on this framework can be found in NIST Special Publication 800-61 Revision 2 at https://nvlpubs.nist.gov/nistpubs/SpecialPublications/NIST.SP.800-61r2.pdf.

# Laying the Groundwork for Your Plan

*There's no harm in hoping for the best as long as you're*
*prepared for the worst.*
~ Stephen King, Different Seasons

Again, you will need to spend a significant amount of time preparing for an incident. This preparation cycle should include your communication planning and should all be captured in your CSIRP. When you're in the middle of an incident, you'll thank yourself that you put all that time into planning a communication strategy, because this is where many incidents go off the rails. Even small incidents can translate to a lot of stress. Waiting for something to occur and then trying to figure out a communication strategy in the heat of the moment can lead to disastrous consequences.

Start by defining what you consider to be an incident and how you will gage incident severity. It's important to have everyone on the same page about what constitutes an incident and how severe you consider the incident. Large companies may have multiple incidents going on at any given moment, so putting a framework around incident response helps to start organizing the chaos.

I think it's helpful to consider at least two categories of incidents: IT security incidents and data breaches. While an incident might involve both an IT security incident and a data breach, it is important to distinguish if data has been exposed or stolen, versus let's say a more standard malware infection. If data has been exposed, potentially exposed or stolen, legal and regulatory requirements will need to be considered. California, New York, and others have requirements of reporting data breaches, usually within a pretty tight timeframe. These rules change and new ones are popping up all the time. It is best to consult with legal counsel rather than trying to figure this out for yourself.

# IT Security Incidents

We'll start with defining a basic IT security incident. An IT security incident is the compromise or malicious attack of a system *or* the violation of security policies or standard security procedures which impacts company operations. While a lot of incident types can fall under this category, the important distinction is that they do *not* involve the loss of data or sensitive personally identifiable information (PII).

Examples of IT security incidents include:

- Detection of malware with remote access capabilities, or multiple hosts infected with the same malware with no evidence of data breach
- A denial-of-service (DoS) attack
- Phishing emails that are not believed to have been opened or acted upon
- General software or system vulnerabilities that were discovered but are not known to have compromised company data
- Harassment delivered via computers
- Unauthorized changes to application/system configurations or application code

- Any other instance where a security notification does not initially indicate any direct threat of improper access to, or alteration, loss or acquisition of company data

IT security incidents can typically be handled by the security group with some assistance from other groups in IT. They are not typically worthy of engaging senior business management unless there's a significant impact.

## Data Breach Incidents

Once it has been determined that the exposure of sensitive data or loss of sensitive data has occurred, it needs to be classified as a data incident or a data loss incident. Why this distinction? Data incidents require input from your Chief Privacy Officer, Chief Compliance Officer, Legal Counsel and many others. An IT security incident, on the other hand, can probably just be resolved and logged without involving many stakeholders. Data incidents will almost always have regulatory and privacy implications, versus IT security incidents which may or may not require regulatory notification depending on your industry requirements. It's always best to check with Compliance if you're not sure if an incident needs notification or not. It's also always best to leave the reporting incidents to regulators to whatever team normally handles this type of communication.

A data incident involves the potential for or suspicion of unauthorized access to, or improper alteration, loss or acquisition of company data. A data incident includes situations where data is either stolen or improperly accessed or maybe even just exposed or *potentially* accessed. A data incident may involve sensitive data that is in any format including hard copy or voice data that has been saved electronically. Examples of a data incident include:

- Bypassing security controls to gain access to sensitive data without proper authorization
- The loss of any electronic storage media (e.g., CD, DVD, flash drive, laptop or desktop computer, hard drive or tablet computer) that contains unencrypted company data, which includes any device where there is indication that the encryption might be compromised (for example, if the password was written down and left in the laptop bag)
- The theft of sensitive data
- Phishing emails that have resulted in loss of or unauthorized access to company data or known to install malware which steals data
- An email containing sensitive data sent inadvertently to the wrong recipient
- Software security flaws that may have given unauthorized access to systems or sensitive data, such as data exposed on a web server
- Improper disposal of sensitive data
- A mailed package with sensitive data sent or delivered to the wrong recipient

Once you have these two categories established and an idea of what constitutes an example of each, you'll want to have a gage for assessing incident severity.

# Incident Severity

Severity is a measurement of impact. How much impact does an incident have? Does it take down the entire network? Does it stop critical business processes? Or does it just irritate users or inconvenience them? If your company already has an incident management process (think system outage more than security incident) then you can probably use those elements in your security incident response process. I would recommend this approach over using your own system so that people don't have to become familiar with two systems or get confused because something seems more important in one system over another.

If you don't have an incident severity scale to leverage, I recommend putting something together as the next step so that you can triage when you have more than one incident in progress. A severity scale helps reinforce recovering the most business-critical systems first, and can be used as a benchmark for measuring the effectiveness of the overall plan.

| SEVERITY | DESCRIPTION |
|---|---|
| **Severity 1: Critical** | An incident that is *critical* has company-wide consequences. A Severity One incident can also be an incident that has severe consequences for any single business unit.<br>Example: on-going exfiltration of data or an active denial-of-service attack. |
| **Severity 2: Threatening** | A Severity 2 incident is *threatening* because it has business-wide consequences but may not be as urgent as a Severity One incident. A severity 2 incident could also be a collection of lower-level incidents that are seen across the company.<br>Example: phishing outbreaks that include malicious attachments. |
| **Severity 3: Moderate** | A Severity Three incident is *moderate* because the vulnerability or exposure which was exploited can usually be remediated with standard methods (such as patch, deployment or other measures).<br>Example: non-critical systems that are not available due to a security issue. |
| **Severity 4: Low risk** | A *low-risk* incident is an event that is identified as being an incident but is more of an inconvenience or minor issue.<br>Example: individual machine malware infection that can be easily cleaned using standard tools. |
| **Severity 5: Informational** | An *informational* event is one where a security incident has not manifest. This might include a system with a known configuration issue that still needs to be fixed but is not creating any imminent danger of exploitation.<br>Example: reconnaissance activity detected with no follow-on attacks |

Of course, some might just opt for a high, medium, low severity scale and use a subset of this rating system. The problem I've seen using this approach is that you can wind up with too many issues being declared high out of an abundance of caution. Using a five-tiered scale allows you some flexibility pushing truly critical issues to the top and less important issues to the bottom. Use whatever works best for you and your organization.

No matter what rating system you settle on, I recommend assigning a high level to any incidents that involve data loss of more than a record or two. If your organization defines service level agreements (SLAs) or service level objectives (SLOs) then you should also assign priorities and time to respond and time to resolve targets. Priority is different than severity because it measures urgency rather than impact.

Again, I would emphasize here that total *prevention* is a great goal but is also not realistic. Anyone can and probably will be compromised at some point, it's just a matter of when. Complete prevention means unplugging from the Internet and putting your information in the proverbial concrete bunker. Even then, there are some theoretical attacks for highly motivated and well-funded attackers. A more realistic goal is to keep the number of incidents reasonably low and minimize the potential impact with a layered controls defense plan and fast incident response.

So far, we've defined what an incident is and what types of incidents are defined by severity. The next element you need to ensure is covered in your CSIRP is roles and responsibilities. Incidents happen when they happen. Key personnel can be on vacation or on business travel. Defining roles and responsibilities of members of the incident response team and the chain of command in the absence of a key person will help everyone be aware of what they need to do and who is responsible for what activities. I recommend going role by role in your CSIRP and have a primary and secondary person outlined for each. If someone doesn't have a plausible backup, be sure to capture this information and start working on training someone to fill the role.

The next step after roles and responsibilities includes the communication plan, one of the most important pieces of the CSIRP. The communication plan establishes relationships and lines of communication between the incident response team and other groups, which might include internal groups or external law enforcement agencies. In the heat of the moment, it's easy to completely abandon the incident response plan and just deal with the issue. However, the most important part of the plan and the one you'll find yourself going to regardless of what the rest of the plan says is the names and contact information of all the key personnel. This includes the first responders and the escalation list.

## First Responder Call List

These are the people in the trenches. You'll want to make sure they have some empowerment to respond first, notify management later. These people need some power to contain and mitigate cyberattacks. Seconds count in incident response, so ensuring that these people have the ability to make decisions in the heat of the moment is critical. You'll want to test this in real-world scenarios like ransomware attacks. Run these exercises as a table-top to gain more comfort that your front-line people have the ability to act fast in the heat of the moment.

## The Escalation List

These are the people who will either be notified after a mitigation has taken place or it will be the "wake them up" call list to notify senior management and get decisions or support for actions taken from the incident response team. It's critical to make sure you have more than one method of communication. I'd recommend a minimum of cell phone and land line phone if they have one. Texting might also be a viable contact method. Because off-hours notifications don't always work, you'll want to have primary and secondary contacts for critical functions (like networking) and senior management contacts for decision support.

### Roles and Responsibilities

Make sure you also detail everyone's roles and responsibilities for an active incident. Typically, you will want to outline roles including the incident response manager, incident team members and management. The incident response manager oversees and prioritizes actions during the detection, analysis and containment of an incident. Once you have roles, definitions and severity defined, your plan is nearly complete. Unlike policies and standards, an incident response plan can be very detailed and you don't have to strive to keep the document short.

## Other Documents You May Need to Be Prepared

In addition to the CSIRP, you will want to make sure that other documents that might be required in an incident are available to you. These include:

- Detailed network infrastructure diagrams
- Backup and snapshot schedules for system recovery
- Web and cloud service connections
- Virtualization systems documentation
- Call tree or contact listing including phone and email details (staff and key vendors)
- A standard incident reporting template. Having a template approach can ensure that whoever is documenting an incident is capturing all the relevant details

## Table-top Exercises

Once you have your CSIRP drafted and finalized, one of the best ways to ensure that your plans are more than just paper plans is to test them in a simulated table-top exercise. But in order for a table-top exercise to be effective, you will need support from all stakeholders in the business, senior management and technology.

I remember running through a table-top exercise once with the business that exposed the fact that no one new how to report a security incident. They really had no idea whom to contact, who the CISO was or even how to start thinking about it. These are the kinds of issues that you don't want to discover the hard way during a real incident.

Table-top exercises are cheap to conduct, but that doesn't make them easy.

Some possible scenarios you should consider testing include:

1) Ransomware or other malware outbreaks
2) Data breach that includes a PII compromise
3) A hacktivist attack
4) Customer accounts and passwords found on the Dark Web
5) A cloud data breach
6) A payroll or other financially motivated compromise
7) Insider threat scenarios
8) Denial of service
9) Natural disasters
10) Business fraud

## Scenario Questions to Consider

Table-top exercises are designed to help businesses think through various scenarios and what issues they might encounter if the situation was real, focusing on key improvement points as you run through the scenario. While some exercises can be conducted in an hour or so, more sophisticated companies may wish to set aside half a day or a full day. They are not going to be effective if decision makers are not available, so plan them way in advance and ask that people keep the calendar clear.

It's best to nominate an individual to run these exercises and to have a notetaker. As a security leader, you probably do not want to run these exercises yourself, as it will be difficult to run them and also monitor responses and behavior of the people involved. The follow-up actions need to be captured and assigned to individuals with completion dates. A simple ransomware exercise is discussed in the following section.

## A Simple Ransomware Scenario

**Scenario**: An employee receives a phishing email that seems enticing. The email discusses employee salaries and promotions data and comes with an Excel attachment. It is actually a spear phishing email that has been sent to an employee with advanced administrative privileges. The employee is curious enough to open the attachment and launch a ransomware program that quickly encrypts all data that it can reach, including sensitive customer information. Before anyone realizes what is happening, a significant portion of the network servers have been encrypted and are now useless. The ransomware program displays a screen that threatens to release all the data if $5 million is not paid in Bitcoin within 24 hours.

**Response Discussion Questions:**

- Who would need to be notified in your organization if this happened to you?
- How would this notification be made?
- If typical communication channels like email have been impacted, how would you make this notification?

- What should management do? Do you pay the ransom? Who needs to approve this decision? This is probably the most important decision to think through in advance. You don't want to start thinking about this when the clock is ticking in a real incident.
- If you are going to pay the ransom, where does this money come from? Do you have cyber insurance, and would it be covered?
- Do you know *how* to pay a ransom in Bitcoin? How quickly could you get money converted and sent if you chose to pay the ransom?
- If you will not pay the ransom, what impact will this have on your finances, your customers and your brand reputation?
- If you will not or cannot pay the ransom, how will you recover this data, or can it be recovered?
- If the data is released, what impact does this have?
- What if you pay the ransom and the data is released anyway?
- How would you prevent this incident from recurring again?

**Security controls**: What security controls are in place or could be put in place to prevent or mitigate this threat? If this phishing email was widespread, could you pull it from inboxes from people who hadn't opened it? Who would you need to engage to do this? How much faith do you have in your backups? Have backups ever been tested? What is the recovery time objective (RTO) and recovery point objective (RPO) for your backups? How much data could be lost and how many days could it take to recover?

**Curve balls**: When running through these scenarios, it's always good to throw a few curve balls by not having certain people be available. Would everyone report an incident by calling the CISO directly? What if that person was on a plane halfway around the world and completely unavailable for the next 12 hours?

Scenarios should be documented, including follow-up actions, notes and results. Don't worry if the scenario didn't go well. If no one knows what Bitcoin is, then capture this as a follow-up and make sure that someone is responsible to figure out how this part of the exercise would work if the event were real in the future. If things went especially bad, it might warrant running through the whole scenario again in the near future. Also make sure to take any significant gaps and build this into your security strategy. Perhaps not enough attention has been given to email security. Email security solutions, additional awareness and phishing tests might all be warranted in the future and may need funding.

# DURING AN INCIDENT: YOU DON'T KNOW WHAT YOU DON'T KNOW

OK, so an incident has happened. People are either starting to become aware of it or you have decided it's time to give senior management a heads-up that something is unfolding. The most important factor to remember during an incident is that you likely don't have the full picture. Yet. For example, you may have detected an intrusion on a single computer. Do you really know that's the *only* system involved? No, you don't. In fact, you only know what you know right at that moment and it is highly subject to change. Make sure you are setting expectations.

Once management becomes aware of an incident, many people will have questions and will be looking for answers from you. This can be stressful, as you may not have answers at this point. Never speculate about any aspect of an incident. Don't promise what you can't deliver and just be very, very factual. Add a caveat that this is what you know *at this time*. Reinforce that this may change, and that what you right know may be subject to change. Ensure that all your written and verbal communication signals that you are taking the problem and its impact on the organization very seriously and you are doing everything you can to manage the situation. Your response, your communication, your body language and your gravitas will all be important to give senior management the reassurance that the company is in good hands and that you are doing everything that can be done to manage the situation.

## A Note on Public-Facing Communications

In the unfortunate event that an incident has happened or is happening, you need to be very, very careful with any public-facing communications. You should have senior management on board with all communications, including *who* is communicating. In most scenarios this is not the CISO. It may be your legal, marketing or public relations department. In a high-profile incident it might even be the CEO. Be careful talking to anyone, including friends, who are asking you about any details.

The first step in even thinking about doing any public-facing communications is understanding your company's policies about public it and if there are teams or people in place that are responsible for handling these issues. Hopefully, this was considered as you put your plan together or conducted table-top exercises.

The best course of action is to designate a point person for external communication and have a single source of the truth. When an incident becomes public, you will not find a shortage of people looking for the inside scoop. These could be friends, the press, business partners or other stakeholders. What you don't want to have is the incident response team fielding requests at the expense of incident response. You also don't want inconsistent responses from multiple people who have no idea what actually happened.

This means that you'll also need to inform employees about not communicating with the press, friends or anyone from the outside who wants to talk about the incident. The key to public communications is to keep them concise and consistent. Be factual, but also understand that you're only obligated to share certain facts with regulators, internal audit and senior management. The whole world doesn't need to know every detail.

## AFTER AN INCIDENT: WHAT DID YOU LEARN?

The aftermath of an incident is important to manage. At this point, it's becoming easier for people to slip back into their day-to-day routines and forget about the incident. This is a lost opportunity unless you act. The most important post-incident task is performing an incident review and root cause analysis. This should be reserved for significant incidents, not every computer that gets a virus alert.

This stated, I hope I have at least got you thinking about keeping a journal. If you don't work with lessons learned anywhere else, focus on yourself and your personal lessons learned. What did you do right? What did you do wrong? How could you improve next time?

For significant incidents, you should spend time to determine and document the root cause analysis or other underlying factors that contributed to the incident and capture the business impact. This needs to be done more formally, preferably in an in-person or video meeting where you can see and understand tone and body language, as well as the words being said. You should do this within 30 days of the incident to make sure you are capturing all the details while they are fresh in everyone's mind.

One of the most important tips for these types of meetings is to keep them open and productive. This is not time to blame the network team for not providing logs fast enough or the security team for having key people on vacation. To keep talks productive, try to capture objective but impersonal facts like how it may have been difficult to pull logs or that key personnel were not available for whatever reason.

Make this a continuous improvement exercise instead of the blame game. No matter how good you are, you can always get better. Create a "safe" environment where people can be candid without being rude or offensive. This means focusing on facts and problems instead of people and personalities.

# THE TEN COMMANDMENTS OF INCIDENT RESPONSE COMMUNICATIONS

OK, we are nearly at the end of this chapter. Incident response is stressful, and communications and personalities play a bigger role than the actual security incident. Here are my ten essential rules for dealing with incident communications.

1. Draft your incident response plan as one of your first actions in a CISO role. Incidents can and will happen at any time. Don't be caught unprepared. Make creating this document one of your top early priorities.
2. Only communicate *facts* during and after an incident. As basic as this is, you need to be crystal clear about what you know as fact and not to guess or speculate on anything. People will make decisions based on what you say or communicate, and it needs to be accurate. Don't be afraid to say, "I don't know." Many elements of an incident may never be fully known. For example, if your data has been exposed to the Internet for months and logs only go back for days, you might never know for sure if the data was actually viewed or copied. And you never will know unless it suddenly pops up on the

DarkWeb. If you're going to provide a professional opinion about what you think, make sure you are crystal clear that this is an opinion only and not a fact. This said, here are some guidelines.

- Don't estimate things that can be counted. If you can provide real numbers instead of estimates, do the work and come back.
- Don't guess something that could be estimated. If you could do some preliminary work and estimate something that you would have otherwise guessed, buy some time and do the work!
- If you do decide to guess, make it clear that it's an educated guess. And that's all that it is.

3. Keep detailed notes and timelines. You will likely be explaining an incident to multiple parties and, who knows, maybe even in court. Your notes are your friend and are so much more trustworthy than your memory. What happened and when? Write it down and keep it where you can find it.

4. Provide regular updates to senior management. Senior management will likely want to be kept fully up-to-date on what's going on with an incident. Make sure you are providing regular updates even if the update is that there's not much new, next steps and when the next touchpoint is. If management is coming to you asking about status, you're probably not being proactive enough.

5. Have a solid communication plan solid before an incident occurs. This should be something that captures names, email and cell phone numbers. Consider how you would communicate in the event of a network denial-of-service attack. Email is not the best answer in this situation. You may even want to consider out-of-band methods like SMS in case a mail server or large portions of the network became compromised.

6. Run a command center during an incident. Open phone bridges. War rooms. Virtual security operation centers (SOCs). Create an open communication environment that you're sure only the "good guys" can access.

7. Update your incident response plan regularly. This is not a "one and done" document, it needs to be updated regularly and stored somewhere where it can be easily found.

8. Document lessons learned and incorporate them back into your response planning. Yes, you need to do a postmortem and be candid about what went right and what went wrong for every single significant incident.

9. Keep talks or updates about incidents to senior management brief. They will let you know when they want more information. Don't go heavy on technical detail and always include next steps.

10. Document your incidents. Keep detailed records, including timelines, responses and people involved. Incident documentation needs to be impartial and factual. Include exactly what happened and who was involved and the timeline. Keep the tone as neutral as possible. There's no blame or fault. Just facts.

# SUMMARY

Establishing an incident response program that includes a plan, table-top exercises and continuous improvement will benefit your organization in many ways.

The stages of an effective incident management plan include:

- Preparation: Establishing and executing a well-thought-out and effective incident response program
- Identification: Discovering intrusion
- Detection: Detecting the presence of a malicious actor
- Analysis: Validating the presence of a malicious actor
- Remediation: Eradicating the intrusion
- Containment: Ensuring any new information systems cannot be infected
- Recovery: Eradicating the infection from the information system
- Mitigation: Ensuring that the information system is configured so that it can no longer be exploited
- Post-incident activity: Evaluating lessons learned and continuous improvement

# REFERENCES AND RECOMMENDED READING

Gartner_Inc. "Fueling the Future of Business." *Gartner*, www.gartner.com/.

IANS. "Institute for Applied Network Security." *IANS*, www.iansresearch.com/.

Kaeo, Merike. *Designing Network Security.* Macmillan Technical Publications, 1999.

Northcutt, Stephen. *Computer Security Incident Handling: Step by Step, a Survival Guide for Computer Security Incident Handling.* SANS Institute, 2001.

Prosise, Chris, and Kevin Mandia. *Incident Response: Investigating Computer Crime.* McGraw-Hill, 2003.

Rajnovic, Damir. *Computer Incident Response and Product Security.* Cisco Press, 2011.

Schultz, E. Eugene, and Russell Shumway. *Incident Response: A Strategic Guide to Handling System and Network Security Breaches.* New Riders Pub., 2002.

Tipton, Harold F., and Micki Krause. *Information Security Management Handbook.* Auerbach, 2008.

Whitman, Michael E., and Herbert J. Mattord. *Principles of Incident Response and Disaster Recovery.* W. Ross MacDonald School Resource Services Library, 2011.

Wyk van, Kenneth R., and Richard Forno. *Incident Response.* O'Reilly, 2001.

# Communicating with Your Team and Colleagues

# 12

*As you navigate through the rest of your life, be open to collaboration. Other people and other people's ideas are often better than your own. Find a group of people who challenge and inspire you, spend a lot of time with them, and it will change your life.*

~ Amy Poehler

Managing a team presents a variety of communication challenges and opportunities but is also one of the most rewarding experiences you can have in your career. A team may consist of consultants, full-time employees or a combination of both. They might even just be matrix employees that you need to engage and influence to support the security program. The days of going it alone in cybersecurity are pretty much over.

One of the biggest mistakes that leaders make is under communicating. They think that everyone has the same level of organizational knowledge and what is going on within a company, without realizing that this is rarely the case. They are overly stingy with praise and sometimes avoid where they need to deliver negative feedback most.

Some CISOs I know literally have hundreds of people working for them. In a big organization, teams of a hundred or more people working in a security function are not that uncommon anymore. But even if your team is only a few full-time people and a contractor or two, there are ways to make communication easier. This chapter discusses adopting the appropriate mindset to managing a high-performance team, challenges with intra-team communication and some difficult topics like when you need to let an employee go.

## MINDSET

As always, it's helpful to start with a good mindset. Having a team is a responsibility. You are your team's advocate and their leader. Not every boss seems to understand this concept. Instead, they blame people for existing problems or try to push everyone out the door so they can bring in a different set of people, often losing the

DOI: 10.1201/9781003100294-14

best talent right away in the process. The most talented people tend to leave first when faced with a bad management situation. They do this because it's easy for them to find something new, often at a raise in salary. Unfortunately, they also take a mountain of institutional knowledge with them. What makes you a good leader is not pulling in all your friends from previous companies; what makes you a good leader is getting the most out of people whoever they may be and wherever they may be with their career progression.

You will ultimately live and die by your team and your ability to attract and retain good people and get things done in an organization. My number one rule with your team is to be nice and give everyone a chance and the benefit of the doubt. They are here to help and support you. Trust them until they give you a real reason not to trust them.

There are many ways to lead a team. Command and control, micromanagement and laissez-faire to name just a few. The way I find the most effective though is by empowering your people and removing the obstacles that stand in their way of doing great work.

Command-and-control managers usually create organizations where no one has authority, and everyone is scared to make any significant decisions without approval from the top. This creates a scalability problem with the manager and is absolutely horrible for morale. When this manager goes on vacation, work grinds to a halt because no one feels empowered to move anything forward without getting permission. Also, I think most people didn't get a job so they can be commanded and told what to do. It's not pleasant working for these managers and the turnover rates are high. I've worked for two of these managers in my career and I couldn't leave fast enough.

Micromanagers get involved with every detail of every project. This is also a scalability problem for many of the same reasons as command-and-control managers. People don't feel empowered to make their own decisions, especially if the micromanager questions every decision that has been made. Sometimes when an employee is very early on in their career, this level of handholding isn't a terrible thing, but I know very few experienced or senior people that enjoy being micromanaged. Micromanagers are often frustrated when someone takes a different approach than the one they would have taken. I think many micromanagers believe they are doing the right thing. But what seems like helpful advice or constructive criticism can sometimes spin out of control into a boss that won't let their team take any action. Micromanagers also tend to be perfectionists, which stifles innovation and slows progress.

The mirror opposite of command-and-control and micromanagers is the laissez-faire manager who barely makes their presence felt. Perhaps they spend their time managing up or perhaps they are surfing the web or running a side business. In either case, they provide minimal leadership and very little gets done. You might still have some star performers that manage to move the needle on a few programs, but otherwise the team is a rudderless ship. This type of manager doesn't get in the way, but they also don't add a lot of value. They are the boat anchor dragging the whole team to the bottom of the ocean.

Poor management styles result in low productivity, poor quality work, high turnover and high absenteeism. The wrong management style also trains employees to disengage or leave, which hurts the entire organization.

My preferred management style is to operate as a team. You are the leader, and you are ultimately the decision maker, but you're also on a team. You have been hired to implement a security program and you will do this through resources that include your team. This means that you're all in this together. It's best to create an environment that provides a vision, gets everyone's buy in and then inspires everyone to do their part. This type of leader is part coach, part visionary and an active working member of the team. They wouldn't ask you to do anything they wouldn't do themselves. If they had enough time, that is.

Trust is key to a healthy team culture. It's better to enable your team to approach things their way, so long as the outcomes are what was intended. Providing just enough guardrails for people tends to result in better autonomy and, when everything gets optimized, the team can become self-managed. This kind of team gets things done regardless of who is in the office and keeps projects moving forward. Everyone has everyone else's back, there's a succession plan and the team is highly resilient.

Developing a team and helping them reach a level where they can largely self-manage is a highly rewarding experience. These are the teams people want to join and don't want to leave without a lot of forethought.

# COMMUNICATION CHANNELS

Communication with teams can be difficult. This is because there are a lot of different communication channels that are beyond your control. A communication channel is simply one person talking to another person or group of people. One of the things I still remember from getting my Project Management Professional (PMP) certification back in 2002 was the concept of managing communication channels. This concept used an equation to describe the total number of possible communication channels. The equation is $n(n - 1)/2$.

Using this model, $n$ represents the number of members. For example, a team with 10 members has $10(10 - 1)/2 = 45$ possible communication channels.

Yikes! A team with 10 people has 45 possible communication channels? Yes, it does. This could be 1-1 communication or 2-on-1 or any combination. Do you have 100 people? Well, that's 4,950 possible communication channels. There is no way to control this many communication channels. But you can take some steps to manage them.

When communication on a large scale gets out of control, you can expect to have problems. Rumors fly. Misinformation abounds. And my experience is that people fill in the blanks with the worst possibilities they can imagine. They imagine scenarios of the company being sold or that they're being fired. If you don't give people some information, you can expect them to make it up themselves and it will rarely ever be correct.

Of course, you should provide accurate, objective updates and not fuel any rumors and speculation. Be transparent and share the information you have and also let your team know what you don't know. Letting people know that there are some gaps in your own knowledge will make you more approachable and trustworthy.

Are you aware that there's a rumor floating around that is causing a lot of disruption? Address it. If it's just a rumor and not based in fact, say so. Let your team know what you know and what you don't know. Sure, they will still fill in some of the gaps through imagination, but at least you can help take some of the interest off the subject and get people focused back on moving the security program forward. Has the company announced it's up for sale? People will be immediately worried about their jobs. You need to address the facts as facts, where what you are allowed to share and try to provide reassurances even when you may not have them yourself. There have been many cases where employees are either asked to stay on and take a bigger role in the new organization or they can get lucrative retention bonuses to stay through the transition. Even bad news isn't always bad news.

# BAD MANAGERS

You can learn a lot from bad managers. I've worked for a few in my career and two who I'd consider absolutely terrible. It's worth a few tips to figure out if you're stuck with that bad boss and maybe also worth reflecting for a moment if you might actually *be* that bad boss.

In his book *How Not to Manage People: The Leadership Mistakes Keeping Your Team from Greatness*, author Mike Wicks cites a few traits that bad managers seem to embrace. These include:

- **Playing favorites**: Look, everyone has some star performers and some laggards. Good leaders stay objective and don't play favorites. Every team member counts.
- **Looking for loyalty**: Bad leaders seem to especially value people who seem overly loyal, but usually in a fake way. Focus on skills and contributions, not fake loyalty.
- **Making examples of people**: If you have a manager that likes to make examples of others, then run. You didn't join a pirate ship, right? Anyone who would try to make an example of an employee in front of others is not acting in anyone's best interest.
- **The passive-aggressive**: Skilled managers listen empathetically and have productive, proactive conversations with their staff members. Manipulative or other passive-aggressive behaviors are a sure sign of a bad manager.
- **The entitled**: I once had a bad manager who delayed the construction of an entire site by months so that his office was slightly bigger than the other manager's offices. These entitled managers prioritize their own desires more than the people they lead. Everyone is there to serve them by taking on unhealthy workloads, generally with the manager taking credit. In contrast, good managers support a healthy work–life balance, teamwork and collaboration.

Hopefully, you are not stuck with a bad boss and you are not the bad boss either. If any of these characteristics resonate with you it is probably time to make some changes.

# BE CONSISTENT

When you're managing a team, be consistent with your communications. No one wants to follow a leader who seems too moody and unapproachable or who changes what they tell you on a daily basis. No one wants to follow someone who plays favorites or someone who allows bad behavior sometimes but not other times. Your goal is to work together as a team, and if your leadership is inconsistent, the team will not have a unified purpose or enough stability at the top.

Finally, set the example you want to see. If you show up late to your meetings, your team will take that as a cue that showing up late is OK for them too. Strive to set the example that you want to see, especially in front of your team.

# BE AUTHENTIC AND CANDID

> *If the content of your speech is not authentic, talking or texting on a device doesn't mean you're communicating with another person.*
> ~ Thích Nhất Hạnh, The Art of Communicating

Again, your team is how you will get things done at an organization and every member counts. That's not to say that you won't' get some problem people or poor performers, but if your team doesn't trust you, I suspect you won't be their manager for long. Either they will leave you or you won't last in the role. Something will eventually give.

Being candid with your team is especially important. If someone on your team is asking for a promotion, but there are reasons they are not being promoted, you need to be truthful and help them define their path forward. If someone wants to be promoted to manager, but they are consistently showing up late and missing deadlines, you need to explain what they need to do to improve and how they can ultimately get to that next level. These are not always easy conversations, as there can sometimes be big gaps between how an employee thinks they're performing and how you think they're performing. But it's best to get it out in the open and talk frankly about it.

Being candid is a little bit of an art. If you go overboard with candor, it can seem confrontational. You'll want to stick to facts, not opinions. Take some time to gather and package your thoughts if you are giving what could be perceived as negative feedback. Keep both the language and your tone of voice civil and professional. If it's an uncomfortable conversation, get to the point quickly. Making an uncomfortable conversation longer isn't serving anyone.

# CASCADING BUSINESS KNOWLEDGE

As a senior manager, you are probably privy to quite a bit of information that your team may not know about or hear from other channels. This might include business strategy updates, new projects or work that is going on in other teams that might have an impact on your area.

If you want your team to be effective, they need to understand some of the same background as you. They need to appreciate the "why" behind projects and what it means for the business. This helps their appreciation for why certain decisions might have been made or why a particular project is important to generate revenue. Give them the "big picture" so they understand the role that security plays in supporting business initiatives.

I always carve some time out from my staff meeting to talk about "general business" updates whenever I can. This helps the team stay on top of information they might not hear elsewhere and gives them more of a sense of community and belonging with the business. While security is there to support the business, we are also part of the business.

# ENGAGE EVERYONE

I never like to be the only one talking in a team meeting. I'm not going to learn anything new that way and I don't pretend that I'm the most interesting person in the room. I generally say my piece and then turn it over to the team to update each other on where various projects stand, what roadblocks are being encountered and where they can help each other. This often leads to pleasant surprises where people can offer help and suggestions to each other. Sometimes it's because they had a similar problem or even had a unique approach when they worked at another company. Having a group of people think about a problem is much more effective than any single person.

If people are not naturally engaged in this process, it's OK to pull them into the conversation with direct questions. Ask open-ended questions though. As you'll recall, this means questions that can't be answered with a simple yes or no.

# LEADING VERSUS MANAGING

*When people say they admire leadership, they're actually*
*saying they value courage, because leaders are those*
*who are willing to go into uncharted territory. They're*
*willing to take a risk without a guarantee of success.*
~ Dan Sullivan

I read a lot of articles about management and leadership that always put the management function in an overly negative light. Leadership is about challenging the status

# Leading versus managing

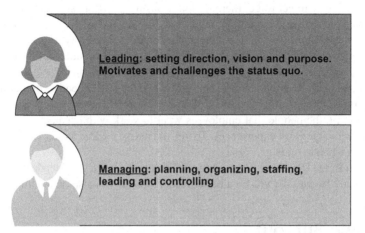

Leading: setting direction, vision and purpose. Motivates and challenges the status quo.

Managing: planning, organizing, staffing, leading and controlling

**FIGURE 12.1**    While many people think that it's better to be a leader than a manager, both leadership and management are important to an organization.

quo. It is about mapping out where you need to go and helping everyone get there together. Leadership refers to an individual's ability to influence, motivate and otherwise enable others to contribute to the organization's success. Leaders set the direction, but they must also use management skills to guide their people in an efficient way. Inspiring people to a greater vision is one thing but enacting the short-term goals and systems to get there is another. Influence and inspiration separate leaders from managers, not power and control.

Managers monitor and adjust work. They regularly look backward to ensure that current goals and objectives are being met. Leadership requires an aptitude to sell, management requires an aptitude to teach. Management requires self-discipline, commitment and accountability to get things done. So, the age-old question of "is it better to lead or to manage?" often requires the answer: "both" (Figure 12.1).

Before you dismiss management in favor of leadership, consider the five basic functions of managers. These are planning, organizing, staffing, leading and controlling. Do you think it's not important to plan, organize, staff, lead or control results then you might be one of those leaders who have their heads too far in the clouds and no eye to the day-to-day requirements of managing a team. They tell a good story, but they don't drive results or help anyone get to their vision.

Of course, not all managers are leaders, just as not all leaders are in management positions. But everyone has the potential to be a leader under certain circumstances. You don't have to be in charge to show leadership. Managers work within constraints and existing conditions. Leaders change constraints and conditions by not accepting the status quo. Challenging the way things are is not for the faint of heart in many organizations. Not accepting that "this is the way we've always done it" takes courage and perseverance. Managers, on the other hand, must understand the vision and drive their teams to do the work necessary to accomplish tasks and get to that vision.

Leadership without management can set a direction that has no method or discipline in execution. In other words, leadership without management can set a direction but not a way to get to get there. A lot of business writing suggests that managers belong in an old-fashioned manufacturing environment and everyone should strive to be a leader instead. But the value in management comes in the ability to keep performance on track, provide high-quality feedback, clarify a strategic vision, motivate and coach others and set some general performance expectations.

Managers can be good leaders, but only when they are already good at managing and are taking things to the next level. When you get both management and leadership right, these are the people that employees want to work with because they create efficient and enjoyable work experiences. To me, the mark of a true leader is knowing when to lead and when to manage.

**Action item:** Do you understand the difference between good leadership and good management? How could you be a better leader? Could you also be a better manager?

## Avoiding Burnout

As a manager, you need to be aware of how your team is really doing. Make it a policy to check in with your employees regularly. As more people are working from home, the lines between personal and work lives can become blurred. The demands of working in cybersecurity often spill into what would normally be personal time. Be ready to offer flexible work hours whenever possible. Encourage your team to take some time off not only in the form of vacations but also some daily "away" time where they can just stretch and go for a walk. A lot of people feel chained to their desk, especially since chat applications seem to demand immediate attention and some may not want to seem unresponsive.

# DIFFICULT CONVERSATIONS

Difficult conversations are a fact of life in cybersecurity. This section applies some of the skills we've learned in previous chapters to some difficult conversations that can arise.

## You Are Giving Negative Feedback

Some managers shy away from giving negative feedback, and it's understandable why. It may feel uncomfortable for both sides of the conversation. This said, negative feedback is a necessary part of managing. There are also ways to make the process less painful for both sides of the communication. Giving feedback is one of the most important things you can do to help your team members. Helping to course correct behaviors can help you avert much bigger problems later.

You almost always want to give negative feedback in person. The next best method would be video, so that you can also capture facial reaction, tone and body language. These nonverbal cues will help you gage how the conversation is working and if the message is being received and understood. You want to be careful that you aren't damaging a relationship, but rather providing feedback in a constructive manner. In fact, it helps to reframe negative feedback as "constructive feedback" as it will help keep the conversation more positive.

Our negativity bias shapes how people hear feedback, so you want to be careful. In the book *The Power of Bad*, authors Roy Baumeister and John Tierney say that it's the "universal tendency for negative events and emotions to affect us more strongly than positive ones." In other words, we really hear and respond to criticism and tend to move right past praise.

## Keep It Specific and Objective

Behaviors warranting feedback should be both specific and observable. You do not want to use vague statements like "your work is sloppy" or "you don't get along well with others." While these statements may be true, they do not make for actionable feedback and may go a little too far down the road of not being objective. You don't want to make feedback personal; you want to keep it simple, clear and objective. A clear statement would sound a lot more like: "Your reports routinely contain multiple errors, omissions and don't appear to have been proofread."

## Make Sure You've Seen the Behavior Directly

Do not provide strong feedback for a situation that you didn't observe first-hand. While there may be situations where you are called in to reprimand someone after the fact, keep in mind that you weren't there, and you may not have all the facts. Use examples that you have observed first-hand.

## Keep Feedback Timely

If someone on your team has an issue in a meeting that you've observed, like speaking unprofessionally to another team member, you need to capture this behavior shortly after it occurs. Resist the urge to provide feedback in front of everyone else, as this will be a more humiliating way to receive feedback. Instead, make sure you get a few minutes immediately after the meeting. I know we're all busy, but do not go about the rest of your day intending to get to feedback later in the day. Feedback needs to be received shortly after the behavior or it will not be fresh in anyone's mind. Delaying giving feedback can lead to a disconnect between the person and the behavior. Delayed feedback will not have the impact you'd like it to have.

## Ask to Give Feedback

I like to use this technique to get people to open and be receptive to feedback. Ask the person if they would mind if you gave them some feedback. Of course, if they are your employee you have every right to give them feedback. But asking makes them pause and think for a moment. Most people will realize that saying no isn't a viable option to your boss and they will be more receptive to hearing what you have to say.

## Situation, Behavior, Impact

A great framework for giving negative feedback is the situation, behavior, impact and action framework, sometimes just called SBI (Figure 12.2). Using this framework, you describe the situation, such as the meeting you were just attending. Add detail, where appropriate, like how there were many senior people and peers attending. Next, describe the behavior observed in detail. For example, describe exact sentences that may have seemed confrontational to others word-for-word or mistakes like not being able to answer questions about their own presentation. Next you want to talk about the

**FIGURE 12.2** Giving negative feedback can feel awkward without practice. Using a framework like Situation, Behavior and Impact can help you quickly get your point across without making feedback personal.

impact this had. For example, treating people in an unprofessional manner may make it less likely that they will want to help in the future and may create unnecessary tension. The final step is determining the action that can be taken to minimize the chance of recurrence in the future. You should ask the person directly how they would handle a similar situation in the future. Don't tell them how they should do it unless they are genuinely struggling to come up with an answer. It needs to be their behavior change, not yours, so it should be their idea on how to change as well.

Let's put that all together in a scenario. You are in a meeting with the infrastructure team to discuss deploying a new security tool. There is some friction because the infrastructure team is busy on other priorities and they don't see yours as important enough to start right now. One of your team members suggests that they would have more time if they didn't take two-hour lunch breaks and surf the web all day. Ouch. If things get more confrontational from here, you may well need to intervene in real time. However, if the meeting continues and concludes your next step is to pull the employee aside in a private area.

> Sean, do you mind if I give you a little feedback? (pause for a response). That was an important meeting with a lot of senior people. We need their cooperation to help implement our new security tool. When you suggest that they are not using their time effectively by taking long lunches, this comes off as offensive and overstepping your authority since they don't work for you. When you say comments like this, it comes off as unprofessional and they are going to be less likely to help us in the future. They will think of us more as enemies rather than partners. Do you understand how they might be offended or annoyed by your comment? How do you think you might handle a situation like this better in the future?

## You Are Receiving Negative Feedback

Not everyone likes to receive feedback. It can seem like an attack on our ego and who we are as a person. But feedback is important to maintain a bi-directional communication loop. If someone cares enough to give you feedback, try to be grateful and listen. Make a genuine effort to reflect on any feedback. Your first instinct might be to feel that feedback is wrong or unfounded but remember that perception is reality. If someone sees you in a certain manner, it's likely that they are not the only one. Feedback is a key driver for good performance and leadership effectiveness. It's important to frame feedback this way, not as a form of punishment. Feedback helps monitor your performance and can alert you to changes that may need to occur. Negative feedback is also constructive, whereas positive feedback doesn't have this benefit.

Of course, if feedback is coming from your boss, then you *really* need to take steps to address it. Even minor feedback can be the sign that something is off track, and you need to make improvements. In fact, I would go the extra step and ask for feedback from your boss on a regular basis. You don't want to wait and get feedback in your performance review. Asking for feedback also shows a level of emotional maturity. It can serve as an early warning signal if something is heading the wrong way.

Feedback can make you defensive and self-conscious. It might even make you angry. It's important to keep your emotions in check and remember that feedback is ultimately to your benefit. When you do receive negative feedback, take a pause and don't rush to react. If the feedback is somewhat vague, you should ask for some specific examples. You can't act on feedback unless you truly understand it. If the feedback is surprising think about if it's factual or not.

If feedback is coming from your boss, you will naturally want to make sure you are following up and taking corrective actions. If it is coming from someone who is not your boss, you can choose to take action or not, but if the feedback points to a disruptive behavior or other major issues, don't be surprised if it comes up again and maybe next time it will be from your boss.

Feedback is so important that I suggest asking for it on a periodic basis. You can ask for feedback from your boss, your team or other colleagues. Remember that perception is reality and if people perceive you in a way that conflicts with your own perspective, you should still take this feedback seriously.

## The Business Accepts the Risk

Another difficult conversation is when the business all-too-readily accepts the risk for an issue that they may be slow to address or may never intend to address. From their perspective, it's fine and they accept that the issue will not be fixed soon. Whatever risk there may be, they accept it. But is your perspective the same?

There's a lot that can go wrong with this conversation. Are you sure that the business *really* understands the risk? Are you sure that there's not some easy steps that could be taken to fix it? If they defined a timeline to fix the issue but it's far out in the future, are you sure it couldn't be done sooner?

While a lot of security people are quick to blame the business, I would flip this around. Are you sure you *explained* the risk properly? Did you explain the likelihood and impact if the risk actually manifests? Are you sure you didn't explain it in a way that you get but nobody else understands? This would be a good time to pause for some reflection. If you didn't explain the risk correctly you need to go back and do it again. I might employ storytelling or other techniques to help personalize the issue and make it a little more tangible.

Sometimes, you explained the issue perfectly. The business completely understood what you said. But it's still just not practical to address the issue. Where do you go from here? While some might continue to push on, these insurmountable problems do arise and sometimes it's just time to document the risk, make sure that very senior people sign off on their understanding and likelihood of the risk and then move on. Even risks that seem likely may never actually happen. The business ultimately owns the data and it's not up to you to prevent them from moving forward. It is up to you to make sure they understand their risks. If you've done that part already then, it's time to document it and move on with life. The best form of documentation for issues like this is a risk acceptance. An example form is provided in the appendix of this book.

# Difficult Conversations: You Need to Let an Employee Go

Letting an employee go is one of the most difficult things you need to do as a manager. Letting someone go has an impact on that person, their mental health, their family and you. You'd have to be pretty cold to just let someone go without it bothering you at least a little bit. That said, cold people are absolutely out there. If you must let an employee go you need to make every effort to do this in person. It's not always possible, but you still need to try. If you must deliver the news via video call or telephone, both you and the person should be in a quiet place that won't be interrupted. Most firms will also want a representative from HR on the call.

Own the situation and absolutely do not delegate letting someone go. Be respectful to the person. The situation sucks for everyone and you need to own your half. Meeting in person is a great way to explain why they're being let go and give them a chance to ask any questions. Every company is different. Sometimes HR drives this process, and you need to go with your company's directive. Also keep in mind that different states have different laws, so you really want to make sure you're following company procedures. Embrace HR during this process, having someone else present with you can be valuable if there are any issues since there's a witness.

Of course, if you are letting someone go for performance reasons, you gave them some indication that things were off track, right? Of course you did. If not, get ready for an unpleasant conversation and shame on you for not being a better boss. Many companies require a formal performance management plan and sometimes this is an onerous process. This is also unfortunate, as sometimes managers take the path of least resistance and just ignore a problem employee completely rather than going through the process. The problem with this approach is that keeping a problem employee can have a negative impact on the rest of the team.

When you have the meeting, don't get sidetracked by small talk. State the purpose of the meeting right from the start, outline the decision that's been made, and explain what it means for the employee. Be specific and honest. Finally, highlight that you appreciate all the work that they've done in the past and wish them well in their future pursuits. There are generally severance benefits that the person is entitled to receive that either you or HR can explain.

Of course, not every scenario where someone is fired is for performance reasons. It's far more painful for everyone if you need to let someone good go because the business isn't doing well and it's not the employee's fault. Of course, they will still wonder why they were chosen over others who might have retained their job. These situations require a lot of empathy on your part.

No matter the reason, keep your explanation for letting them go simple and direct. These conversations aren't fun, and you want to get through them quickly for everyone's benefit while still being respectful. Try to end on any high note you can find, but don't force it. Sometimes, there are no high notes in these situations. However, don't discount the fact that the relationship may not be working for either of you and you might both be grateful to part ways.

# THE LANGUAGE OF INCLUSIVE LEADERSHIP

*Inclusivity means not "just we're allowed to be there,"*
*but we are valued. I've always said: smart teams will*
*do amazing things, but truly diverse teams will do*
*impossible things.*
~ Claudia Brind-Woody

Non-inclusive language is any language that treats people unfairly, insults or otherwise excludes people. I am not an expert on this topic, but I think it's an important one to cover. We haven't really covered communication unless we acknowledge that this problem exists because it has a lot of impact on effective communications and leadership.

Diversity, equity and inclusion (DEI) has become a critical subject in the modern workforce, and for good reason. Diversity is about the presence of differences among different people. It may include race, gender, language or any other number of attributes. It's what makes us all different from each other. Equity is promoting justice, impartiality and fairness regardless of people's differences. Inclusion is an outcome that ensures that everyone feels included and welcomed. This section focuses on inclusion because it is the most critical to clear communication. You can't have communication if you are not including everybody.

All language is built from a set of rules. Inclusive language is similar. These principles will hopefully help you enter conversations mindfully of your audience's preferences. To be clear, by inclusive, I don't only mean race or gender. I also mean geography, culture, age and all the things that make one person different from another.

I remember one time having a conversation with my overseas team. I was going on vacation and told them I would be "out of pocket" over the next week. Of course, I knew what this meant. But no one else did. Finally, someone had the courage to fess up and ask what the heck I was trying to say. Oops. That's my fault. Why would I assume that everyone knows what an American slang term means? Especially one that, when you think about it, makes absolutely no sense. These terms are called idioms, and while they are OK for casual conversation with friends, they may not be appropriate in every setting. Idioms, industry jargon and acronyms can unintentionally exclude a lot of people and will prevent a clear message from getting through.

You are not going to connect with people by using terms they don't understand. You are also going to shut down listeners when you alienate them with language that may make them feel left out, insulted, confused or worse. Here are some general principles for being inclusive in both written and spoken language.

## People First

Always put your audience first and try to understand their needs. I like to think that in general, people are people. I don't personally like making distinctions based on gender, race, sexual preference or other characteristics unless it's directly relevant. Hint: it's almost never relevant in a business setting. What you need on your team is multiple perspectives. You need team members who feel like they're on a team that cares about them as individuals. What you don't need is people who feel that they are less important than others or that there's some kind of secret handshake or club where they weren't invited.

Consider different perspectives on your team and people's backgrounds. I once listened to a senior leader complain how long it had taken to get in and out of a location using the corporate helicopter, where most people had no choice but to drive in a very busy area of New Jersey. Not the best tactic to relate and build rapport with your audience. In fact, by failing to take the audience into account, this casual comment served only to create resentment and anger rather than unity and teamwork. If you're going to connect with your audience, they need to be able to relate to you.

## Don't Let Casual Language Get too Causal

Obsessive-compulsive disorder (OCD) and attention deficit disorder (ADD) are real conditions. So are many other conditions you might use to describe people in casual, inconsiderate ways. Consider that the person you talk to may have had a personal experience with intellectual disabilities and many other risky terms that you may throw into a conversation. In an office setting it's best not to go anywhere near these subjects. Is there really a need to talk about personal characteristics of any of your employees? It's not very likely.

Especially on your own team (where you have direct control), you need to treat everyone equally with no exceptions. Yes, some people are more senior, skilled, etc., but this has nothing to do with anything other than their knowledge and experience. We are all on a continuum, so don't play favorites. Don't make anyone feel left out of the conversation. Maybe I'm fortunate that this came naturally to me, but in my experience it's clear that it doesn't come naturally to everyone. If you have a problem in this space, please, please, take the time to work on it. Think about your language and the way you treat people and how it might leave others behind.

I don't want to catalog every "politically correct" term, because I question if this matters at all. Everyone on your team is a human being and you need to treat them with respect and not worry too much about what makes everyone different. It's what makes us the same matters a whole lot more to a functioning team. Respect different perspectives and different backgrounds. Be the change that everyone wants to see. If you do this, you won't need to worry too much about figuring out what words you can and can't say in a business setting.

Unfortunately, personal biases can manifest in strange ways that even you are not fully cognizant of. Do you only say hello to certain people on your team? Why is that? Is there someone you sort of avoid on your team or in the company? Why is this?

## Show Zero Tolerance

The behavior you accept is the behavior you get. Don't accept behavior or language that leaves part of your team behind. I wouldn't want to work for that manager, and neither should you. This means you need to stand up to bullies, correct political incorrectness and sometimes just take it all offline and try to educate people on how inclusive leaders behave. Just don't let bad behavior on your team go. That's a slippery slope.

## Don't Let Expert Knowledge Get in the Way

Sometimes, expert knowledge can alienate people who are not at the same level of expertise as you. Technical jargon and acronyms can leave people feeling that they aren't as smart or just aren't part of the conversations since you might as well be speaking Latin to them. Be conscious, even when you're speaking to your own team, that overdoing it with tech talk may leave some people feeling they are not part of the conversation.

## View Multiple Perspectives

Understanding multiple perspectives, even if you don't agree with them, will help build trust. If employees feel their experience and perspective is heard, they will feel more included. Sometimes, it will be necessary to pull people into conversations. Some people don't participate in conversations because they don't feel that they're really part of them in the first place. Make every opportunity to pull people into the conversation and help them feel part of the group.

## Try to Use Gender Neutral Language

Use gender neutral language whenever possible. Using "guys" to address everyone is gendered language that may imply that men are the preferred gender at your organization. I must admit, I sometimes catch myself using terms like guys, but I'm trying to get better at it. You may not be perfect at avoiding terms like these, but it all starts with being conscious that you are doing it and then striving to improve. Hey, nobody's perfect.

Think about the terms you use that may seem non-inclusive like "guys" or "manpower." That said, I also hate the use of words like "s/he," which just sort of strikes me as silly. It seems awkward to say things like: "everyone must log in to the system, regardless of if he/she wants to take the awareness training or not" versus the more neutral: "everyone must log in to the system, regardless of if they want to take the awareness training or not."

## Avoid Language with Implied Connotations

The English language is quite complicated for people to learn. Some words may mean different things based on their context. For example, a crane is a large, beautiful bird in nature, but it's a totally different meaning if you're talking about a construction site.

These words are called homonyms. Homonyms are words that have the same spelling and pronunciation, but entirely different meanings based on their context.

Other words in the English language have implied meanings, especially when applied to a certain demographic. An example might be the word "bossy," which is skewed to be a negative term describing a woman that is a bit too direct in communicating their expectations. There are a lot more of these terms to watch out for and I'm not going to try to catalogue them all here. Just be sensitive to your audience and choose your words wisely. You're trying to communicate, not offend, alienate or otherwise hurt anyone's feelings. This takes some self-awareness and of course some empathy. What terms might you be using that some might find offensive, or otherwise non-inclusive to others?

# Be Authentic

Most important of all is to be yourself. While many rules and ideas in this section are meant to help your language to be more inclusive, I'm not asking you to be an entirely different person overnight. I think if you show a little empathy, most of the thoughts in this section will come naturally. Be sincere and actually care about people and I think most of this will take care of itself.

# Lose the Ego

I've previously mentioned that there are some big egos in the security industry. Too many CISOs (I've worked for some) come in, take over and take credit for every good thing that happens and blames all the bad things on everyone else. They clean house and then repeat the cycle again when things still aren't going their way until they eventually run out of excuses and are shown the door.

You're not getting much done without your team, so include them. "I" have never delivered a world-class security program. I have led a team of people that have though, and "we" did a hell of a job as a "team." Catch the difference?

# Some Notes on Virtual Attendees

In the post-pandemic world, we are likely to move from being mostly virtual to having some form of hybrid office work where some people are in the office and some people are virtual. Maybe your organization was already like that before the pandemic. In either case there are some limitations when everyone is in a conference room and there are a handful of online attendees. Take steps to include online attendees in the conversation. Remember, online attendees can't just start brainstorming on a whiteboard and they likely won't be able to see that if you do. Take the time to explain things that can't be seen, like whiteboards. Share handouts or presentations. Remain cognizant that people are on the call and it might be helpful to pull them more into the conversation by asking them questions or their opinion proactively.

# RUNNING EFFECTIVE ONE-ON-ONE MEETINGS

Strong leadership teams are critical to your success. Holding one-on-one meetings with your team gives you an opportunity to help build that leadership team. These meetings are essential conversations for career progress, updates and general employee morale. They help build rapport and trust and can greatly improve team and employee performance.

One-on-one meetings are typically 30–60-minute meetings on either a weekly or sometimes bi-weekly basis. Some over-extended managers hold monthly one-on-ones, but I think this is a bit too infrequent to have value. A good one-on-one meeting doesn't only focus on status and project updates. These meetings are an opportunity for managers to directly improve employee performance and for employees to have their voice heard in a private setting by management.

An agenda for a one-on-one meeting should be driven by your direct reports, not by you. This makes life a little easier, as it's one less meeting to plan and manage. Your part in making a one-on-one meeting is still critical though and there are several things you can do to make sure these meetings are productive. Note that we will cover your own one-on-one meeting with your boss in the next chapter.

## Make It a Safe and Candid Environment

When I have a one-on-one meeting with someone this is a private meeting where I want open dialogue. This means that you need to create and enforce a "safe" environment where ideas and opinions can be shared openly. It's important to connect on a personal level with your direct reports, so you want the free exchange of ideas and opinions without judgment or punishment. I like to start these meetings with casual chat and some open-ended questions like "what's your most important topic today?"

## Your Role in a One-on-One

Your role in a one-on-one meeting is to be a good boss and manager, a good mentor and a good coach. That said, you probably want to default to coaching. The reason for this is, quite frankly, so you do less of the talking. Instead, you should be asking questions, actively listening and challenging them to get out of their comfort zones. Coaching should be something that all managers do with their direct reports. Coaching helps you understand how people think about their work, their careers and their relationships with the organization. Informal coaching can also help deal with issues before they become problems.

## Ask Great Questions

As I mentioned, you want your reports to do most of the talking in their one-on-one meetings. This meeting is for them to help them grow, become more comfortable with the organization and get help in any areas where they are stuck. Part of being an effective coach in these meetings is by asking great questions. Questions like where your employee wants to go in their career or what areas of the company they would like to learn more about will give you insight into their personality and help them think through where they are taking their career.

# SUMMARY

Managing a team and working with other colleagues represents a great opportunity to show leadership and also be a better boss. Having a team is a responsibility. You need to get things done through your people and will ultimately succeed or fail not on your own abilities, but your abilities to lead a group of people to accomplish organizational goals.

- The wrong leadership style alienates team members by being either too controlling or too lax with people. The best way to lead is to help every member of your team by removing obstacles and valuing each individual.
- The number of communication channels on your team or in any large group can be represented by $n(n - 1)/2$. A team with 10 members has 45 possible communication channels. These can't be controlled but should be managed.
- No one enjoys inconsistent leadership. Don't play favorites and try to be a source of stability. Be candid without crossing the line and being rude.
- As a security leader, you are often aware of business details that your team would never be aware of otherwise. Make sure you are conveying the "big picture."
- Leading and managing are both important. Don't emphasize one over the other, you need to do both.
- Giving negative feedback and letting people go are difficult communications that need to be handled carefully.
- The situation, behavior, impact, action framework can help people understand the impact of poor behavior and choose better courses of action in the future.
- Inclusive leaders try to keep everyone on their team feeling engaged and included. Be conscious that your words or actions may alienate some people if you are not mindful.

# REFERENCES AND RECOMMENDED READING

Cannon, David L. *CISA Certified Information Systems Auditor Study Guide*. Wiley, 2009.
Cole, Brent. *How to Win Friends and Influence People in the Digital Age*. Simon & Schuster Paperbacks, 2012.

Covey, Stephen R. *The 7 Habits of Highly Effective People*. Simon & Schuster UK Ltd., 2020.

Doke, DeeDee, et al. *Building a Team: The Practical Guide to Mastering Management*. Dorling Kindersley, 2011.

Fountain, Lynn. *Leading the Internal Audit Function*. Auerbach, 2020.

Gantz, Stephen D. *The Basics of IT Audit: Purposes, Processes, and Practical Information*. Syngress, 2014.

Kegerreis, Mike, et al. *IT Auditing: Using Controls to Protect Information Assets: Using Controls to Protect Information Assets*. McGraw-Hill, 2020.

Patterson, Kerry, et al. *Crucial Conversations: Tools for Talking When Stakes Are High*. McGraw-Hill, 2012.

Pickett, K. H. Spencer. *Audit Planning a Risk-Based Approach*. Wiley, 2006.

Pompon, Raymond. *IT Security Risk Control Management: An Audit Preparation Plan*. Apress, 2016.

Slater, Robert. *Jack Welch and the GE Way: Management Insights and Leadership Secrets of the Legendary CEO*. McGraw-Hill, 2001.

# Managing Up
## *Finding Your Boss's Communication Style*

**13**

One of the most important people you will be communicating regularly with is your boss. It's critical to master this communication path, as ultimately your boss is the main person who will determine your success or failure in your role. You are also going to spend a lot of time with your boss, so don't leave this relationship to chance.

Many people complain about working with their bosses. While there certainly are some relationships that are doomed to fail from the start, I think the main reason for difficulty is because people don't prioritize building a good relationship with them. A good relationship gives you a solid foundation for executing the security mission and building the trust needed to help you resolve issues quickly.

I recommend having a strategic communication plan to work with your manager more effectively. No matter how good or bad your manager may be, you are going to need to make the relationship work for your duration in the job. Putting together a plan forces you to think about how you will handle communications for one of your most important relationships.

One of the keys to effective communication with your boss is to first understand their communication style. Refer to Chapter 1 and the DiSC model. Do you know what your boss's personality type is? This is a great starting point and can drive how much detail they want to receive or how to best interact with them. But because this is such a critical relationship, you need to go deeper and get a better understanding of how to best operate together.

Without getting into the philosophical arguments about *where* security reports in an organization, you still likely have a boss, and that boss probably likes to get information in a certain way. Do they want every detail? Do they want to be CC'd on every piece of email? Do they only want to be made aware of potential surprises? Does your boss work evenings and weekends? Does your boss prefer email instead of phone calls? Do they respond to SMS text messages? Do they expect you to answer emails on weekends? Every person has a different style, and you need to understand how your boss prefers to receive their information. It's OK to just ask these questions outright. It's important to get these preferences right, not what you *think* is right. While I'm sure you'd figure it out over time, why start off on the wrong foot?

DOI: 10.1201/9781003100294-15

Since I manage a team as well, I'll share my communication style with you.

- I am an early morning person. Like 4 or 5 AM early. If I send you an email at 4 AM, I am not expecting a response until business hours. That doesn't mean I want to see your response 72 hours later though. This is simply a quiet time that I can catch up with messages.
- I respect vacations. I will *only* call your cell phone or text you (SMS) if I really need help. Emails can and should be completely ignored by me during this time. I might be impressed if you respond to something that's mildly important, but if you respond to everything, I'll have concerns that you're not disconnecting from work in a healthy way.
- If I am texting you or using corporate chat applications like Microsoft Teams, it is probably moderately important, and I am at least hoping for an answer within an hour or so if you're in the office. I am not, for the record, trying to figure out if you're supposed to be working from home but are really outside cutting the grass. Some managers seem to use chat as a way of making sure their people are tethered to the computer screen all day.
- I don't use the phone a heck of a lot. If I am *calling* you on your cell phone it's probably pretty important. If I really need to reach you on an evening or weekend, this is the method I'll choose.

People have all kinds of different styles based on the many ways they prefer to communicate. Getting to know their preference will put less friction on getting things done. And getting things done is what we all want, right?

I have had bosses who travel extensively. This can introduce some challenges in that they are in different time zones, on planes or otherwise generally unavailable. You will still need to push for whatever face time you can get in person, but never default to complete silence. If your boss isn't big on email, but that's the only way to send updates or communicate issues, this is still a better course of action than delaying important information. In other words, don't let your boss forget that you exist and that you're doing great work moving the security program forward. While of course you want to let them know about potential issues, make sure you at least send a status report on a weekly basis. I provide a quick and easy format later in this chapter. Here are a few suggestions to start building a good relationship with your boss.

# TECHNICAL DETAIL

You'll want to get a feel for how much technical detail your boss can handle or would like to have. If you are reporting to a CISO, it's likely they can handle a good amount of technical detail and they may want to hear every detail. If you are the CISO reporting to a CIO it may depend. If you are a CISO reporting to a Chief Risk Officer, Chief Financial Officer or Chief Operating Officer, technical detail may not be something they want to see. Figure out their preference and tailor your updates appropriately.

# OFFER TO HELP

Your boss is likely to be one of the busiest people in the organization and one of the few people that might be busier than you. They will appreciate it if you can offload some of their burdens and help them meet their deadlines as well. Offer to cover meetings, get involved in projects outside of your scope or just about anything that can make their life easier. This will also help you be seen as more versatile than just being the "security guy" and it will be appreciated by your overworked manager.

Honor your manager's time. Be on time to their meetings, answer their phone calls and make sure your meetings have an agenda ahead of time, so your boss knows what points you'll be covering. This is a two-way street. Don't just provide updates on your programs, always ask if there's anything they need from you.

# OWN YOUR AREA

Strive to do excellent work and manage your area relentlessly. You own your functional area and there are no excuses. Your boss doesn't want to have to worry about the details or feel like you aren't on top of things. Demonstrate that you are in control of your area and ready to handle any problems that arise. I've been fortunate enough to report to bosses who are genuinely concerned with all the things that can go wrong with information security. Not every boss is like this and you might even have one who is *overly* concerned or paranoid that something will go wrong. In either case, you'll want to make sure that they feel they have the right person on the job to handle all issues that may arise.

This also means you own your success and your failures. Things will go wrong, and you need to own the problem and do everything you can to make things right. As former Navy Seal and all-around badass Jocko Willink says: "Extreme Ownership. Leaders must own everything in their world. There is no one else to blame."

# CONNECT ON A PERSONAL LEVEL

You don't have to be best friends with your boss, but you do need to make a personal connection. While I have generally been friends with most of my bosses over the years, there have been a few that were, well let's just say, difficult. In this case, you do your best to win them over. If the relationship will never work, make the call and move on to other opportunities. A bad relationship with your boss will eventually result in your failure. There are very few other possible outcomes unless your management chain changes.

When building a relationship with your boss, you don't need to be all business. Certainly, if you are working for a new manager, they may want to get an indication that you have mastered your area before they want to share with you how their daughter's college progress is going. You need to build your credibility first, but after that you should strive to connect on a personal level. Asking how their weekend went doesn't require a close bond; it only requires a genuine interest in hearing the answer. Remember, your boss is a person too, just like you. They have all the same insecurities and awkwardness that we all have.

# NO SURPRISES

Just like you, your boss probably doesn't like to be blindsided with an issue they didn't know about in advance. I like to give my boss an early heads up on developing issues and describe how I'm handling it or if further action is warranted. While this might mean that my boss is aware of minor issues that never manifest into big problems, I'd rather err on the side of caution. Some bosses can get too involved trying to dictate what your response should be with a potential issue. This is OK if they are providing useful feedback, but you don't want to create the impression that you don't know what to do about a problem. That's why you generally want to include what you're already doing to monitor and control an issue from escalating, not just the fact that there is a problem brewing. We all get a ton of email. Unless your boss is really on top of email, you might choose another medium like chat, SMS or the good old-fashioned phone call to give them a heads-up that something important is developing. You don't want them to read your email update after *their boss* has already asked them a surprise question about it in the hallway.

# SET EXPECTATIONS

Your boss is not likely to know the details of how long certain efforts may take or how difficult it might be to pour through logs to investigate an incident. Make sure you are setting expectations and putting some parameters around your work, especially if it is something they asked you to pursue. You want to under promise and over deliver. And of course, keep the commitments you make. If your boss asks you for something, even something that may seem trivial, make sure you follow up and get answers. Don't wait until they remember that you never got back to them.

# READ THE NEWS, BECAUSE YOUR BOSS AND THE BOARD DO

As a security leader, you are expected to keep tabs on major items in the news on a daily basis. Don't just use technical sources either. Stay on top of business sources like the

*Wall Street Journal* and the *New York Times*, which are typically what your boss and the board might be reading. You want to know way in advance if there's a developing security issue that may raise questions.

Be proactive when you see something that might relate to your business. If you read in the newspaper that there's a major security issue or problem like the SolarWinds hack in 2020, put together an email communication for your boss and senior management. One framework I like to use for this is:

- Use BLUF to create a summary of the problem. Keep this short, like a few sentences, right at the top of the email.
- What it means for us: Is this going to kick off a big patching exercise? Enhanced monitoring? No impact at all, it's just for awareness?
- Mitigation strategies: if the problem can't be fixed right away, what steps will you be taking to mitigate the risk?
- Tracking: if there is going to be a patch deployment, how will it be tracked?
- Source: where you saw the story or add a link to the source for anyone interested in more detail.

A quick word on sources too. As a student of journalism, one of the first things I learned was to question sources. Is a problem being overblown because a vendor is trying to push their product? Is the source you're using known for quality alerts that don't just say that the sky is falling? Make sure you're dealing with a real issue before you go propagating it further. Of course, if you got it from a quality source like the *Wall Street Journal*, you can count that it's been vetted. Still, you might want to correlate it with more technical sources and make sure you understand the problem enough that you could put it into simple terms for others. Think in terms of composing an elevator pitch if you ran into your boss in the hallway. How could you provide a brief summary of the issue in just a few sentences?

When examining issues like this, set expectations early that this is what you know *right now* and that things are subject to change. The SolarWinds issue started out as the FireEye breach and then got much worse. Remember, in a security incident you don't know what you don't know. You only know what you know right now.

Obviously as information changes, you'll want to provide updates. To keep things clear, I like to simply reply-all to the original email message (if that was your initial medium) rather than starting a new message. That way, everyone can clearly see what's changed since the original message all in one place.

# PROGRAM STATUS REPORTS

Security programs have a lot of moving pieces to track. You are probably running multiple projects, operations, managing incidents and more. Don't assume your boss knows all the great work you're doing. You need to produce a status report that shows where projects and programs stand and showcase the work you're getting done and talk about where there are issues that may need management intervention. Ultimately, your boss is supposed to help remove some of the roadblocks you are encountering. You want to make sure there's an open and ongoing discussion about any issues.

Status reports have the added benefit of demonstrating to *anyone* where the program stands and how projects are tracking. This could be an auditor, an external regulator or the board of directors. It could also be other business stakeholders. Having a regularly updated status has a lot of benefits. Done correctly, a good status report will answer the questions everyone is always bugging you about before they even ask. A status report should be a high-level tracking document and not include every detail and milestone. Consider using only a page or two. Ideally, you can get everything onto a one-page management dashboard, but big programs with large teams will likely need more. Please don't get them down to one page by using six-point font though. If you must, go to two or even three pages. I bet if you worked at it, you could get it all on a page though.

If you have a need to track more details, you might want to have one page for overall program status and individual pages for projects. This would be one page for program status, like a dashboard, and no more than a page for each significant project. PowerPoint is an ideal format for these reports, but you can get away with Word or even Excel. Pick whatever tool you're the most proficient using.

I recommend covering each of the following elements.

1) **A title of the report**, for example Company X Cybersecurity Program Status
2) **Date**, so it's clear how old the report is or if it's the current version
3) A general **project name** for each project, such as Firewall Upgrades
4) For each project name, add a **project description** which can be a few sentences. Even if something seems self-evident, like firewall upgrades, it might add a lot more color to say "upgrade hardware and software on 36 Palo Alto firewalls and verify functionality by January 15, taking into account business outage windows"
5) For each project, indicate an **overall status** if the project is currently on track, at risk, or off track. I like to have an actual color associated with this, such as red, yellow, green. If it's an informal status, there's no need for qualitative metrics, red, yellow, green is fine. A quick note on colors. Security people tend to be overly pessimistic. Be careful marking every single project red or taking the opposite side and calling them all green and then something goes off the rails later. Try to put a candid rating that you could defend if questioned. Are you going to miss a date? It's red if there's no chance in making it, but yellow if you could double down efforts
6) For each project add a **status summary**. For example, "10 of 36 firewalls have been upgraded without incident. Remaining 26 have been scheduled for weekends over the next four weeks"
7) Make sure you are highlighting any **risks and issues** that may impact project deliverables. These are the items you will want to discuss in person with your manager. Ideally, you are also explaining how you are planning to address these risks and issues. Be sure to highlight where you could use their support, if necessary

A sample status template is included in the Appendix. This is a living document that you will want to update regularly. I like to update my status weekly, and I meet with my manager to discuss open items, issues, challenges and progress regularly. I think it's a good idea to create a copy of last week's status and archive it, so you can go back to a specific date. It's been rare that I need to do that, but every time I did have to do it, I was grateful that I had a copy.

Do you think this might be overkill for your boss? No problem, here's the short version that you can send weekly as a simple email. To be fair, I recommend discussing status in person as often as possible to create a better connection. However, some bosses travel quite a bit or have a lot of direct reports and this isn't always practical. At a minimum, make sure your boss is kept aware of the following every week, even if it's just an email:

- What I completed
- What's in progress
- Issues, incidents and risks
- What I need from you (if anything)

In the "what I completed" section, you want to add significant progress, project milestones or any other achievements you and your team may have accomplished. "What's in progress" should focus on the next week or two, not the next year or two. And definitely not every single project under your area, just the important ones. Don't include trivial items like how you attended meetings; try to include fewer items that move the program forward, not every little thing so you can prove how busy you are.

In the issues and incidents section, I like to point out any security problems or incidents that may be either developing or underway. This is *not* the time to surprise your boss with a big incident that they haven't heard about yet. We went over giving an early heads up in Chapter 11. However, you can certainly provide updates on incidents that your boss already knows about in this section.

Finally, in the "what I need from you" section, you want to add any issues, approvals or support you need from your boss to keep the program moving forward. If you have it all covered, then no problem. Put an N/A or "nothing at this time." Alternatively, cut it out completely and only add it back when there's something of note.

Don't forget that you oversee an important area that is prone to problems and issues. These problems may not be your fault, but keeping an open line of communication going at all times will make your boss more comfortable that you are on top of your area, understand the issues and are making progress with them. It's also an opportunity for you to remind your boss where they can help. Don't be shy about taking that opportunity.

## MANAGER ONE-ON-ONE MEETINGS

In the previous chapter, we discussed holding one-on-one meetings with your direct reports. If you recall, a one-on-one meeting agenda should really be driven by the employee. This means that when you're meeting with your manager for your one-on-one meeting, you should come prepared with an agenda and topics for discussion.

A lot of people like to default to status updates in one-on-one meetings. While it's certainly OK to do this, you are missing a giant opportunity if this is the only thing you talk about in these meetings. One-on-one meetings should also cover your own career development, areas where you are having difficulty, problems and obstacles you're encountering. A good format for a one-on-one meeting can include:

- **Status and metrics on projects**. Again, while this isn't the only subject you'll want to cover it can still be an important one to include. Feel free to leverage other program updates you may compile for other stakeholders, but make sure you're only discussing the important ones in this meeting.
- **Updates from the last meeting**: Highlight any issues, challenges or progress that have happened from the last meeting.
- **Personal and career updates**. I like to also include any significant topics that are going on in your life and career. Studying for a new exam or certification? Moving? Getting ready for vacation? These are all good topics to include to help build a personal rapport and possibly get some advice.
- **Issues and roadblocks**. Struggling somewhere? Stalled with procurement? Your job requisitions are still pending HR approval? These are issues to discuss, as a well-placed email from your boss might help move the mountains in the organization.
- **Other topics**: A few other topics for consideration include talking about recent work challenges, getting better insight into a business strategy, getting help with any tricky communications or managing a difficult personality in the organization.

Your manager one-on-one meeting is an opportunity to strengthen the relationship and take advantage of your boss's knowledge and experience with the organization. Make sure you are properly leveraging these meetings by having clear agendas and taking advantage of time beyond just providing status updates.

## Checklist for effective manager communications

1) My manager is a _____ from the DiSC model in chapter 1.
2) My manager's preferred communication channel (e.g., email) is: _____
3) My manager wants what level of technical detail when I communicate with them? How many details should I provide overall?
4) I will provide regular program updates to my manager through what medium?
5) My boss and I share what things in common?
6) What are some items we can discuss in our one-on-one meeting that go beyond typical status updates?

# SUMMARY

It's critical to establish a good working relationship with your manager. When this relationship is strong you will have a lot of support you need to execute on the security program. When it is difficult or strained relationship, you need to take steps to immediately improve it.

- Your manager has their own communication preferences, including how they like to receive information, on what frequency and providing a preferred level of detail.
- Offer to assist with issues that are outside of the subject of cybersecurity. You'll be seen as a stronger leader overall.
- Take extreme ownership of your area. Don't make excuses or blame others. There is no one else to blame.
- No one likes negative surprises. Make sure you are giving your manager an early heads-up if there are developing issues.
- Be proactive. Keep on top of the news and security developments that may impact your company. Just make sure the source is credible.
- Keep a running program status report. This can serve as a way to brief any senior manager or auditor on where the security program stands.

# REFERENCES AND RECOMMENDED READING

Baldoni, John. *Lead Your Boss: The Subtle Art of Managing Up.* AMACOM, 2010.
Bing, Stanley. *Throwing the Elephant.* HarperCollins World, 2003.
*Managing Up.* Harvard Business Review Press, 2014. https://store.hbr.org/product/managing-up
   -hbr-20-minute-manager-series/16863
Scroggins, Clay. *How to Lead When You're Not in Charge: Leveraging Influence When You Lack Authority.* Zondervan, 2017.
Watkins, M. D. *The First 90 Days.* Harvard Business Review Press, 2012.
Willink, Jocko. *Extreme Ownership.* Pan Macmillan Australia, 2018.

# The Board of Directors

# 14

*A modern cybersecurity program must have Board and Executive level visibility, funding, and support. The modern cybersecurity program also includes reporting on multiple topics: understanding how threats impact revenues and the company brand, sales enablement, brand protection, IP protection, and understanding cyber risk.*

~ Demitrios 'Laz' Lazarikos

## PRESENTING TO THE BOARD

Not that long ago, it would have been unimaginable that a security executive would be presenting to the board of directors. Now, it is becoming routine that the CISO has a standing agenda item with the board to provide an update on the state of cybersecurity within the organization. Sometimes this is annually and sometimes more frequently.

While this can be quite intimidating, the board of directors is ultimately a collection of individual people. And like all individuals, they will have their own communication styles and preferences. According to a recent survey conducted by the Ponemon Institute, only 9% of security teams feel as if they are highly effective in communicating security risks to the board and to other C-suite executives. This number needs to change. Less than one in ten CISOs feel like they have this area under control.

Frankly, there is no bigger communication disconnect than the CISO and the board of directors. CISOs come in discussing tools and technology and some APT group and board directors ask, "so are we secure?" This is usually followed by an awkward silence or an answer that is less than satisfying for everyone involved.

The key to making this process more effective is by planning a communication approach well in advance to a board discussion. It will be crucial to use simple business language that still gets the point across. You will need to become a master of compressing the right information into the right amount of time, no matter how complex the topic or how short your timeslot you've been given. This is high-stakes communication, don't leave success to chance.

DOI: 10.1201/9781003100294-16

# First Seek to Understand ...

Let's start by considering why CISOs are being brought in front of the board of directors in the first place. Cybersecurity has become a massive business issue and the board needs to know how this can impact business operations and the bottom line. They also want to hear it directly from the source, which is why CISOs are personally presenting on cybersecurity regardless of their reporting line.

Board members are potentially personally liable for failing to "appropriately monitor and supervise the enterprise" according to the 1996 Caremark decision made by the Delaware Chancery Court. You need to put your conversations in this context. The board might well be held accountable to a function that they don't understand. Your job is to help them understand real risks and how they are being addressed. You'll also need to convince them that you are the right person to lead this effort. In a sense, every board presentation is like a job interview. If you lose the board's trust, you won't be in the role much longer.

Depending on your business, technology is either critical to the business or it may actually *be* the business. Corporate boards face a new set of challenges in understanding the risks that this presents. Yet many directors do not have the skills and experience needed to provide an adequate level of oversight. Cybersecurity weaknesses represent a very new existential threat for business. But board members typically don't understand cybersecurity or even technology to any great level of detail in most cases. While this is starting to change, there's a long way to go.

The board is typically worried about how a business makes money, what are the risks and threats to the way they make money and how do they protect their assets and operations from problems and competitors. From this perspective, cybersecurity is one of many problems they need to consider. While cyber problems might magnify the risk exponentially, these are really the same risks a board member might have been focused on in 1970: how do we stay in business, keep our competitors at bay and run successful operations?

Hopefully this gives you a little context on where most board members are coming from. They have a lot of concerns to think about and you are one of many. A useful way to think about this is to use is the Amazon threat versus the cyber threat (hint: this is probably not that useful if you currently work at Amazon). Which threat is bigger, a cyber incident or Amazon moving into your business marketspace and cutting your margins by ¾ and crushing you like a bug? Just like life itself, a little perspective goes a long way. Cybersecurity is important, but it's on a list of important topics, not the only topic on the list.

The more you prepare for a board presentation, the better it will go for you. Even seasoned CISOs don't try to wing this conversation. Preparation is key. Here are a few considerations for you to get ready for this presentation.

# Get to Know Your Board, If Possible

Most board members have a biography published if you are a public company. They might have a profile on LinkedIn and you might even get a chance to meet with them 1-1 in an informal setting. Learn their background and it may give you some insight into their communication preferences. Board members have backgrounds in finance, operations or other disciplines that might dictate if you should focus on spending, operational metrics, etc. Again, board members are just people like you and me and they all have different styles and communication preferences.

Not everyone can get time with individual board members privately. That's OK, and don't push for it if it's not happening. But if you have the chance to do it, jump on it. Even getting a close relationship with a single board member can be that inside feedback loop that makes it easier to deal with every other member.

It's a fact of life that board members are also human beings with all the same shortcomings as everyone else. The National Association of Corporate Directors (NACD) published *A Field Guide to Bad Directors.* One of the great examples in this publication is what they call the "fifteen percenter" director which they describe as:

> They pre-read and understand about 15 percent of the board materials, are about 15 percent attentive in committee or board conversations, are 15 percent knowledgeable on any given issue, and wing it magnificently based on long-practiced shallow-and-wide preparation for all board work.

Unfortunately, the fifteen percenters definitely exist. This is your lowest common denominator and the person you should strive to keep engaged. If you succeed with this person, the stronger directors will fall in line.

## Prepare Relentlessly

Many CISOs spend days preparing for a board update. This sounds like a lot of time for a simple meeting, but without the presentation being properly tailored, it is unlikely to be effective. Again, presenting to the board is a high-stakes communication scenario. If the board is not confident in your abilities to protect corporate assets, one well-placed recommendation could land you on the unemployment line. Prepare your presentation and rehearse it in front of a mirror or camera or neutral live audience. Then do it again and again. Don't make yourself crazy or nervous, but you do want to make yourself prepared.

## Know Your Business Risks

Everything you say to the board should reflect something about business risk. This means you need to understand your risks first. This includes industry risks, risks to your specific business and anything that relates directly to how your business operates and makes money. This might be a good time to review your organization's 10-K or to speak with peers on the audit committee or enterprise risk department. You are going to use these risks to provide context and discuss potential financial or reputational impact. Once you are crystal clear on these risks, you want to make sure you're really focusing in on the security controls that directly address these risks and drop any discussion on the ones that don't. In other words, you might not want to focus on things like firewall upgrades that are table stakes for a general secure network. What are you doing to protect the company's crown jewels and reduce *business* risk? You are going to want to articulate both best and worst-case scenarios to leadership and help them understand potential risk impact.

## Have a Clear Agenda

The purpose of your presentation is not to talk about technology, but the business. Remember, this is about them, not you. Many presentations to board members continue to focus on technology and data points that are only relevant to IT security operations personnel. Executives will not be interested in hearing about this, and you won't make any meaningful connection. So, don't go in with a long list of tools and toys that have been deployed. I bet most board members won't understand even what you might consider a common acronym like DLP (data leakage protection). Don't use these terms without a simplified explanation that accompanies it.

## Be Data Driven with Meaningful Metrics

You want to be data driven with the board by using metrics that have meaning and relevance to them. You don't want to include a high volume of operational metrics that provide no strategic insight. Provide enough context for your metrics to be meaningful and business friendly. Refer to the metrics chapter of this book for additional guidance on this subject.

Think about what your board needs to know and then come up with the supporting data and metrics. Board members are looking for insight into the state of the organization's cybersecurity program and the business implications of cyber risks. This message needs to be loud and clear, not buried in a mountain of metrics.

## Be Concise and Stick to the Main Points

If the board invites you for a 20-minute presentation, don't take 30 minutes. In fact, your best bet would be to allow at least some time at the end for questions. This means, given a 30-minute slot you need to allow at least five minutes for questions and your presentation should be 20–25 minutes. Don't stray from your main points and make sure your agenda is clear and concise. Don't raise unnecessary details that don't support your main points.

It also means that you should plan on no more than five minutes per slide. Some simple math means that if you are given a 30-minute slot, do not go in with more than five slides (5 slides × 5 minutes = your 25 slot + 5 minutes for questions). Some CISOs go in with 40–50 slides and a 20–30-minute slot to get through them. This isn't effective and you'll never get through the material in any meaningful way. Now is also not the time to try and impress the board with how much work you and your team are doing, but rather how much risk you've reduced for the company.

## Use a Framework

Consistent use of a framework will help demonstrate how the security programs approach their mission and will provide some consistency and a method of organization in presentations. I personally like the NIST Cybersecurity Framework (CSF) that can be found at https://www.nist.gov/cyberframework. You may find that other frameworks better suit your business needs. Perhaps COBIT or ISO 27001 may be a better fit for you. Frameworks provide a great way to organize and communicate the security program. A framework also provides an assessment mechanism that enables organizations to determine their current cybersecurity capabilities and show paths to enhancement and improvement.

## Talk About Business Risk, not Fear Tactics

Using fear, uncertainty and doubt is not effective with boards. While this technique may have had some traction in the past, the reality is that it has lost all impact. Boards are constantly being told to be afraid of cybersecurity events, and many now tune out the warnings. To remain credible, speak in business terms: risk, impact and likelihood. Also, mind the number of threats you're discussing. There will always be an ocean of vulnerabilities, exposures and threats that you could discuss. Highlighting all of them as equal priority is not productive. On the other hand, if you are overly optimistic about the state of the security program, you may be setting yourself up for some difficult questions in the event of a breach.

## Use Storytelling

We covered storytelling in Chapter 6. If you are a CISO, you might only address the board for a few minutes once a quarter or even only once a year. This means that there is more pressure to keep presentations crisp and effective. In these types of situations, storytelling can be more powerful and effective than just reviewing pages of PowerPoint slides.

## Be Consistent

Make sure your messages remain consistent over time. The more you can make presentations on cybersecurity consistent in their message, tone and even format the more the board will hear the message. Remember, messages need to be repeated many times before they start to sync in. Every time you repeat something consistently, you help enforce the message and make it more likely to stick. If you vary the same message too much, it will seem like a brand-new message.

## Read the Room, not Your Presentation

If you have prepared and rehearsed, you will not need to spend a lot of time looking down at your notes. Instead, you'll want to look for the nonverbal cues that indicate things like "go deeper on that point" or "speed it up." Don't take it personally if everyone wants you to wrap up, the average board meeting covers a lot of different topics and your part is still a very minor part of what might be a very long day, or days, for them. Board meetings may be two days of 5–8-hour meetings covering a huge array of topics. Sorry for the reality check, but you're a minor part of that. Your objective should be to get in, get out, and live to fight another day.

Be ready to pivot fast based on what the feedback is telling you. Also, resist the urge to focus only on the most expressive face in the room. Research shows that people tend to focus on the most expressive faces in the room, no matter if it is positive or negative feedback. While it's good to know that someone may feel more strongly one way or the other, it can offer a skewed view on how the entire room is reacting.

## Anticipate Favorite Topics

It feels like most boards are interested in the state of the program, how we compare to our peers and any incidents and insight into how they happened. Incidents can be either at your company or a competitor. These are some of the "favorite" topics in my opinion. Remember, board members are highly interested in knowing the level of acceptable risk, and peer comparison is a great way to do that. Make sure you have some benchmarks ready and that you are current with incidents and breaches that have happened in the news, especially if they are in the same industry as yours.

## Be Strategic

You'll want to stay high-level with the board, not get down into program details. Show them where you're going and get their buy in to the program. Remember, that unless you are a security company, security is likely not a revenue generator, but table stakes for companies that have a large digital footprint. Show the value your programs have in business terms and supporting revenue-generating initiatives.

## Follow Up

If the board asks you a question that you can't answer, just say so. However, you also need to follow up on their question. You need to make sure you are demonstrating that you are taking concerns or questions seriously. Make sure you know *how* to get back with them as well. Timely feedback to a question or concern is best rather than waiting for the next board meeting. Usually, there is an organizer for board meetings, and you can get answers back to them through this channel.

## Anticipating Questions

Are you going to talk about a competitor's breach? Anticipate that the board may ask you how they are safe from the same type of event. Was there a recent big headline regarding cybersecurity? Stay current and be ready to address it. If you're talking about your program, be ready to field questions on what benefits the company is deriving from the investment. While you can't anticipate every possible question that could be asked, going through this process will keep you more prepared for the questions that do come and you'll have more agility coming up with answers.

Anticipating questions isn't always easy, as board members are becoming more educated on security issues. This means that questions could range from the somewhat ambiguous "how secure are we" to more difficult questions like "why aren't all of our systems fully patched at all times?"

It is best to run through some potential questions in advance and make sure you can answer them. Here are some potential questions a board might ask:

- How do we compare to our peer organizations? Are we spending more or less than them on security?
- How are we monitoring emerging threats and staying ahead of the curve?
- Do we know what our crown jewels are and where they're located?
- What are our greatest cybersecurity risks, why are they considered our greatest risks and what are we doing to manage them?
- Who are our adversaries are and what they are most likely to attack?
- What are our breach detection and response capabilities?
- Are we compliant with required cyber regulations?
- How do we know if we have NOT been hacked?
- How will you prevent breaches from happening in the future?
- How are we integrating cybersecurity risk into our enterprise risk-management program?
- Are we putting enough resources against the security program?

Be careful with that last question. Board meetings are not the time to say that your management hasn't given you the resources necessary to run the program or to break out a wish list for millions of dollars of security spend. Have you made it clear to your management what resources are required? Are they supportive? This is the right channel to discuss funding, not in the board meeting unless you've been directed otherwise.

# Boardroom Failures

I think it would also be worthwhile to consider some of the main reasons that CISOs fail to resonate with board members.

This isn't a comprehensive list; these are just the most common problems:

1) Providing too much technical information to board members who probably have no idea what the average mean time to patch (MTTP) is or why they should care
2) Providing too much information in general
3) Not making a personal connection with anyone
4) Not making the subject interesting or engaging
5) Selling too much fear, uncertainty and doubt

6) Painting an overly pessimistic picture without any corresponding action plans
7) Painting an overly optimistic picture without being realistic about the threats
8) Failing to convince the board you are the right person to lead the program
9) Not making enough measurable progress over your tenure
10) A significant breach under your watch that wasn't handled well

Notice that the last item doesn't just stop at "a significant breach under your watch." I think we've all moved on from hiring a CISO and thinking that the company is now bullet proof. But you may live and die based on how you handle a real incident. That's why those table-top exercises we went over in Chapter 11 are important to both the business and your career longevity.

## Board Scenario: You've Been Breached

These days, there is a high likelihood that you may be called to discuss a breach in front of the board of directors. In fact, you might even be asked to present a competitor's breach and what can be learned from that as well. The board needs to understand risk to the company, and you need to help educate them.

Needless to say, these are high-stakes conversations. I have known a few CISOs who are no longer on the job due to board presentations that didn't go well. Make no mistake, these are the top executives of your company and if you can't demonstrate knowledge and competence, you might be looking for a new job a lot sooner than you thought.

Hopefully, most of your presentations to the board will be uneventful walk-throughs of metrics decks showing how much progress the program has been making since you joined the firm. This is your best case. But what if you've experienced a data breach? If it's a big enough breach, you are likely going to be called in front of the board to explain the situation. This can be a very tense conversation. While it is certainly possible the board is looking to attribute blame to someone, don't go into this situation with any preconceived notions. But definitely prepare.

## Strategy

Because breaches can take quite a while to unfold, it is also possible that you will be asked to discuss a problem that isn't fully resolved. If so, you need to be very careful about what is communicated, but also be responsive to the questions that are likely to arise. You need to give the board the assurance that you have the situation under control and are the right person to lead the response.

**Tactics:**
- **Know your role**. Security's role in board reporting during breach may differ from standing board presentations. The CISO should seek to partner with the broader crisis management team (e.g., Legal, HR, Privacy,

Heads of Business Units, etc.). If you have a crisis team, be prepared to play more of a supporting role during the board presentation. For example, if the primary focus of the board is about dealing with customer inquiries, this is probably not your conversation to drive.

- **Avoid speculation or promises**. Present and discuss facts only. Avoid speculating or making promises that may prove incorrect or impossible to meet. For example, if the board asks how a breach occurred, but your investigation isn't complete do not speculate on how you think something happened. Instead, explain that the investigation is still underway and then outline what you know to be facts as of today. I would also explain that facts may change as the investigation continues and that you will keep them informed.
- **Minimize documentation**. Be careful what is documented and communicated to the board as this information may be legally discoverable. Seek guidance from legal counsel for handling any communications around an incident. This includes being careful sending privileged and confidential information via email. Lean heavily on the legal team for guidance.
- **Establish a communication cadence**. Tell the board when they can expect updates throughout the life of the incident. Never let the board feel it must proactively seek information from you, as this can interfere with your team's response and create mistrust between Security and the board.
- **Anticipate questions**. Understand the board's fiduciary role and how that affects questions the board is likely to ask. In particular, the board is often focused on: (1) has the incident been contained, (2) how big is the incident and (3) how does the incident impact regular business activity?
- **Signal a serious tone**. Ensure all written and verbal communication signals that Security takes the breach and its impact on the organization very seriously and is doing everything it can to manage the situation.

**After the Breach:**
- **Minimize documentation**. Again, be careful what is documented and communicated to the board as this information may be discoverable. It may be advisable to keep the presentation sparse and rely on verbal discussion. Seek Legal input when you are in doubt.
- **Anticipate questions**. Be prepared to answer questions board members, in their fiduciary duties, are likely to ask. Common questions include: (1) what are the costs and implications for the organization; (2) was the breach due to security negligence; (3) was the breach appropriately handled once discovered and (4) what is Security doing to protect against similar breaches in the future?
- **Focus on the future**. When the first board presentation after a breach is fully resolved, this can be an opportunity to link the breach to Security's strategic plans moving forward. Linking a past breach with a future plan helps provide assurance to the board and keeps the conversation future-focused rather than focused on the past and what could have been done better.

- **Practice, practice, practice**. In some cases, the board may be evaluating whether they still have confidence in the CISO after the breach. The first board presentation after a breach has been fully resolved is an opportunity for the CISO to regain trust and provide assurance. It is advisable to practice the presentation with a trusted peer, peer coach, etc., to perfect the message and anticipate tough questions.
- **Lean on an outside expert**. In some cases, it may be helpful to get presentation support from an outside expert. This can lend additional credibility to the CISO's message, particularly if the expert was involved in Security's response and can speak knowledgeably about the breach. This voice can also lend credibility to the message that breaches are a matter of when, not if, and that the breach was not caused by Security negligence.

# Board Scenario: A Competitor Has Been Breached

In the case of a competitor's breach, ensure that your analysis also includes what your team is doing or has done to avoid a similar situation. Outline investments in relevant controls. Discuss ways that your incident response team is prepared for a similar incident. This should go without saying, but also don't speak in absolutes: "Our defenses guarantee this couldn't happen here."

**Situation**: A competitor experiences a data breach and you have been asked to present the situation to the board of directors.

**Strategy**: It's also common for the board to want to speak with their own CISO when a competitor has been breached. This is obviously a little less tense than dealing with your own breach. In this scenario, your strategy should be about knowing as much as you can reasonably know about the competitor's breach and be ready to explain how that could or wouldn't happen at your firm. If you don't have a good answer on why it couldn't happen at your firm, be prepared to explain how you are addressing any issues and fortifying the defenses.

**Tactics**:
- **Gather information.** Maybe you can reach out to the CISO at the breached organization, or others in your industry with some insider knowledge, to get a better understanding of the breach. Having these conversations before presenting to the board lends better credibility to the presentation and signals Security's understanding of the breach and its ability to protect against (or detect) a similar attack. Of course, you will want to read and digest all the public information about the incident as well.
- **Avoid disclosing privileged information**. Be careful to avoid disclosing any privileged information about the competitor's breach to the board, as this information may be permanently recorded and subsequently discoverable. An easy rule of thumb is to put a stronger reliance on publicly available information to establish known facts.

- **Focus on what is relevant**. Focus on aspects of the competitor's breach that are relevant to your organization, rather than summarizing everything that is in the news. The board has access to news reports and typically is looking for an expert analysis on what happened at the competitor and what it means for your organization. Not a rehash of what's already in the paper. Try to add value to what everyone has likely already read.
- **Provide (appropriate) assurance**. Ensure that analysis of the competitor's breach is coupled with an explanation of what Security is doing in response. This includes outlining past and future investments in relevant protect controls, describing monitoring efforts to detect similar attacks in your organization, and discussing ways that your incident response is prepared (or preparing) to respond to a similar incident if it were to occur at your organization.
- **Never overpromise**. Never make statements or provide assurance that may not be met (e.g., "this can't happen at our organization," "we're positive we haven't been breached by these adversaries," etc.). They'll hold you to your word.
- **Take advantage of teaching moments**. Breaches in the news can be an opportunity to remind the board that breaches are a matter of when, not if, and that incident response is an important investment.

# WHAT'S IN A GOOD BOARD PRESENTATION?

What goes into a good board presentation PowerPoint deck will vary from industry to industry. However, here are some basic elements I would consider for every presentation.

## A Summary from Your Last Meeting

Were there any follow-ups from the last board meeting? You absolutely want to make sure any questions have been addressed and that the issue or question is formally closed. If there were no follow-ups or actions, simply provide a summary of what was previously discussed.

## Business Risks

Next, you will want to have something on any significant changes in the threat environment to your industry or specific business. This should be risk-weighted and cover the most significant threats that would have a genuine impact. This is not the time to try and impress the board with how many risks and issues are out there. They only care about significant, high-impact risks that have a high probability that they should be worried about.

Any big risks should have a program element that maps to a mitigation strategy. For example, you don't want to go to the board and explain that their biggest risk is something that you aren't currently covering in your program. Make sure you can line up programs to risks. Worried about increasing ransomware attacks in your industry? Great, what are you doing about it? What programs are you running that map to reducing ransomware risk? Make sure this is clear and that it's clear how much longer these programs must go before they are fully in place. Is there anything the board can do to help? This isn't a terrible time to highlight what kind of support you need to make this happen.

I personally like covering breaches in the news even if a company isn't a direct competitor. What can be learned from other breaches? How do we know it wouldn't happen here? You'll want to demonstrate that you are on top of the threat environment and being proactive.

If you are far along enough to have a roadmap, you'll want to include a slide on progress and where your programs stand. It's OK if there's a lot more to do, so long as you're demonstrating progress from presentation to presentation.

If there's time, I also like adding a deeper dive into a particular security topic. There are a lot of options here. Business travel to hostile foreign countries? Social media? The list of topics is endless, but you might also ask what topics they would like to see covered in future presentations.

## The Threat Environment

I always like having something that details updates to the threat environment and what risks the industry may be facing. Make sure your risks match your industry and are directly relevant. You can figure this out by past incidents, competitor breaches and other industry news. You need to be able to answer simple board questions like "why would they attack us?"

Input for this section should be real-world events, but also developing threats. If nation-state attacks are common in your industry, try to make sure you have the most current information from sources that might include paid threat intelligence services, DarkWeb reports or industry insider information. This section really needs to be cutting-edge to be credible. If you can't make it credible, don't bother including it at all.

## Program Trends

With program trends, you'll want to cover where your security program stands, what you've done to make improvements and where you must still make progress (Figure 14.1). Maturity models can sometimes help. However, when using a maturity model, you need to accept the fact that some companies are perfectly content being a C student. Not every business wants to be best in class in cybersecurity. They want to focus on core strengths and be able to keep up with industry peers. Don't take it personally. You

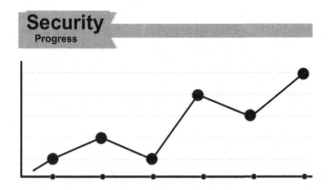

**FIGURE 14.1**   While it's useful to report how metrics change from month to month, it can be even more valuable to show trends over longer periods of time. Consider showing trends for six months or more so that you can show where the program has been and where it is going.

should try to figure out what makes sense for your organization and aim for the appropriate level of maturity. Seek feedback from your boss and a few trusted peers on where your targets should be, but it's rare that you should strive to be best in class across all categories and even rarer that you could actually hit those targets anyway.

## Keep It Professional and Polished

Make sure there are no grammar or punctuation issues, but also pay attention to the overall layout. Make sure everything is crisp and clean. Make sure headings are straight. Make sure the presentation prints well, but also looks good on an overhead. Needless to say, this is not the deck that you should include Dilbert cartoons or anything humorous. Keep it 100% professional.

# BOARD METRICS

While I've included a whole chapter on metrics, please keep the following guiding principles in mind when preparing board-level reports:

- Make sure you have actionable and relevant data. Ask yourself *why* this is relevant to the board and why they should care.
- As always, avoid all technical jargon and be concise. Edit relentlessly to make sure.
- Be relevant to the business. Translate technical issues into business impacts.
- Be strategic. You should be enabling dialogue about risk and management's response to risk issues. Don't just throw data at them.

- Think fewer, but better metrics.
- Remember that the overall metric may matter a lot less than the trend.
- Do you know what your organization's crown jewels are and where they are stored? From a data perspective what is the most important thing? Customer data? Financial data? Trademark or intellectual property? The secret formula for Coca Cola? Make sure you can put those crown jewels in the context of your security controls and how they are being protected.

# EDUCATING THE BOARD SLOWLY OVER TIME

If you have a regular cadence with the board, use this opportunity to squeak in a little education every time for them. There are certain things that would make everyone's life easier if they started to understand these few basic points. These include:

1) Even the best technology will have problems. All software has flaws no matter if you built it yourself, bought it from a security company or got it from a large company like Microsoft.
2) If it can run code, it can run malicious code. Macs, printers, cell phones are all in scope for security problems.
3) Even with the best security controls in place, bad things will still happen. The NSA, CIA and just about any high-profile company you can think of has been hacked before. No one is immune.
4) There's no such thing as perfect security. Even given unlimited budget, people and time there will be flaws along the way. Security is a risk-management tradeoff.
5) The security problem is large, and a lot larger than it was 20 years ago. There are datacenters, clouds, third parties, mobile devices, Internet of Things (IoT) and a growing list of problems. The risk-based program focuses efforts where they will have the greatest impact. Securing every device from every threat at all times is an unsolvable problem.
6) The board is responsible for understanding security risk. That doesn't mean that they really understand it though. That's why you're there as an advisor and educator, so keep it all in simple terms. You wouldn't want your doctor to tell you that you have *orthostatic hypotension* when all they really meant was "you got dizzy from standing up too fast."

# SUMMARY

Presenting to the board takes practice and preparation. Don't leave these high-stake communications to chance. A lot of what the board wants to know is simply:

1) Where are we?
2) Where do we want to be?
3) How do we get there and what will it take?

- Understand your audience. Get to know individual board members if possible. Make sure you think about what kind of information a board member would be interested in understanding.
- Relentless preparation is key. Board meetings are high-stake communications and should not be left to chance.
- Do not go over allotted time or bring more detail than you can possibly get through in the allotted time. Plan for questions.
- Read the room, not your presentation. Be sensitive to body language and empathetic that board meetings are long and somewhat tedious for members. You are only one topic being discussed.
- Make sure any and all questions raised are answered either on the spot or with follow-ups if you don't have the answer on hand.
- Many people fail by providing the wrong detail or too much detail. Make sure what you are presenting has some relevance and avoid overuse of metrics that don't resonate.
- Presenting your own or a competitor's breach can make for some stressful communications. Make sure you are sticking with facts and letting them know where there is a lack of certainty.
- Following a similar general format is helpful for board members to follow. Some elements to consider are business risks, program progress, threat environment updates and any breaches either at your own firm or in the news.
- If you have a regular meeting cadence with the board, use this as an opportunity to try and teach them or raise awareness on one thing in every meeting.

# REFERENCES AND RECOMMENDED READING

Fitzgerald, Todd. *CISO Compass: Navigating Cybersecurity Leadership Challenges with Insights from Pioneers.* Auerbach, 2020.
Gentile, Michael, et al. *The CISO Handbook a Practical Guide to Securing Your Company.* Auerbach Publications, 2006.

# Working with Auditors

# 15

*Two-thirds of the Earth's surface is covered with water. The other third is covered with auditors from headquarters.*
~ Norman Ralph Augustine

The audit process is one of the most misunderstood and hated processes in the IT world.

Working with auditors represents a special kind of communication situation for most. Audits are a fact of life in the financial services industry, but most public companies also have an internal audit function, and more are starting to turn their attention towards cybersecurity issues. The threat from cyberattacks is significant and could have a tremendous business impact. Many audit committees and boards have asked auditors to help them better understand this risk in the form of performing audits on the business.

Clear and accurate communication during an audit is critical. Communication channels need to be addressed early in an audit and issues should be understood by both sides before being documented and addressed with senior management. Audits can come in the form an internal audit, an external regulator, a Payment Card Industry (PCI) audit or any other number of other sources. This chapter covers some dos and don'ts about handling this kind of communication and aims to change the way that most security leaders are approaching auditors and audits.

An auditor's job is to review the actions and duties of all employees and departments within a company to verify that controls are in place and operational. Many people interpret this as being questioned on their abilities, but the audit function performs a critical role in protecting and improving a company. Sometimes, auditors are just gathering information and trying to understand how your part of the organization works. In other cases, they are verifying that controls are working as stated, usually by either sampling or obtaining evidence of functionality. Because auditors pinpoint weaknesses in systems and recommend corrective actions, many people dread audits. It also takes some upfront work to help auditors do their job. You will have to walk them through different aspects of your area, gather evidence and then take any corrective actions necessary if there are issues. Since most people have enough to do already, audits are seen as an unnecessary burden. Rest assured that this is not the case. Audits can actually help drive the security program and you should see auditors as partners, not as enemies.

Working with auditors does take some balance. Auditors are people and have all the same communication challenges that everyone else has. An additional challenge is that an auditor who has been asked to review cybersecurity functions may not have a strong background in the subject at all. This means that it will be up to you to make

DOI: 10.1201/9781003100294-17

sure that they understand what they are reviewing and that they're reviewing the right things in the first place.

# THREE TYPES OF AUDITS

Audits typically fall into three categories: compliance audits, system discrepancy audits and process assessment audits.

Compliance audits are probably one of the most predictable, as they generally have clearly defined objectives and criteria for achieving a satisfactory rating. PCI is an example, as the audit objectives are pulled straight out of the PCI Data Security Standard (DSS) document. Compliance audits are either regulatory in nature or industry specific. Regulatory audits are the result of legislation and may carry significant penalties for non-compliance. Some industry audits, such as the Health Insurance Portability and Accountability Act (HIPAA), are a bit open to interpretation, which can lead to conflict during an audit.

System discrepancy audits are audits where problems or discrepancies are noted in an audit, but the root cause of the problem may be unknown. For example, an auditor may flag that duplicate transactions have taken place in a system, but they don't know how this happened. It is up to the owner of the system to determine root cause and fix the underlying issue. Because of the vague nature of system discrepancy audits, a lot of friction can arise while the root cause is being tracked down

Process audits generally review existing policies, standards and procedures to ensure they are being followed. These audits are generally pretty straight forward unless your documentation is out-of-date or you don't have documentation for processes that an auditor would consider mandatory, like change control.

# PREPARE IN ADVANCE

Audits will go a lot more smoothly if you are prepared. Most audits are not a complete surprise. Auditors typically operate on annual or semi-annual cycles and they generally publish their schedule in advance. In other words, there's no good reason not to be prepared.

A few things to know in advance include:

1) **Mindset**. As always, we will start with your mindset. If your attitude going into an audit is negative or confrontational, it will come out in all of your communications and set the tone for what will likely be a painful audit. Try framing auditors as a group that is here to help manage risk and make the cybersecurity program even more effective. The wrong attitude interprets issues as a personal attack on you, your program or your team or as a

complete waste of time. Auditors can make your life very difficult, but they are also providing an important function. Try to treat the engagement as a partnership and it will go much smoother.

2) **Audit scope**. What is the purpose of the audit and what, specifically, are auditors there to examine? For example, a PCI audit focuses on credit card-holder data. This means that systems that have nothing to do with cardholder data should be considered out of scope. In fact, keeping a tight scope is the key to passing a PCI audit. Make sure you steer the conversation and deliverables to only items that are in scope and make sure you review the scope well in advance to audit fieldwork beginning.

3) **Review previous results**. Review previous audit results well in advance of a new audit, because you want to make sure that previously identified issues have been fully addressed. Repeat findings can be very serious in some environments like banking. But in any environment, repeat findings create the impression that you are not taking the issue seriously. At a bare minimum, reviewing previous results will give you an idea of what to expect on the next audit, so it's always time well spent. If there were previous issues that were not addressed, you need to either have a plan to address them or create a risk acceptance that has been signed off by senior management if it's simply not feasible to fix an issue.

4) **Program documentation**. Having up-to-date security program documentation will greatly help an audit go smoother. Create or obtain relevant network diagrams. Have a written security program in the form of policies, standards and standard operating procedures. Having this documentation on hand will facilitate the whole audit process and help the auditors get an immediate understanding of your area. If you are required to comply with specific standards, like PCI or the HIPAA, make sure you can show how your program is meeting these requirements. All documents need to be up-to-date though; refresh them if necessary or if they're hopelessly out-of-date simply get rid of them. Do not leave old documentation on intranet sites or SharePoint servers. These documents are sitting ducks to become your first audit issues. The most important document to have is your security policy. Without a written policy, there can easily be disagreement over what data should be protected or what services should be allowed. Without a written policy there is no way for an auditor to know whether or not the organization is following its policy because there is nothing to audit against.

5) **Create formal communication protocols**. How do you expect communications to be handled during an audit? Will there be weekly touchpoint meetings? Are meeting minutes going to be captured? Will all audit requests be submitted in writing? This greatly reduces the chance of confusion with a request.

6) **Have a point person**. You can facilitate communication protocols by establishing a focal point for all audit requests. This has a lot of benefits if your organization has a lot of audits. Having a single, dedicated point person can help with duplicate requests and also make sure that requests are handled promptly. If auditors are making requests to individuals, you can't track

centrally who has responded and who is not responding in a timely manner. In addition, a point person can provide guidance to users who have never participated in an audit. This person can educate users about appropriate ways to answer auditor questions and best interact with auditors. It will also reduce the burden on the operations staff that may need to compile evidence that a control exists and is functioning as expected. This person needs to be relatively senior, but many successful audits have shown that the more that senior management is involved, the more likely the process is to go smoothly.

You should also familiarize yourself with the standards, regulations or laws that you are being audited against. Read Pub 1075 for IRS audits or the Federal Financial Institutions Examination Council (FFIEC) documents for financial audits. Many auditors will audit outside the bounds of these documents and that's OK, but they will certainly include material from the PCI-DSS on a PCI audit. In some cases, auditors may even provide their audit "program" that they are using to conduct the audit. There's no harm in asking for a copy of this prior to any audit fieldwork.

Of course, you should also be familiar with your own policies and standards and be following them. See the policy section of this book. As a rule of thumb, it is better to not have a policy than to have one that no one is following. The logic of this statement is simple. Lack of a written policy is a documentation issue. Not following a policy that is documented is a breakdown in governance. I'd rather be written up for the first issue, not the second.

Here are some of the standard program documents that many auditors would want to see in place.

- An overarching Security Policy
- Change management procedures that cover how code is pushed into production
- Patch management policies and procedures
- Personnel onboarding/offboarding procedures
- Disaster recovery and business continuity
- Access control documentation

# RISK AND CONTROL SELF-ASSESSMENT (RCSA)

Rather than let your auditor discover your system's vulnerabilities or a process that is not documented accurately, it is far better to discover them yourself prior to the audit. In addition, some laws such as the Health Insurance Portability and Accountability Act and PCI's Data Security Standard require a periodic assessment of your systems and networks. Performing an assessment allows you to discover areas of weakness (i.e., vulnerabilities) and determine what to do about them. Performing a risk assessment can help prepare for audits, detect problems before auditors do and demonstrate a high level of program maturity.

One method that can be very helpful is called the Risk and Control Self-Assessment. RCSA is a process through which operational risks and the effectiveness of controls are assessed and, in some cases, tested. If your organization has an RCSA process, review the security material to ensure it is current and relevant. RCSA typically covers all operational risks, not just security.

A typical RCSA process starts by identifying risk to a business entity. What does the business do and what kind of inherent risks apply to it? Controls that mitigate the identified risks are then documented. Finally, after identification, the controls need to be assessed to see that they are working as intended and are suitable for the purpose. If there is any lapse in the controls, corrective action needs to be taken.

When vulnerabilities are discovered, they should be analyzed to determine the risk level they pose to the organization. Once that's been done, three options exist:

1) For high-risk vulnerabilities, issues should be remediated (fixed) soon if not immediately.
2) For vulnerabilities whose risk is lower or whose risk can be mitigated, a risk acceptance statement should be written, documenting why the vulnerability does not constitute a significant risk to the organization, why the impact to the business fixing the issue outweighs the risk, or what mitigating factors are in place or being put in place to lower the risk to the organization. A sample risk acceptance is included in the appendix of this book.
3) For vulnerabilities that are too risky to accept but cannot be fixed right away, a work plan should be created. If the vulnerability is also discovered by the auditor, demonstrating that it was identified in a self-assessment process and already has a work plan shows a greater level of maturity and will potentially reduce the risk rating assigned by the auditors. This is one of those instances where you want to be on top of identifying your own issues, not surprised by auditors with issues that you had no idea were present.

What should the RCSA measure a system against? Typically, they use security "best practices." One example of best practices is the PCI Data Security Standard. Other examples include COBIT and ISO 27001. I also like the material found at the Center for Internet Security (https://www.cisecurity.org). There are plenty of best practices to draw on, but you're best off picking a framework that's comprehensive and focusing on it until everything is in place.

If you don't have a formal RCSA process, ISACA.org is a great place to start. Their COBIT framework is a great starting point that already has a lot of documentation, checklists and other useful information.

# DURING AN AUDIT

Security people are a busy bunch. We don't have a lot of time for uninvited guests, especially the ones who can make more work for us like auditors or regulators. This stated, you also can't afford to have these interactions go poorly. I'm proposing a few rules for when auditors and regulators show up that can help.

## Be Cordial

Audits are not fun, but they are necessary to ensure that a company's controls are in place and running as expected. An auditor is simply doing their job. Treat them professionally and with respect.

I worked at one firm that had an especially difficult relationship with their primary regulator. The auditor assigned to the firm had a long history with the company and didn't think that they took security concerns seriously. These audits were off-track the minute they started because the lead auditor was looking to send a message and was working with a confirmation bias that everything was going to be a mess. Sometimes, there is existing baggage like this in place and you just have to do your best to build rapport and trust. Treating them professionally is the first step in that process.

## Be Accurate and Honest

Auditors don't always have the same level of experience and depth as you do. Make sure that if they are asking you to produce a report or information that doesn't make sense to you, you are well within reason to ask clarifying questions or even why the information would be helpful to obtain in the first place. You can feel free to suggest alternatives to getting the answers that they want. You also don't want to provide the wrong information because you didn't understand the question in the first place. This can lead to a lot of confusion.

This may sound obvious, but never lie or mislead auditors or regulators. This can get both you and your company in a lot of legal trouble. Auditors also have an uncanny knack of knowing when something doesn't add up. Don't give them any reasons to doubt your credibility. Unethical treatment of auditors and regulators will almost certainly end with your termination.

## Answer Clear Questions

If you don't understand what an auditor is asking you, don't guess the answer to the question. Instead, ask for further clarification on anything ambiguous or help them rephrase the question in a way that's clear. Don't answer questions that would be best answered by someone else in your organization or questions where you don't know the answer. If you don't know the answer to a question, just say so. If someone else is in a better position to answer a question, don't answer it for them. Instead, refer the auditor to the right person and don't assume that person is already in the room. Finally, if something is not under your direct control, don't answer questions about the process. Refer them to the process owner for an official answer.

## Assume Auditors Know Nothing About Your Area

Don't assume auditors know your area as well as you do. While auditors are generally sharp people, they also tend to have fairly wide responsibilities across an organization or industry in the case of regulators. In a large company, you might find an auditor or department dedicated to technology or even cyber audits, but this is not always the case. You will likely need to educate them on the specifics of your area. Try to gage their experience and understanding of cybersecurity, but it's likely that they are nowhere near your own level of expertise.

This means you should plan on making sure they understand your various program elements and your general roadmap. Be prepared to explain why you are doing certain things and why you are prioritizing certain efforts over others. Again, having a documented security program is helpful here, as the auditor can take this away and come up to speed on their own time.

## Be Concise and Specific

While I mentioned that you should never lie or mislead an auditor or regulator, you also don't want to provide them with more information than what is asked for. Providing too much information can confuse everyone and lead to investigations of other areas that might have otherwise been considered out of scope.

Answer every question the auditor asks to the best of your ability, but you don't need to elaborate outside of the specific question being asked. Another important point is to answer the question and then stop. Too many people keep talking and wind up raising unnecessary issues. And of course, all the rules of good communication still apply like avoiding technical jargon and acronyms.

## Passing an Audit Does Not Mean You're Secure

I'm sure I don't need to tell you that passing an audit doesn't mean your systems are secure. Auditors and regulators have a limited scope and limited time to complete their work. They also have nowhere near the level of expertise that you and your team have. Just because you pass an audit doesn't mean that there's not more work to do. The security of your data, systems and networks is still very much your responsibility.

## Create a Partnership

Once the audit has started, ask the auditors to bring any major issues to your attention as soon as possible. There is no reason to wait for an audit to conclude to address issues, and in a best case, an issue could be addressed before the audit closes which can be noted in the management response. Better yet, you can sometimes discuss an issue or risk before it is formalized in a report. This way, you'll still have some influence over the language and impact being described.

Sometimes a problem that didn't get funding can gain traction when an audit issue is attached to it. Having a good relationship can be quite beneficial to the security program and you can work in harmony to help get the right management attention on a problem.

## Address Issues

If you are going to have a partnership with audit and regulators, you need to take their findings seriously and close issues in a timely manner. Avoid repeat findings, which can make it seem like you're not taking findings seriously.

This means you need to put some project management formality around audit items. Someone on your team, or elsewhere in the organization, should be flagged to track and manage issues to completion. Often in a cybersecurity audit, issues are flagged that span multiple areas. Auditors might assign issues that are not under your direct control, but instead fall in areas like infrastructure, architecture and the service desk. This means you need to manage your issues, but also keep tabs on issues that fall in other areas. You'll need to make sure that owners are defined, action plans are drafted and that progress is being made on issues over time.

Unfortunately, security issues often span multiple areas. It's not enough to simply address issues under your direct control. This is one of the more difficult aspects of being a CISO. You are often responsible for problems that are outside of your direct control. But security issues are security issues. Management will likely not make the distinction about what is or isn't under your direct control. And a failed security audit will reflect poorly on the security group regardless of how many other groups may be involved.

## Audit Action Plans

When you do have an audit issue, you will be required to respond and write an action plan to address the issue. There's a little bit of an art to this, because you need to make sure that the audit finding is specific, and that the response is tailored to meet every point of the audit finding. Avoid general responses that are open to interpretation. For example, if you have access control findings and write a response saying that you'll develop a comprehensive identity and access management (IAM) program, how are you going to demonstrate that you've met the audit item in a timely manner? An IAM program can take years to get up and running and even then it's a work-in-progress. This response is also bad because it would be open to interpretation on what elements belong in an IAM program and when you could call the audit issue resolved. A better response would be if ten systems were found to have access control issues then you will address those ten systems. This isn't the time to over-commit and talk about a comprehensive IAM program that covers every application and server in the company. Go ahead and create that IAM program anyway, but don't make it a dependency of closing that audit issue in a timely manner. Providing a practical and easy-to-achieve action plan is always the way to go versus the gold-plated idealistic plan that has little chance of success.

## Don't Argue

When an issue is identified, your one and only defense is if the issue is factually accurate or not. An example might be that they were looking at a wrong server and found an issue that didn't have any relevance to the production environment. That is, so long as no sensitive information was exposed.

If the issue is factually accurate, there is nothing more to discuss than the overall risk rating of the issue (generally high, medium or low) and when you are going to have the issue addressed. Don't be surprised if issues are reported to the company's board of directors. Auditors are generally required to communicate in writing issues found that are relevant to the governance of the company. This is all the more reason to perform some sort of risk assessment in advance and fix as many issues as possible before the auditor arrives.

## Auditors and Regulators Don't Run Your Program, You Do

Finally, remember that auditors and regulators do not run your security program. They should not tell you specifics about *how* to implement security controls or programs; they should only communicate some high-level requirements of what those programs should include. If you have an auditor who is trying to dictate implementation details, I'd recommend finding that person's manager and discussing this problem openly. Senior audit managers know the proper role of audit, which is not to dictate implementation steps but to advise on risk.

# AFTER AN AUDIT

When an audit winds down and issues have been resolved, I'd recommend going right back to preparing for the next audit. This means making sure that issues have truly been closed and anticipating what the new problems or focus areas will be, especially for regulators. I know when I worked in finance we kept close tabs on what issues regulators were raising with peer institutions. If there was a new topic being brought up at JP Morgan, it was bound to work its way to Citigroup eventually and vice versa. Keep informed and be proactive. Audit issues and regulatory problems get all the wrong kind of attention, so it's best to stop them from happening in the first place.

# SUMMARY

The best way to work with auditors and regulators is to treat it as a partnership. Creating a partnership with auditors requires crystal clear communication, trust and honesty.

The more work you do before an audit begins, the more likely it is that the audit will go smoothly.

- Clear communication is critical during an audit. Establish clear communication channels and protocols. Try to nominate a point person to field all requests and ensure a timely response.
- Don't approach audits with a negative attitude or it will affect every step of the process negatively. Never argue with or lie to auditors.
- Audit objectives may vary depending on whether it's a compliance audit, system discrepancy audit or a process assessment audit.
- Make sure you understand the audit scope and objectives and agree with them prior to fieldwork or control testing begins.
- Review all previous audit results and make sure there are no repeat issues.
- Documentation is key to passing an audit. Policies, standards and procedures all play a critical role in facilitating the audit. Having a documented security program is a requirement.
- Having an RCSA process in place can help proactively identify issues before they are discovered in an audit.
- Always be cordial, accurate and honest. If you don't understand a question, clarify it rather than trying to force an answer that may lead to other problems. Be concise with your answers and don't volunteer a lot of unsolicited information.
- Remember that you run the security program, not auditors or regulators. Don't let them overstep their bounds by dictating implementation details.
- Your only "defense" in an audit is if an issue is factually incorrect, but you can work with auditors to negotiate risk ratings for issues.

# REFERENCES AND RECOMMENDED READING

Cannon, David L. *CISA Certified Information Systems Auditor Study Guide.* Wiley, 2009.
Fountain, Lynn. *Leading the Internal Audit Function.* Auerbach, 2020.
Gantz, Stephen D. *The Basics of IT Audit: Purposes, Processes, and Practical Information.* Syngress, 2014.
Kegerreis, Mike, et al. *IT Auditing: Using Controls to Protect Information Assets: Using Controls to Protect Information Assets.* McGraw-Hill, 2020.
Pickett, K. H. Spencer. *Audit Planning a Risk-Based Approach.* Wiley, 2006.
Pompon, Raymond. *IT Security Risk Control Management: An Audit Preparation Plan.* Apress, 2016.

# Your Next Job

# 16

*Nothing in the world can take the place of persistence.*
*Talent will not; nothing is more common than*
*unsuccessful men with talent.*
*Genius will not; unrewarded genius is almost a proverb.*
*Education will not; the world is full of educated*
*derelicts.*
*Persistence and determination alone are omnipotent.*
~ Calvin Coolidge

Even for seasoned security professionals, interviews and the job search represent one of the most difficult and stressful communication scenarios you will face short of a major security incident. Some people even consider interviewing more stressful, and it's easy to understand why. The stakes are high and every word you say is being judged as you are weighed against other candidates. You will be communicating through your résumé and online profile, on the phone and in video meetings and face-to-face. The fear of rejection is very real and to a degree maybe even probable, as even great candidates may lose out to someone who's just a "better fit" or less expensive than you.

With the average CISO tenure being 18–24 months, the odds are good you may find yourself in an interview more than once in your career. The goal of this chapter is to discuss the ways to make this process less stressful and increase your chances of success.

## PREPARATION

If you are either employed and looking for a new role or unemployed and looking for a new role, you are going to need to work on your mindset first. Looking for a new role can be stressful. You are going to need to treat a job search as either your only job if you are out of work or as a second job if you are still in a role but looking to make a change. Either way, looking for a new job is a full-time job in and of itself and you should treat it that way. You need to be organized and methodical in your approach.

Hopefully, you already have a résumé, but if not then putting one together is your first step. If you do already have one, it may not have been updated in some time. I strive to do an annual update of my résumé regardless of if I'm considering another position or not. This way, when and if the time comes, you're not starting with material that hasn't been touched in ten years.

DOI: 10.1201/9781003100294-18

# YOUR PERSONAL BRAND

It's important in your career and especially in your job search to establish a personal brand. People who know me know that I'm pretty modest. Tooting my own horn and pushing a personal brand doesn't come that naturally to me. But building a personal brand is becoming mandatory.

Personal branding is an ongoing process of marketing yourself to make you stand out from everyone else. Some people find this to be a very uncomfortable process and others seem to spend *most* of their time marketing themselves and getting little else done. Building a personal brand is not something that you need to spend a lot of time doing. In fact, I personally think that spending a little bit of time over a long period is the best way to build and manage your personal brand.

Branding yourself requires an understanding of your career goals, strengths, interests and specialties. In other words, you want to focus on what makes you unique and avoid things that could detract from your professional image, like having an AOL email address. Unsolicited job offers may come from your personal brand since recruiters may see your profile and reach out to you. If you still don't think personal branding is for you, think of it as how other people see you and talk about you and how you want to be seen. We all have a personal brand, it's best to put some conscious effort into yours.

While there are a lot of tools and platforms out there to build your personal brand, one of the most important ones in the business world is LinkedIn. LinkedIn has taken on a bigger importance than Twitter and Facebook for job searches, networking and in professional settings. We will discuss managing LinkedIn later in this chapter.

# AN EXECUTIVE RÉSUMÉ ISN'T WHAT YOU THINK IT IS

One of the challenging documents you will write in your security career is your résumé. Your résumé is something that many people don't spend enough time honing and improving and others spend far too much. Your résumé is the document that will get you the interview when there isn't another way, like an inside track. Even if you do have an inside track, you will likely need to produce a résumé for the people that don't know you in the interview chain.

Once you have an interview lined up, the importance of your résumé goes down. This is because the hiring manager has probably already determined you are qualified to be at the interview. From this point forward, I'd wager that very few people involved in the interview process will spend more than a few minutes looking at your résumé. If you've ever hired someone, you've been guilty of this as well. Can you honestly say you've read every word of every résumé that's come across your desk? I know I haven't. I've even been in situations where the interviewer didn't "get a chance" to read my résumé and they either try to fake something quick while you sit and wait, or they just skip it altogether and open with the "tell me about yourself" question.

Recruiters and hiring managers sometimes spend 10 seconds or less on your résumé. That's a real number, by the way. This means you need to ask yourself if your value proposition is clear. Will your résumé entice someone to pull you in for that interview? Is your résumé differentiated from your competition? Does it provide tangible results from your previous roles? These days, it is also important to make sure your résumé is applicant tracking software (ATS) friendly as well, so that you come up in applicant online searches, which we will cover in the next section.

Trying to get all this information in a document that may only be read for 10 seconds can seem like a daunting task. But it helps to think of your résumé as a sort of marketing sheet, since that's exactly what it is. Your résumé is supposed to help the hiring manager distinguish what makes you better or more unique than other candidates. Anything that doesn't help market why you are the perfect hire for the particular job you're applying for is working against you.

As you get more senior in your career, it can be tempting to have three or more pages in your résumé. Don't do it. No one is going to read it and it makes your résumé more likely to be filed straight into the garbage can. I've personally received résumés that are 7–8 pages. This alone makes me likely to screen it out because I think it shows a lapse in judgment.

For most people, you should strive for one and maybe two pages. Rarely, if ever, would I advise going over two pages. I have worked at many firms for a quarter of a century and my résumé is still just two pages, with all the best material on the first page. The key is to be concise, relevant and easy to read, with points that really highlight what you can or have accomplished.

To help you limit the size of your résumé, you can try:

- Limiting the description of each role to 3–5 bullet points. Use more bullet points for your most recent jobs and fewer for the older roles.
- Only include your most impressive achievements and be quantifiable, wherever possible. Avoid nebulous "responsible for …" items. I can make a monkey "responsible for" leading a team of 50. That doesn't mean he did a good job.
- Only list job experience going back 10–15 years. Some people disagree with this or think it creates the perception of gaps, but the reality is that experience older than 15 years is barely relevant in the information security industry. If you are in the latter portion of your career, I don't think anyone cares about that brief stint you had doing something completely unrelated to security. If you do go back more than 10 years, the detail should be less and less as you go. The most detail should appear in your current position, since this is the most relevant.
- Adjust the margins and fonts, but don't go nuts. Use a proportional font that has varying widths to display each letter, like Times New Roman. Fonts like Courier are not space efficient. I wouldn't go for anything too unusual; err on the side of being conservative. You are applying to a senior executive job,

so showing a lot of creativity and fancy formatting probably doesn't help you as much as it would in a liberal arts job. Résumés that seem to succeed in capturing the most attention feature simple layouts, with clear sections, easy-to-read bullet points and heading titles.

# "MUST HAVE" RÉSUMÉ ELEMENTS

There are a few items that pretty much every résumé needs to include. These are the standard fields:

1. **Your name and contact information**. This doesn't have to include your home mailing address, but it should include your email address, a phone number (mobile is best) and a general location. City and state are usually enough. Make this one line to conserve space.
2. **Your work experience**, of course. You want this written with a strong emphasis on quantifiable achievements. There are two schools of thought here, as this section should be the majority of your résumé. One school believes you should include every job from college. Another says the last ten years is fine, as experience prior to that is largely irrelevant. In either case, the further back you go in your career, the less detail you want to include from more than ten years ago. No one cares that you were an OS/2 wiz back in the late 1990s.
3. **Education**. This should include degree and major, school name, location, graduation date and GPA if above 3.0. I would also include relevant industry certifications, such as CISSP.

If you stopped here it would be fine. I'm not a fan of including hobbies and other elements that have nothing to do with the job you're pursuing. Your love of hang gliding and parasailing is probably not going to land you that next big security role. Also, in a more senior position, your "technical skills" section is sort of a take-it-or-leave it section. Do you really expect that the CISO should know JavaScript? You can probably save some space by leaving this out and focusing more on your accomplishments and leadership skills.

## Objective Title or Branding Statement

You can either put the title from the job posting at the top of your own résumé or update the title of your résumé to closely match the title of the job posting and required skills. For instance, if you are running incident response for an organization looking at a job titled "Director of Incident Response" you could simply put that exact title at the top of your résumé. You can also show the field and industry expertise in your title in the résumé such as "High-Tech Sales Executive."

## Professional Summary Statement

A professional summary demonstrates your unique value to a potential employer. You want to show you have as much of the desired experience and skills as possible. Review the job posting of interest and highlight any key phrases. Next, weave these key phrases into the professional summary section of your résumé. Don't make this section too bulky, but make sure to show off your top qualification.

Have you avoided typos, punctuation mistakes and overused terminology (team player, seasoned, go-getter)? Remember, your résumé is essentially a marketing flyer. It needs to address a few basic questions like what makes you qualified, why you are better than other candidates and why we should hire you instead of someone else. Every word or element on your résumé that doesn't help make these points is working against you. Focus on achievements and how you personally helped projects get to the finish line.

# TAILOR YOUR RÉSUMÉ CHECKLIST

Tailoring your résumé to the role is a great way to make sure that the hiring managers and HR departments see you as a fit for the job. Here are some helpful tips:

- ☐ Review the company, position and industry. If it is a public company, review SEC filings like the 10-K or the investor relations portion of their website. If they're a private company, try to find out as much as you can from other sources like Crunchbase (https://www.crunchbase.com/) or Bloomberg. Do you know anyone at the company or maybe have a friend-of-a-friend? Check on LinkedIn for second or third-degree connections.
- ☐ Quantify any achievements wherever possible. Did your centralization project result in a savings of two million dollars annually? Say so! Again, you want an accomplishment-oriented document. Statements liked "served as the head of incident response" are nowhere near as effective as "defined a new incident response process leveraging a team of 20 that resulted in 30% faster response time."
- ☐ What is the company looking for and how are you a good match? Is this captured in your résumé? Go line-by-line if you have a job description and make sure that it's clear where you meet all requirements. Often times, first screens are done by HR departments or recruiters. Don't assume that a security person is the first one reading your résumé. This said, if you don't meet a particular requirement, don't let this deter you. Supply and demand is still far off in this industry and just because you miss a requirement or two doesn't mean you won't be considered for the role.
- ☐ If you have a professional summary section (and I recommend that you do) tailor this section by key phrases from the job specification into this paragraph. Make sure this section is not too long!

☐ Have you figured out what the company is *really* looking for? Plan how you are going to market yourself to meet these needs.

☐ Review your entire résumé to see if there are other items that should be highlighted to underscore your fit for the role.

# BEATING ONLINE PARSERS

We should spend a moment talking about the applicant tracking software system. ATS is a résumé parsing tool that handles the automated storage, organization and analysis of digital résumés. Résumé parsing software provides companies an efficient way to identify keywords and skills in order to sort through many applications. Many companies rely on an ATS or online job boards, which use matching technology to compare an applicant's résumé to a job posting.

I won't spend a lot of time on this subject because senior security professionals will very likely not find their next job by applying to jobs on Monster.com or Indeed. Still, it's useful to understand that résumé parsers are out there and how to work with them.

Many people think that PDF provides the most control over how your résumé displays and therefore submit their résumé in this format. The alternative is MS Word or a compatible editable format. Unfortunately, with online parsers, PDF is not an ideal format. This is because PDFs can be read in two different ways, one of which is essentially a photo of your résumé which can't be indexed. When you can't be indexed, you won't be found in a search.

An important point to remember is that companies are increasingly relying on parsing software. If you know there's a likelihood that you're going to be going through an online parser (like Taleo) you should write your résumé with parsing software in mind as your first audience, not a hiring manager or person on the other end. Here are some other tips to keep in mind:

- Use your name in the filename (e.g., jimbob_cv.doc).
- Don't use PDF, use MS Word format.
- If you feel you must use PDF, make sure it can convert to a format that can be read and indexed.
- Avoid headers and footers, which tend to never work out right.
- Pick a single font for the whole document.
- No tables or columns. Keep it simple.
- Absolutely no WordArt, funky spacing or trying to be different just to be different. This is not a creative profession and you'll probably just be seen as weird rather than unique.
- Don't export your LinkedIn profile. This is lazy and sloppy. I hear that LinkedIn has also taken some steps to make this less-than-effective for parsers, since they technically compete directly with them. In either case, it's best to use a properly formatted résumé that you control, not a social network site like LinkedIn. That said, you do want to make sure that the factual

information in your résumé matches what appears on LinkedIn. Yes, people do compare the two and will note discrepancies.

# MANAGING LINKEDIN

I'm sure most or all of you have at least heard of the business social networking site LinkedIn (https://www.linkedin.com/). I confess that I've become somewhat disenchanted with LinkedIn lately. It's overrun with marketers, random connection requests and people looking for your time, money or attention in one way or another. Microsoft, who now owns LinkedIn, has actually taken several steps backward with the service such as now allowing you to mass accept connection requests without running special scripts.

I still think LinkedIn is a necessary evil and I do recommend using the service overall. A lot of companies are attempting to bypass the sometimes-sketchy world of recruitment by doing their own searches via LinkedIn. LinkedIn has over 30 million companies with 20 million open job listings. Even CISO jobs, traditionally the property of retainer-based search firms, are starting to show up more on LinkedIn. This is not a market you want to completely ignore. I personally had a few jobs approach me through LinkedIn and, in fact, the origin of this book opportunity came to me through a LinkedIn connection. Sometimes, you need to take the bad with the good.

In addition, in the new world of remote working, LinkedIn is one of the few professional online networking sites that I would recommend. It is far more business and employment oriented than Twitter, Facebook or other common social media platforms. Because of its business focus, you won't find a lot of the nonsense topics that plague other social media sites.

LinkedIn provides a good venue to connect with like-minded professionals in the industry and keep track of former colleagues. The platform is part business card, part résumé. The positives of using LinkedIn are keeping in touch with friends and possibly finding new opportunities. Unfortunately, in an industry that is so driven by products, you will also likely be bombarded by vendors, conferences and other business owners trying to make a sale. The signal-to-noise ratio is almost unbearable. Compounding this problem is that the tools that help manage this platform are generally geared towards the mass-marketer, leaving the recipient of these campaigns defenseless to manage the onslaught.

With LinkedIn, your profile deserves the most attention. You're not on the job market? Well, if you are a hiring manager people are looking you up trying to understand your background and preparing for their own interview. You're going to want an up-to-date profile almost every way you look at it. Fortunately, there are only a handful of important components needed to create a good profile. Let's go through them.

## Use a Professional Profile Picture

You need to have a photo for your profile. This is your calling card and how people will picture you as they engage. It is literally your first impression, so you might want to get a professionally taken photo. Your photo should be relatively recent and should look like a professional, office-friendly photo. You can also consider a background picture to help enhance the overall look. This isn't the place to have that picture of you on the dirt bike. Save that for Facebook or Twitter.

## Write a Good Headline

Your LinkedIn headline is one of the most visible sections of your LinkedIn profile. It is also one of the most read sections of your profile, since it shows up first in search results. You want to expand on the default headline. You will also appear in more LinkedIn searches by using strategic keywords in your headline.

A lot of people simply use their current job title as their headline. This is squandering an opportunity. Your title is already covered in later sections of your profile and it does nothing to sell your value or differentiate you from the competition. You also want to avoid things like "seeking new opportunities" since very few people who are looking to hire are looking for people "seeking new opportunities." This is doubly so in an industry like cybersecurity where we are still near zero unemployment.

## Don't Skip the "About" Summary

Similar in importance to your headline, do not leave your summary blank. This is your opportunity to have your "elevator pitch" about what makes you unique. It may also be one of the few items that get a careful read from anyone.

## Watch for Junk Words, Focus on Action Words

Just like in your résumé, you want to avoid all the trite phrases like:

- Motivated self-starter
- Action-oriented
- Strategic
- Passionate
- Enthusiastic

All of these may be true, but they are not helping your cause. These words are table stakes and don't explain what you achieved. Try to find something a bit more original and maybe measurable. Better terms might include:

- Created a ...
- Launched a ...
- Started a ...
- Redesigned ...

- Centralized ...
- Expanded our ...

Obviously, these terms should include what was created, launched, etc. Focus on concrete achievements, not personal attributes, and quantify wherever possible.

At the end of the day, LinkedIn is a search engine just like Google. When people are looking for something on LinkedIn, they type some words into the search box and LinkedIn serves up the most relevant results. But how does LinkedIn decide what's relevant and what shows up first? Keywords are one of the biggest factors.

## List Your Relevant Skills

It's one of the quickest wins working with LinkedIn—scroll through the list of skills and identify those that are relevant to you. Doing so helps to substantiate the description in your Headline and Summary and provides a way for other people to endorse your skills. However, the key here is staying relevant. A long list of skills that aren't really core to who you are and what you do can start to feel unwieldy. Take time for a spring clean of your skills list every now and then.

## Publish and Share Content

It's up to you how active you want to be on this platform. If you are really trying to build your brand, LinkedIn does offer a good place to post original content, comment on posts from others and build a reputation. The more active you are on LinkedIn, the more likely you will come up in searches. The platform favors active users over lurkers.

The good news is once you get your profile set up, the general maintenance isn't that bad. Active participation also doesn't have to take up that much time. LinkedIn messaging is another issue for you to manage though. For me, I check these messages very quickly and will not take the time responding to everything. Ninety percent of my messages seem to be either vendors or conferences. Hopefully, LinkedIn will start making life easier for the recipients of messages rather than making it easier to mass-message a lot of people.

## Keep Your Résumé and Linked In Roughly in Sync

Unlike your résumé, you can feel free to add details to your LinkedIn profile. While I would still keep details relevant and use your best material, there is no reason to limit the number of words or try to fit everything into a page or two. This stated, employers do compare your résumé and online profile. Make sure that what's in your online profile matches up with your physical résumé. Titles and timelines should all match, as well as the basic job description and accomplishments.

## Let Recruiters Know You're Open to Searching

Finally, make sure to adjust your settings to let recruiters know that you're open to new opportunities. If you are still in a current role, there are settings to make sure that only recruiters can see your status, though I'm not sure how much faith I'd put in this if I really didn't want anyone in my current organization to know.

# SOME THOUGHTS ON FINDING YOUR NEXT OPPORTUNITY

I won't spend a lot of time discussing exactly *how* to find your next opportunity, but it is worth mentioning in passing. Your most effective methods are going to be (in order of preference):

- **Personal networking**: This is the hidden job market and is the most effective way to find a new role. Spend the majority of your time on this, not on searching job boards.
- **Referrals**: Similar to networking, references from friends make you something of a "known quantity" and carry a lot more weight than a completely unknown candidate.
- **Targeted recruiters and headhunters**: There are retainer-based searches for senior executives and commission-based recruiters for everything else. Note that a commission-based recruiter will likely want you to take a job so that they get paid, but a retainer-based recruiter gets paid by the hiring company to find the best candidate for the role. Retainer-based firms are great, but they also tend to have a limited number of roles at a given moment compared to a commission-based firm that has experience with cybersecurity. You'll need to register your interest with many retainer firms rather than depending on one or two.
- **Company websites**: If you already have your dream employer in mind, you can try the company's website. Many of these jobs do make their way onto job boards, but if a company can hire you outright without going through a commission-based recruiter, this saves them recruiter fees and is preferred.
- **Cold calling and job boards**: A lot of job searchers spend most of their time here, when it's actually the least effective method. Cybersecurity has a lot more people than it did a few years ago and you will be competing with potentially hundreds of other candidates.

There are other methods of course, but these are the most common and, in my view, the most effective. Once you have someone interested in talking to you about a role, it's time to prepare for the interview. This is one of the most important steps, as it is what will ultimately lead you to a job offer.

# THE THREE QUESTIONS OF JOB INTERVIEWS

Once you have someone interested in either meeting or speaking with you about a role, there are several steps you should take to make sure you're prepared (Figure 16.1). You need to know who you will be meeting with and their background, understand the company and the position you are targeting. This sounds basic, but a lot of people don't do the prep work and wing it instead.

The first step is to understand the company. What do they do? Read their website, press announcements, earnings announcements (if the company is publicly traded) and analyst reports on the company. I like spending time with the 10-K, which will also outline some cybersecurity elements.

The next step is to look up the LinkedIn profile of the people you are meeting. You might find an interesting connection, e.g., having the same alma mater, a shared social cause or other common ground. (But don't fake it—you can't "pretend" that you belonged to the same fraternity.) Bottom line: do your homework!

A philosophy that has served me well as both a candidate and a hiring manager is the three questions a company needs to answer during the interview process. These questions include:

1. **Can you do the job?** Are you capable and qualified for the role? Are you a security architect applying to a CISO role? Do you have enough skills and knowledge to do this? A lot of people are great engineers and fail at the CISO role because they don't realize the success criteria for the role.
2. **Do you *want* to do the job?** Is this something that interests you and you want to pursue? If you're a CISO applying to an architect role, will you be able to make this transition?
3. **Can we stand working with you?** Are you going to fit our culture? Will you be able to work with our people, your colleagues and possibly a team you'll inherit?

Let's look at each of these in detail.

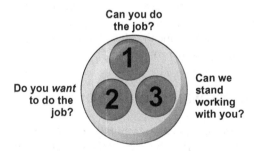

**FIGURE 16.1**  The three questions of job interviews can help you gain an edge over the interview process. These questions not only drive if you will hire or be hired by someone but also how that person performs in the role.

## Can You Do the Job?

The answer to this question comes in two parts. Part one is found in your résumé. Remember, your résumé should be a one- or two-page document that details the most important accomplishments in your career. Resist the urge to focus most of your job search crafting the perfect résumé. Why? Because your résumé doesn't matter that much. Note that I said, "that much." It matters, but its main purpose is to get you in for an interview. People will spend minutes and maybe only seconds reading your résumé before making a snap decision on bringing you in for further interviews or not. In some cases, it is a machine making this decision. Also, if you have been recommended by someone, most people will barely bother reading the résumé since a recommendation goes a long way.

That doesn't mean you shouldn't spend time making sure your résumé is accurate, grammatically correct and representative of your skills. It needs to be all of those things. And you need to have a solid document, just don't spend *all* your time on it.

## Do You *Want* to Do the Job?

As strange as it may sound, some candidates get part-way into the interview process and realize that this job isn't going to be the right fit. Maybe the budget that you thought would be there isn't there. Maybe the job is a CISO job that has no team, funding or support from the business. Maybe the company culture just isn't going to work for you. This is a two-way process as well. If I were interviewing a CISO-level candidate that was applying for a more junior job reporting to me, I might be skeptical that this is really going to be a good fit since you'll really want to be in my role, not the one you're applying for.

In either case, the best thing to do in scenarios where it's not going to be a fit on one side or the other is to get through the first set of interviews the best you can. If you're not going to be a fit on the hiring side, you likely won't be invited back. If you, as a candidate, don't like the fit it's time to bow out at this point. Cut losses quickly and move on to the next opportunity. Whatever you do, don't keep pursuing an opportunity you don't want just to keep busy with interviews or to play it all the way through. This is a waste of everyone's time if you have no intention of accepting the role. Remember, you interview to get a job offer and that's the only reason. If you don't want that offer because it's not the right role, it's time to pull the plug even if the hiring manager or HR wants to bring you back for another round.

## Can We Stand Working with You?

The final question is if the hiring manager thinks they can stand working with you or if you'll fit the company culture. By now, the hiring company thinks you can do the job. After all, they brought you in for an interview, didn't they? From here, your résumé takes a distant back seat, and everything becomes a personality and culture fit. To a large extent, this is driven by "gut feel" from the hiring manager and key interviewers.

Soft skills are critical here, but even superb soft skills don't guarantee you're going to be a great culture fit. Coming from an extensive Wall Street career and applying to a software start-up staffed with mostly 20-somethings? I'm not saying it won't work but proceed with caution on both sides of the interview table.

Guess what? These three questions will be used to continuously evaluate you even after you get the job. If you fail technically at the role, you'll be removed. If you don't want to do the job, it will come out in your performance and you will either move on or be pushed out of the organization. And even if you are capable and you do want to do the job, if you don't integrate into the culture you're not going to succeed.

Keep in mind that the more senior you get in your career, it is possible that the hiring manager may know very little about cybersecurity. That's why they're hiring you, after all. The style of questions that will come from a hiring manager at this level may be much more geared towards behavioral questions, such as asking you to walk through a challenge you had at work and how you overcame it. A lot of situational interview questions sound like:

1) Tell me about a time when ...
2) Tell me about a challenge you overcame at work.
3) Tell me about a time you had to choose something else over doing a good job.
4) Describe a situation where you saw a problem and took steps to fix it.
5) Describe an incident or problem that didn't go your way.
6) Describe your biggest work failure and how you handled it.
7) Tell me about your biggest career accomplishment.
8) Describe a situation where your boss was completely wrong. How did you handle this?

These situational questions are either great opportunities to drive the conversation towards your strengths or they are catastrophic landmines where a candidate can literally talk themselves out of the job. You need a method of organizing these ambiguous, open-ended questions in a way that won't undermine your chances of success. This is where the STAR framework really shines.

# MANAGING SITUATIONAL INTERVIEWS WITH THE STAR FRAMEWORK

Once you are senior enough in your career, you won't get a lot of the "tech-out" questions where the interviewer is trying to verify your technical skills. Instead, you will likely be presented with behavioral questions or situational questions like the ones listed in the previous section.

A great framework to answer behavioral questions like this is the STAR framework. The STAR framework helps you tell a brief story and answer behavioral or situational questions. It also provides you with a platform to walk through some of your prior experiences in a way that highlights both hard and soft skills (Figure 16.2).

**FIGURE 16.2** Leadership interviews will not go deep into technical questions, but favor "situational" questions to assess your leadership abilities and soft skills. The STAR framework can give you an edge in these types of interviews.

Before we get into the nuts and bolts of the STAR framework, you'll want to think about your career and some stories you can draw on that can help illustrate who you are and what makes you who you are. If you recall how to journal from Part 1 of this book, this can be a wonderful tool to help you get your stories on paper and put them into a logical format.

Why do you need this step? Because you have no idea what questions will be asked of you in an interview, so you'll need to be able to draw on several types of stories and different experiences from your career. In other words, when you're thinking of your personal stories, think about your successes and failures. You'll probably need examples of both. Think of the difficult people you've worked with and how you made the relationship work. Think about your strongest highlights and make sure the sequencing and detail is clear in your head. While you'll also want some difficulties to draw on, you don't have to pick the worst parts of your career that don't have a happy ending.

Once you have an idea about what highlights you'd like to draw on, we can start putting them into the STAR framework. STAR is an acronym that stands for situation, task, action and result. Let's look at each element.

**Situation**: Describe the scene or situation, providing enough detail and background. This will serve as the context of your story. You want to describe a challenge or situation you were faced with and why it was important. It could be a project, an incident or an event. You want to provide just enough background, don't get bogged down in details. Just include important, relevant detail.

**Task**: Describe what your role or responsibility was in the situation. Were you the leader? Was this your project or someone else's? Who else was involved that is relevant?

**Action**: Explain exactly what you did and how you showed leadership or drove a result. If it was a problem, what did you do to resolve the issue? If it was a team effort, make sure you are focusing on your part and what your specific actions were.

**Result**: Share specific outcomes that you achieved. Are there any metrics or data that support why this was a success? Did you learn something you can apply to future situations?

The STAR framework is relatively straightforward and resembles the corporate storytelling framework from Part 1. Now, let's look at an example of how to put STAR to work.

Situation: You have been asked to explain about a time when a previous effort or program didn't go as expected.

Strategy: Use the STAR framework to provide some structure around an answer that shows how a project didn't go according to plan and how your efforts and leadership helped fixed the situation.

# Tactics

As I mentioned, it's best to have these examples thought through way in advance of an interview. You can draw on an example that shows strong leadership and other qualities that might be required for the job. Remember, at senior levels you don't have to be the only hero of the story. You'll want to demonstrate teamwork, leadership, crisis management and other qualities that a hiring manager would be looking for.

Let's use these strategies and tactics to craft an answer using the STAR framework:

**Situation**: Explain enough background (don't overdo it) that provides context. "I was leading an effort launching a Data Leakage Prevention tool across 50 countries including many in the European Union ... when I suddenly got a heated call from our chief privacy officer (CPO)." Perhaps you could explain how you had not fully understood some privacy requirements of your new system and that somehow the CPO heard about the project and called you up agitated that they had not been consulted.

**Task**: Next, you would explain the challenge or the task in front of you. You could say something along the lines of how your program had specific milestones, but now there were data privacy concerns putting the whole project timeline in jeopardy. This was a board commitment, but you couldn't break EU privacy rules in the name of hitting your targets.

**Activity**: Using this framework, you could detail what you did to fix or recover the situation. I would strive to use examples that had clean resolutions. In this case, you can hopefully detail how you engaged, HR, Legal, Privacy and senior management to attempt to get things back on track. Perhaps a few concessions were made along the way, highlighting your negotiation skills.

**Result**: Here is where you detail your results. Using this simple example, an ideal result would be how you got the project back on track, addressed privacy concerns and learned to be much more sensitive to privacy issues in the future.

There are a few important things to note using the STAR framework. Use specifics and quantify your success where possible. Make the story tangible and personable. Keep your answers concise to convey the maximum achievement in the minimum amount of time. Don't provide too much detail, especially irrelevant details. Some people keep talking well beyond they've made their point and this usually doesn't end well. Aim to keep these answers short, like five minutes or less. Finally, finish on a positive note so that the overall impression is strong. If there was a lesson to be learned, say what it was.

# FOCUS ON YOUR ACHIEVEMENTS

A lot of candidates will put their focus on every detail of what they did in previous jobs. There are a lot of big egos in security and it probably makes a lot of sense to most people to go through all the great things you've done in previous roles. I managed a big team. I had a large budget. I ran both operations and engineering. There's only one problem with that. The hiring manager only really cares about what you are going to do for *their* company in the future. As they say in the investment world, past performance is no guarantee of future results. So what should you do instead?

Rather than providing every detail of everything you did prior, speak to your accomplishments. Prepare and rehearse specific examples and tell a brief story that shows how you've made a difference to your company and the customers it serves. Of all the qualities I'm looking for in a candidate, motivation tops the list. I want to know what gets people excited about their work, culture fit and authenticity.

## Culture Fit

The more senior the position being filled, the more the interviewer will emphasize culture fit. Technical skills at a senior level are assumed. The rest is about fitting in and being able to lead and motivate others. Culture fit matters at more junior levels as well. Studies show that nearly half (46%) of new hires fail within the first 18 months, largely because of cultural incompatibilities. Understand what suits you best, from your ideal work environment to the type of boss you want to work for and learn from. The more you know about yourself, the better you can demonstrate culture fit to your prospective employer.

# AFTER THE INTERVIEW: THE LONG SILENCE

A lot of your interviews will lead to black holes. Here are some tips to increase your chances of not falling through the cracks without overcommunicating.

## Write a Thank You Note or Thank You Email

A lot of people think a thank you note should be a personalized, handwritten note sent through the physical mail. My problem with that method is timeliness. An email can be sent and received in hours (or less), whereas a physical letter will take more like 2–3 business days at a minimum. Which one you choose is your call, but when you do write a follow-up note (which you should) be sure to:

- Keep it brief.
- Restate your interest in the job and any relevant details on why you're qualified. This is a good time for your elevator pitch.
- Thank them for their time. Both sides of the interview process are kind of a pain. Acknowledge that they took time out of their day to meet with you and thank them.
- The thank you note is also an excellent opportunity to add any significant information you may have forgotten to say in the interview.
- Stay enthusiastic about the role.

Here is an example thank you note that you can tailor for your own purposes.

Dear [interviewer]

Thank you again for meeting with me yesterday to discuss CISO role at [Company]. I enjoyed our conversation and am very excited about the possibility of joining the team and building a world-class security program.

I would like to emphasize my experience presenting complex concepts in simple terms to diverse audiences including boards, the audit committee and business stakeholders. I also I appreciated hearing about your company's journey to the cloud and some of the new regulatory and privacy challenges that come with that. I believe there's a great deal I can offer to help solve these problems and I look forward to continuing the conversation.

Thank you again for your time and consideration. If there is anything I can do to assist with the decision-making process, please do not hesitate to contact me.

Regards,

[your signature]

## Check-In

After you send the initial thank you note, you can check in after a week or so. Despite the risk of seeming annoying, following up is a reasonable thing to do and it will remind the company of your interest. Also, when you do follow up, make sure that you keep the communication positive and assume you are still in the running. I have been on both sides of the hiring equation and you really don't have any idea what's going on or why they haven't jumped on you as the perfect candidate. Just accept it.

## Keep in Touch

If you don't get the job, it might still be worthwhile to keep the employer in your contacts. You'd be surprised even at the top level how many CISO jobs either fall apart during the deal phase or of course within that first 18–24 months. Sometimes you might have been a runner-up candidate but lost out on a minor detail that made the other candidate more attractive. Never take it personally and don't burn any bridges just because you didn't get the role. You want them to think of you first if anything changes or even down the road.

# NEGOTIATING JOB OFFERS

Building on what you already learned with negotiating skills, when it comes time to review a job offer, you are going to want to strive for a win–win outcome. This means that both sides will hopefully come out of this feeling positive. It is definitely possible to blow an offer in this phase by being overly demanding or difficult, so you had better start this process knowing what a successful outcome looks like to you. Presumably, if you have gone this far in the process you want the offer to be successful. There have been instances where candidates change their mind at the last minute and decide to take "moonshots" with salary demands and other perks, not really expecting to get them but maybe willing to change their minds if they did. It's a win–lose strategy that they're OK letting become a lose–lose strategy.

Assuming, you are trying for a win–win strategy, you'll need to first isolate your non-negotiables and your negotiables. In a job offer situation, it's easy to focus on salary only, but this is not the only thing you can or should be negotiating. Figure out what's important to you. Maybe it is more money, but it could also include perks like:

- Additional vacation time/paid time off
- Your job title
- Working from home
- Tuition reimbursement
- A flexible work schedule
- Professional dues
- Paid for conference attendance
- A sign-on bonus
- A guaranteed first-year bonus

What you don't want to do is shoot for all of these plus a huge bump in pay. You'll be seen as difficult, and I have definitely seen situations where the offer is rescinded because the candidate was too demanding. Strive to strike a balance and make sure you are using your active listening skills. If a company says that corporate policy won't allow something, it's probably useless trying to pursue it further. Instead, focus on a different item on the list or maybe it's just time to accept the offer as it stands.

# SUMMARY

Finding a new job can be stressful. All of your communication skills will be put to the test. Understanding how to write a good résumé and respond to situational interview questions can help you stand out from other candidates.

- Finding a new job is a fact of life with cybersecurity professionals. Average tenure of senior roles is not long, and you will likely find yourself looking for a new role a few times during your career.
- Three common questions an interviewer is trying to answer during an interview are if you are qualified to do the job, if you want the job and if you fit the company culture. Your résumé is important for the first question, but the interview is critical for the last question.
- Expect situational questions instead of technical questions as you get higher up in your career. The STAR framework offers an easy way to frame answers by defining a situation, task, action and result to talk about previous experience and why you are the best candidate for the job.
- Résumés should be kept short (1–3 pages at a maximum) and should focus on achievements instead of responsibilities. They should factor in that online parsers might be the first reviewer, not an actual person.
- Tailor your résumé to the specific job by including details from the posting or from the company's needs.
- You should establish and maintain a personal brand using tools like LinkedIn.
- Searching for a job, whether you currently have one or not, is a stressful process that takes a lot of time.

# REFERENCES AND RECOMMENDED READING

Bolles, Richard Nelson. *What Color Is Your Parachute?: A Practical Manual for Job-Hunters and Career-Changers.* Ten Speed Press, 2020.

Cuddy, Amy Joy Casselberry. *Presence: Bringing Your Boldest Self to Your Biggest Challenges.* Little, Brown Spark, 2018.

"Mind Tools: Online Management, Leadership and Career Training." *Management Training and Leadership Training—Online*, www.mindtools.com/.

Norman, Jeff. *LinkedIn: Tell Your Story, Land the Job.* Tycho Press, 2015.

"Outplacement & Career Transitioning Services." *Challenger, Gray & Christmas, Inc.*, 6 Jan. 2021, www.challengergray.com/.

Storey, James. *The Art of the Interview: The Perfect Answers to Every Interview Question.* CreateSpace Independent Publishing Platform, 2016.

Tom Maddison—Former Fortune 500 Company CHRO. "RiseSmart Outplacement Services." *RiseSmart*, www.randstadrisesmart.com/.

# Consultants and Sales

# 17

## *Building and Maintaining Client Relationships*

> *Pretend that every single person you meet has a sign around his or her neck that says, "Make me feel important." Not only will you succeed in sales, you will succeed in life.*
> ~ Mary Kay Ash

I promised that this book wasn't only for CISOs but for senior security leaders of all kinds. This chapter covers communication in the product sales and consulting professions. Even if you think you are never going to be a security consultant or product sales professional, I recommend you read this section anyway. In a way, we are all consulting and selling information security. We join a company to understand their problems and offer advice and solutions on how to fix them. We offer a service to our companies in the form of creating and implementing security programs that meet business objectives and integrate with the rest of the company. Some of the strongest security programs I've seen treat cyber as a business. I also know many former CISOs who pivoted into consulting or product sales in the later stages of their career. They choose a competitive salary and none of the operational responsibilities that come with running a program. Don't count these areas out as a potential turn in your career at some point. I sit on several product advisory boards and wouldn't rule this out myself in the future. So read on, hopefully you'll learn something or at least gain a different perspective.

This chapter discusses the sales process consulting and the product side of the business. Whether you are selling products or services, the approach and communication challenges are very similar. But first, it helps to understand the sales process in general.

DOI: 10.1201/9781003100294-19

# THE SALES PROCESS

*It is not your customer's job to remember you. It is your*
*obligation and responsibility to make sure they don't*
*have the chance to forget you.*
~ Patricia Fripp

Consider this for a moment. Don't you have to "sell" your program to a variety of stake-holders? Don't you have to "sell" yourself as the right person to be leading the program? Don't you have to "sell" security controls to people who may not be receptive to them? At a minimum, maybe this section will help you recognize when you're working with someone on the sales side that actually "gets it."

And if this still doesn't resonate with you, think of it this way. We are all providing security services. We serve companies and they pay us for those services. Maybe we shouldn't call it sales, but use the word service instead. As Sir Winston Churchill said: "We make a living by what we get, we make a life by what we give." When you start to think of sales as serving others and not as making money, you are heading in the right direction.

To provide a little context about my background on these topics, I have spent a lot of time understanding businesses in my career. When I go into a company, I like to really learn how the business operates and what makes it run. I've probably spent as much time reading *BusinessWeek* and the *Wall Street Journal* as I have reading about the latest issues in cyber. I've advised investment banks in the product space and worked with the Cyber Investing Summit in New York City for several years. I've also served on several product advisory boards. There are many product companies out there who think they have amazing products that "sell themselves," but the sales don't come because the businesses don't have the problem that the product addresses. As much as business and cyber leaders don't always understand the sales side, the sales side doesn't seem to understand the business side.

If you are in a sales or sales support role, you will be faced with some unique communication challenges. Vendors pushing for a fast sale cause tension with security people on the other side of that sale. Even when it is a product that we want to buy, the whole process sometimes just feels "icky" for almost everyone involved.

Consider this as an important principle for this whole chapter. No one likes being "sold" to, but people like being buyers. There's a big difference. The word "sales" conjures up some horrible images of people who are more interested in commission than people or solving problems. But good salespeople are problem solvers, relationship builders, strong communicators and great listeners.

Let's deconstruct the sales process by understanding a standard simplified model of how it works. While there are several sales models to pick from, I think a good way to understand this as it relates to security is by the following seven-step model (see Figure 17.1):

1) Prospecting
2) Warm up

3) Qualifying
4) Presenting
5) Overcoming objections
6) Closing
7) Follow up

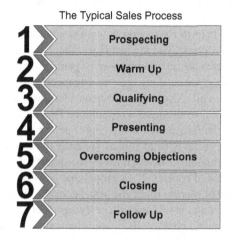

The Typical Sales Process

1. Prospecting
2. Warm Up
3. Qualifying
4. Presenting
5. Overcoming Objections
6. Closing
7. Follow Up

**FIGURE 17.1** The sales process typically comprises seven steps. Sales professionals that rely too much on the hard sale and closing rarely do well in the role.

Let's discuss each step to give you a flavor for how the sales process works in practice.

In a sales context, prospecting is simply finding someone who might be interested in purchasing your product or at least *likely* to be interested. Most products have some sort of key audience and should be marketed to the most likely industry, department and function lead that might purchase your product. Trying to market an HR system to an IT manager might not be the right venue. Trying to market a fixed-income management product to a hospital is definitely the wrong industry. Your product fits a demographic that includes certain industries (finance, manufacturing, etc.) with companies of certain sizes. For example, governance, risk and compliance (GRC) products will not be a likely purchase for a company with only 40 employees. On the other hand, if you are a small startup company, you probably won't be effective by starting off trying to sell to JP Morgan or Citigroup, as they are likely to choose more established vendors. Language barriers, support requirements and procurement obstacles will make you wish you had never started down this path. Start small and work up to that big fish.

Since all salespeople are generally under at least some pressure to make sales and reach quotas, they should not waste time with unlikely prospects. You should also be able to answer simple questions such as:

• What makes your product or service unique?
• What problem do you solve and how big is that problem?
• Who has this problem that might be looking for a solution?

- How strong is their need?
- Do they have the budget?
- Do they have the authority?
- How likely are they to seek your help?
- Why would they choose you or your product instead of someone else's?
- Is this urgent enough that they would act? Is this the right timing?

These questions should help in narrowing down a small set of likely prospects. If you haven't gone through this phase, you are just marketing randomly to anyone and everyone who will listen. This is not an efficient way of generating leads and will waste a lot of people's time, including yours, along the way. Prospects with no need and no urgency are not really prospects at all. Also, before you consider pushing unnecessary products, consider that this will weaken the seller's reputation. This means that prospects will be less likely to engage in the future even if offerings become more relevant.

Once a likely prospect has been identified, we enter the warm-up phase. This is also sometimes called the initiating contact phase or the qualifying phase. In the warm-up phase, you are making initial contact. The channel that you choose can be a deal breaker for many. There are a lot of ways to make the initial contact. This could include cold calling, sending emails, reaching out on LinkedIn or trying to get an in-person or video meeting. A general rule is to avoid phone calls. Studies have shown that 80% go straight into voicemail and only about 2% of these calls result in a next step. I worked at a company where they removed employee voicemail altogether. While this caused a bit of concern at first, most people realized that it was only vendor calls most of the time anyway.

Social media may seem like a good way to reach out, which for professionals usually means LinkedIn. But not everyone manages LinkedIn closely and some people have abandoned the platform altogether even though their account remains active. Given the choices, email tends to be one of your best options. Or maybe better yet, use a combination of options. Just keep in mind that this is an introductory phase. Don't jump right in to trying to make a sale.

In the qualifying phase of the sales process, you are attempting to learn more about the prospect including goals, challenges, budget and other factors that will increase your chances of success in making a sale. Most of my friends in sales suggest that only 50% or so will actually continue to be good prospects, so don't be discouraged if your numbers are similar.

The presenting phase is where you will prepare yourself to answer key questions about your unique offers and benefits, as well as what problems you can solve for them. This is sort of a nurturing phase. Nurturing leads involves educating them about your product or service and personalizing your communication to address common challenges. In this phase, you want to be very helpful and responsive to questions by either answering them directly or working quickly to get answers and follow up. You'll find that some things that seem like good fits in this phase may simply not have the right timing or a strong enough interest to continue.

Stage five, overcoming objections, is not always necessary if there's strong need, existing budget and the timing is right. Otherwise, you'll need to address any obstacles to the sale in this phase. These might include common objections like price, contract terms, deployment concerns, product roadmap or any other items that could be a roadblock to making a sale. In this phase, you'll want to use patience and empathy with their concerns. Added pressure or pushy sales tactics are not going to help the sale or your reputation. I've had instances where I've requested a different sales representative because we actually had interest in the product, but the rep's soft skills were so bad and so pushy that I wouldn't continue the conversation. Make sure that any objections in this phase have been fully addressed. Ask clarifying questions if you haven't managed to hit the mark.

Stage six, closing the sale, is where everyone is finally in agreement on solution, price, product, service and what is being offered. You might find yourself handed over to a procurement department from here, but you shouldn't lose contact with your prospect. If there's a little hesitancy in this phase, light incentives like price discounts, free trial periods or other perks can be offered. Just be careful not to stray into overly aggressive tactics.

The final phase, follow-up, ensures that your prospect is happy. For your very happy prospects, you might ask if they'd be willing to be a reference client. Follow-up is an important step, so don't skip it. Follow-up helps ensure that your prospect doesn't feel like you ran the minute the check was cut and don't care about if they're satisfied with the end result or not.

This seven-step process has hopefully given you a little insight on how the sales process works and what can go wrong along the way. Now, let's go a little deeper into both the product and services side of sales.

## But Cybersecurity Sells Itself, Right?

Everyone gets that security is a big problem now, though this was not always the case. However, if you are in sales you are going to have to come to terms with the fact that a lot of companies consider cybersecurity a cost center, not a strategic investment. This means that there might be interest, but not budget for your product. There might be budget but too much skepticism to move forward.

Here are some actual points I ask myself before I'll consider buying a product:

1) What problem does this solve?
2) How much do I care about this problem?
3) What new problems will this product introduce? This one is important, as every product is likely to introduce new challenges like alerts that need to be chased.
4) How much effort will it be to implement, manage and maintain this product?
5) Am I going to need more staff to run this thing?
6) Am I ready for this product *now*? Is this really a current priority?
7) If I choose to move ahead with this product right now, what opportunity cost am I missing on the security roadmap? In other words, what am I *not* going to be working on because I'm working on this? I want to take a risk-based approach, so does this really address a big business risk?
8) Is this the most appropriate product for my needs? It might be too much or too little in terms of features and benefits. For example, with a small staff and budget, maybe the built-in but already paid-for version built into Microsoft or Cisco is enough for my needs. You don't want overkill or underkill.

These are all important questions. If a product doesn't solve a real problem for me, why would I buy it? If it does solve a problem, is this a big problem in relation to the product cost? If a solution is not "practical and sustainable" then it can quickly become a management burden. Am I going to have to hire new people just to manage your alerts? Sometimes, we're just not ready for a product either. If I don't have vulnerability management under control, I'm not going to be that interested in focusing on GRC since it's at a different level of the maturity scale. So I might be interested but not right now because it's not the right time on the strategic roadmap. I want the most appropriate product for my team, their skills, our budget and the company's risk profile.

There are a lot of considerations before I'm willing to pull the trigger on a new product. I've worked in companies that have purchased products they never bothered to deploy because the problem wasn't really that big or the complexity of deployment outweighed the benefits. This doesn't benefit anyone in the long run. Even if a sale was landed, there's an overall bad view left behind of the company that made the sale.

### Common Sales Mistakes

Salespeople get a bad rap because a few of them are really doing a very bad job and spoiling it for everyone else. Following are some common problems I see watching how sales operate in the field when they are marketing to security leadership.

## Assuming the CISO Is Your Only Prospect

I see way too many sales reps go straight to the CISO trying to market something. At first, this seems fine. The CISO is likely the final decision maker, knows if there is budget, and has some view of where things fit on the strategic roadmap. The problem is that about 50,000 other vendors also know this and are all trying the same route.

Might there be a champion a little lower in the organization? Why not try to aim a little more directly where the problem lives in the organization. Selling a tool to help with incident response? Why not start with the head of incident response instead of always going straight to the CISO. I can tell you that when my team comes to me explaining that they need a tool to be more effective, it means a lot more than any one of thousands of vendors telling me that I need their solution.

## Making Assumptions That We Already Know Everything About Your Product

This one is tricky, because sometimes the salesperson doesn't seem to know very much about the product either. This isn't really forgivable, but it's a fact. So, before you start explaining your solution, make sure how much the person on the receiving end knows about your product, or in some cases even cybersecurity in general. Don't make assumptions about the customer's knowledge. Don't talk down and be conscious of coming off as overly arrogant. Also don't waste slides telling us how important security is and how scary it is out on the big, bad Internet. We know that already. Are these stock slides in your marketing deck? Great, skip them for the experienced prospects.

## Being Insulted by the Word Vendor

There are certain things that will be very difficult to change after a certain point. There was a time when cybersecurity was known as information security. Then a shift in the industry came about and it became known as cybersecurity. This was being communicated all the way up to the board of directors. To me, the term "cybersecurity" implies a lack of concern for physical paper, datacenter physical security and many other elements of a good security program. But after a while you are fighting against the tide, so I use the term cybersecurity too. But I still make sure to cover the important elements of an information security program like physical paper.

I mention this story only because there are a lot of companies referred to as "vendors" that would rather be "business partners" or "partners" or "trusted advisors" or some other term, any other term other than vendors. Don't fight the tide. A customer is a customer and don't try to correct them or change a term that the entire company might use just to feed your ego. You're selling and they're buying. And that's OK. Names are labels, don't worry about it. And certainly, don't make a big deal about it.

## Selling Fear

Don't come in to established security professionals with more than a year or two of experience and try to sell fear, uncertainty and doubt (FUD). We all get that security is important and that's why we have a job. Pushing this tactic too aggressively makes you lose instant credibility. Also don't think that you understand the risks to our businesses as well as we do. If you think something is a risk in a specific industry, don't waste your time disagreeing if your prospect doesn't agree with you. You probably don't know better than they do.

## Selling Miracle Solutions

Another misguided tactic is coming in after someone's big breach saying, "our product would have prevented (fill in the threat)" and having no evidence whatsoever to demonstrate how this is true. This kind of sales pitch is almost as bad as the FUD sales tactic. You think it's true that it would have stopped an attack? Well, you better be coming in with a whole lot more detail than you probably have. My bet is that you're just assuming that it would have stopped whatever threat and you have no clue if it's true or not. Even if you're sure it's true, using this method makes you come off as arrogant, which is never a good sales tactic.

Cybersecurity is full of mediocre products that barely work or that add so little value as to be barely worth purchasing. These seem to be the same products that claim to be the silver bullet cure to all your security problems. Be realistic about what your product actually does and maybe even have a few "real world" success stories instead of the works of fiction.

## Don't Make Your Sales Targets Their Problem

Having worked in some very large corporations and a state government in my career, they all had one thing in common: they did not move quickly. Not only that but the snail's pace seemed to turn into a full stop whenever something went into the procurement department. Typical salespeople look to close sales at the end of monthly, quarterly or yearly cycles. On the buyer side, this is not my problem. CISOs typically have zero influence on how fast procurement will or will not move something through the cycle. Just be aware that there is a procurement cycle and that it will take time to get through. Smaller companies, with correspondently smaller sales dollars, may be nimbler at the purchasing process. Don't walk into a large, global bank on June 15 and expect to seal the deal on June 30 just because you're in a hurry to make a sales target. You would be dramatically underestimating the procurement process and how long it can take.

## Do More Listening than Talking

I've had a few vendors ask me to get on a call not to buy a product but to offer advice. If this part of the request was genuine, these were truly not sales calls. But they also talked all over each other and they talked over me. For every bit of information I offered, they defended that I must be wrong and their product design was right. If you're going to ask for time, make sure you're doing like 80% listening and 20% talking. Otherwise, don't hold the call in the first place.

But this is true all the time with sales. You're not going to know the right things to say if you haven't listened enough to define the problem. Active listening, once again, is a top communication skill.

# BETTER SALES STRATEGIES

I'm conscious that we covered a lot of what goes wrong in sales cycles. So if you work in sales and want to be more effective, what's the best tactic? I recommend this simple, four-step process.

1) Don't sell products; solve problems and provide service.
2) Have your value proposition down cold.
3) Keep the message concise and consistent.
4) Know when it's time to move on.

My first bit of advice is that while you are ultimately there to sell products and hit sales targets, your higher calling should be to solve the customer's problems first. Make sure that the prospect you are working with actually has the problem that your tool solves and that they think it's enough of a priority to investigate deeper. Again, this means doing more listening than talking. You may think your tool is amazing, but if the customer doesn't have the particular problem that it solves you are best moving on to someone who does.

You also need to have your value proposition down cold. Why does your company exist? Why should the customer choose your product over the competition? Why is your product, better, faster, cheaper, whatever than everything else that's out there? Be clear who your competitors are and why your product is the right choice. To be clear, I sincerely hope that management, not front-line salespeople, have determined a lot of this in advance and communicated it down the ranks.

Keeping the message concise and consistent across all channels is critical. One thing I hope you have learned so far from this book is that communication seems easy because everyone does it. But that doesn't mean that everyone is good at it. It's also worth noting that if you are a salesperson and can't figure out your company's value proposition and neither can anyone else then it's probably time to find a new company. You will fail at your sales targets and honestly there are better things you can be doing with your time than being part of the "bad security product" space.

Finally, know when to move on to more promising prospects. There's no point in trying to force a sale when it's not happening. The last thing you want to do is annoy a potential customer or worse yet gain a bad reputation as a pushy salesperson.

Finally, once you are actively engaging in the sales process, you should consider the following tips.

1) The security team and CISO are extremely busy. We are all understaffed and have multiple competing priorities. Mind how much time you are asking for and be on point to use that time wisely. Start on time, end on time and don't ask for 90 minutes if you could really cover it all in 30. And don't expect to take 40 minutes if you asked for 30.
2) Know your customer. The same way you should read up on background for an interview, read up on the customer and company ahead of time. In other words, do your homework. Dealing with a security newbie? It happens, so tailor your pitch to be a little more basic. Long-time multi-decade CISO? Cut the fluff and get to the point. As a state CISO in the public sector, I wouldn't expect to see the same presentation you might have brought me when I was working at Citigroup. Your comprehensive General Data Protection Regulation (GDPR) coverage probably doesn't have a lot of relevance for a state government. Know the industry, know your customer, tailor your pitch. Again, do your homework.
3) Avoid the "let's hop on a call" approach. I know that I certainly avoid it when someone tries this with me. I don't have time to hop on a call for every random product that wants to get in front of me. This is not a sales strategy; it's just a nuisance. How about trying to convince me that it would be worth my time and why I'd want to know more.

4) Don't bother with the green-field pitch. Unless a company is literally brand new, they probably already have an SIEM. And authentication. And compliance and a bunch of other tools. It's great that you have a portfolio of products covering the entire enterprise. But we also have an existing mess to deal with. You're going to have to integrate into this, not expect me to scrap our existing investments and start over with your comprehensive product suite.

5) Don't name drop all the big clients that you're "working with." This is a very small community, and we all talk to each other. If you really got the product deployed at a big client, you better be ready that I might pick up the phone and confirm that it's going well. And if you haven't made the sale yet? Don't even bring it up. Just because you got in the door at General Electric doesn't mean they're serious about you and a non-buyer has zero clout in making a new sale. Focus instead on peer companies or industries of similar size and scale where you *have* deployed and where it's going well.

6) Never talk about low-skilled security staff and how your product solves for that problem. You have no idea what kind of team I'm running or what kind of talent I have. In either case, my job is to protect and advocate for my team. Don't expect me to throw them under the bus for the sake of a product sales pitch.

7) Listen from the start to the finish. Can you re-phrase what problem I'm trying to solve in your own words? Do you know why that problem is important to me or what other challenges I'm having in implementing my security program? It's important to capture these details, as they likely all came out at some point in the conversation. The better you understand my problems, the easier it will be for you to take at least one of those problems off the table with your solution. And again, if you don't solve a problem that I actually have then why are we talking?

8) Avoid relentless follow-ups. Avoid the daily follow-up calls and emails. Every security person is juggling multiple projects and tasks. There's a good chance that if you haven't heard from us in a while it's because there's an incident, competing priority or other issues that need to be addressed first. It's OK to follow up but keep track of how aggressive you're being and maybe even suggest that we get back to you with a more suitable time if we indicate that the calendars and priorities are in play.

# SELLING SECURITY SERVICES

Being a security consultant usually involves assessing software, computer systems and networks for vulnerabilities. Then comes designing and implementing the best security solutions for an organization's needs. Cybersecurity consulting can be a very lucrative business, but strong communication is critical.

In a large firm, it is common for most of a consultant's time to go towards generating new business. This is a shame because new projects don't have to always come from new clients. Time spent looking for new sales might be better spent strengthening existing relationships, getting repeat business or referrals to other clients. Building these relationships takes strong communication skills. In fact, the best consultants are also some of the best communicators in the business.

No matter if you are an independent consultant or working at one of the big security consulting firms, there are some communication skills you'll want to perfect. These include keeping a positive mindset, exceptional communication skills, problem solving, a strong understanding of your client and exceeding expectations.

# Keep a Positive Mindset

Is keeping a positive mindset a communication skill? You bet it is. If you fail to keep positive, this will manifest itself in your communication in many ways. Everything from your body language to your "general vibe" will echo what you are thinking. Consulting can be stressful and it's important to show a positive face to your clients no matter how you are personally feeling. I'm reminded of the Maya Angelou quote: "I've learned that people will forget what you said, people will forget what you did, but people will never forget how you made them feel." Make sure your clients feel good after they interact with you. Don't complain or be negative and try to keep a positive demeanor.

# Exceptional Communication

Timely, efficient and clear communication is your priority. Consulting requires very clear communication skills, projecting an image of expertise and building client relationships. You'll want to be timely with communication because you need to stay top-of-mind with your clients, but balance that by not being a pest or over-communicating. You will build strong relationships by making your clients feel comfortable being open and honest with you. They should feel that their ideas and concerns will be taken seriously and that you will bring expert solutions to help solve these problems.

# Know Your Audience

All the general rules of strong communication apply with consulting, but especially understanding your audience. This means not only knowing your primary company contact on a personal level, but you'll also want a view of your client's industry, it's problems, compliance requirements and challenges. If you are typically dealing with one point-of-contact over and over again, you'll want to know as much as you can about them and remember details like their interests, the names of their kids and any other details that can help you connect on a personal level. Remembering these details can really help you connect, and ultimately communicate much better.

## Don't Make Sales, Solve Problems

Make sure you don't just see your clients as a paycheck, but as people with problems. You are here to serve them, not to make sales. Adopt the service attitude and the sales will take care of themselves. While ultimately your company may hold you accountable for making a certain number of sales, don't make this your only focus. Understand your client's problems and be ready to show your value by solving problems, not by making sales. The sales will follow.

This is a tough concept for some people to get. It can be stressful being measured to sales targets. It's a specific number that you either hit or you didn't hit. If you didn't hit your numbers, you may already be on the path to the exit door. Very few sales professionals miss targets more than twice in a row. It's easy to see only this side of the situation and therefore spend most of your time looking for new business.

While new business can, indeed, lead to new sales, being a problem solver will lead you to repeat business, word-of-mouth recommendations and referrals much faster. In other words, sales will find you, not the other way around. Build a reputation as a problem solver, and in an industry crowded with people just pushing for sales, you will stand way out from the crowd.

## Exceed Expectations

The consulting industry is crowded with lots of people and companies trying to do exactly what you're trying to do. You'll need to exceed expectations and do so consistently. Develop a reputation for excellence. Under-promise and over-deliver. Don't oversell yourself or promise unrealistic results or timeframes, but make sure you're consistently doing better than expected. Always look for where you can go above and beyond in delivering value. This is a small industry and people will talk about you if you're good or bad. So be excellent and let them talk!

---

# SUMMARY

---

I think a lot of sales roles are made worse by people focusing on sales targets and dollar transactions rather than solving problems and serving others. When you reframe sales as solving problems and serving people, sales are the natural result.

- Sales roles are difficult and require exceptional communication skills. Sales require reading nonverbal cues, building relationships and solving problems. But there is a natural friction when there is an over-emphasis on completing a sale.
- The same rules of communication apply for sales, but especially knowing your audience. Don't pitch the importance of security to established CISOs

and don't focus on technical details with people who may not have that background.

- Solving problems will take you much further than simply trying to land sales.
- The simplified sales model includes prospecting, warming up, qualifying, presenting, overcoming objections, closing and following up on sales.
- Cybersecurity sells itself, except when it doesn't. Make sure that a prospect actually has the problem that your product solves or else move on to someone who does.
- Common sales mistakes make the overall process much more difficult. These include marketing only to the CISO, presuming what level of knowledge a prospect has of your product, selling fear or miracle solutions and not listening to what real problems or objections are being discussed.
- Better sales strategies include solving problems, keeping your value proposition simple and clear and knowing when it's time to move on.

# REFERENCES AND RECOMMENDED READING

Adamson, Brent, et al. *The Challenger Customer: Selling to the Hidden Influencer Who Can Multiply Your Results.* Langara College, 2018.

Bookhabits, Paul Adams. *Summary of the Challenger Sale by Matthew Dixon and Brent Adamson: Conversation Starters.* Blurb, 2018.

Holden, Jim, et al. *The New Power Base Selling: Master the Politics, Create Unexpected Value and Higher Margins, and Outsmart the Competition.* Wiley, 2012.

Konrath, Jill. *Agile Selling: Get up to Speed Quickly in Today's Ever-Changing Sales World.* Port Folio Penguin, 2015.

Schultz, Mike, and John E. Doerr. *Insight Selling: Surprising Research on What Sales Winners Do Differently.* Wiley, 2014.

Signorelli, Brian. *Inbound Selling How to Change the Way You Sell to Match How People Buy.* Wiley, 2018.

Weinberg, Mike. *Sales Management Simplified; the Straight Truth about Getting Exceptional Results from Your Sales Team.* Amacom—American Management Association, 2016.

Ziglar, Zig, and Kevin Harrington. *Secrets of Closing the Sale.* Revell, a Division of Baker Publishing Group, 2019.

# Conclusion and Key Takeaways

I sincerely hope you have enjoyed reading this book. I can say that I have learned quite a few things along the way and have increased my own appreciation for how hard it is to communicate. If I had my own "ah-ha" moment it was how all communication is an abstraction of something more tangible. A thought, idea or event that I must get out of my head and into yours. That's not easy! Now take a complex topic like cybersecurity and you can see why communication is so hard in our industry. We all come from very different backgrounds and have different technical knowledge. Trying to get everyone on the same page is extremely difficult.

Writing this book has increased my own mindfulness about the subject of communication and the importance of soft skills. They are both intertwined, but the good news is that they can both be improved with some targeted and deliberate practice.

They say there's no better way of learning a subject than by trying to teach that subject to someone else. Just taking the time to think about communication and how you can improve it puts you far ahead of the game. I encourage you to share what you've learned with others and with your team. It will help reinforce your own understanding. There's enough room for the whole world to get better at communication, so don't keep information that can help to yourself.

While I already knew that communication skills were important, writing this book really made me think about some of the people I interact with regularly and how I might be picking the best form of communication based on my preferences rather than theirs. We all do it, and that's OK. Sometimes it's late in the evening and I just fire off an email, when I know that a phone call the next day would probably be better for the recipient and their communication style. Set yourself up for success by keeping your messages simple, communicating them through channels that the receiver is comfortable using and then watching the feedback to make sure your message was received and understood.

Remember, learning about communication is a process. You're never "done" learning about this subject and the more you invest in this skill, the further you will go with your career. And just because something is a "soft skill" it is still a skill. It can be learned, practised and taught. You can get better at it with deliberate, focused practice on the right things.

Let's keep the feedback loop going! Do you have any comments or thoughts about the book? What are your big communication challenges? You can reach me on LinkedIn at https://www.linkedin.com/in/jeffreywbrown/ and on Twitter at @packetsniff. Maybe when we are all back in person, I can meet you at a conference and we can discuss it over a coffee or a glass of Malbec.

DOI: 10.1201/9781003100294-102

I wish you all the success in your lives and careers. May all your conversations be meaningful. May you touch lives and connect with many people. And may you finally get that big security project approved and funded because you explained how important it is by putting it in business terms that everyone understood.

# Appendix

- The Expert Communicator's Cheat Sheet
- 10 Common Communication Mistakes
- Sample Security Charter
- Sample Program Status Report
- Sample Risk Acceptance Form
- Incident Response Communications
- Job Search Communications
- Executive Job Search Checklist

---

## THE EXPERT COMMUNICATOR'S CHEAT SHEET

---

**Mindset**: Get your mindset right first by reviewing these fundamentals about communicating.

- Senior security leaders will be faced with a wide range of communication scenarios. They need to master verbal, written and listening skills to be successful.
- Mastering basic communication skills will ensure that the security program is understood at many levels of the organization and will make it easier to demonstrate progress and get funding.
- Communication starts with understanding your audience and what kind of information they care about hearing.
- Listening is as important as communicating. Use active listening skills and stay engaged in two-way communications.
- Speak in business terms, not technical terms. Avoid acronyms and "tech talk." Are you making your message easy to receive?
- Virtual meetings are here to stay. Learn to use them to your advantage by looking good, sounding good and leveraging tools like chat.

**Communicate**: When you are communicating, think about the following elements.

- Did you plan and think about what you are trying to say as the first step? If you are going to say something is it clear in your mind first?
- Did you consider the audience you're communicating to and what they already know? What about what they would *like* to know?

- Are you sharing *insights* or just information?
- Did you consider the most appropriate channel to use to communicate to this audience?
- Did you double check the Seven C's of Communication? Are you being concrete, clear, concise, correct, complete, coherent and courteous?
- Did you make sure that what you're saying is factually correct?
- Did you make sure you didn't select *your* favorite way to communicate, versus the most appropriate way for your audience?
- Are you putting your bottom-line up front (BLUF)?
- If you wrote something, did you read it out loud before you sent it? Did you proofread it for punctuation and other errors?
- Do you have a feedback loop, and can you pivot the conversation based on audience feedback?

# THE TOP TEN COMMON COMMUNICATION MISTAKES

Everyone makes communication mistakes from time to time. However, you'll protect your reputation if you avoid the ten most common errors.

1) **Neglecting your mindset**. Emotional responses like yelling at a colleague or sending angry emails can damage your reputation and make it harder to connect with people again in the future. Use emotional intelligence skills to better connect with people and keep your own emotions in check.

2) **Using the wrong communication channel**. Sending bad news in email or detailing complex directions verbally are both going to be ineffective. Make sure you are choosing the best communication method factoring in your audience preference and the complexity of your message.

3) **Choosing your own preferred communication channel**. Remember, you are communicating to connect with an audience. Using a "one-size-fits-all" approach to communication doesn't take people's different personalities, needs and expectations into account.

4) **Not editing your work**. Spelling and grammar issues make you look careless and sloppy. Check your work for tone, punctuation, spelling and grammar. Don't rely on spell-checkers: they won't pick up words that are used incorrectly, such as "affect/effect." It can be hard to spot errors in your own work. If something is really important, consider having a colleague give it a review before sending it.

5) **Completely avoiding difficult communications**. Just because something is difficult doesn't mean it should be ignored. Performance issues and other negative messages still need to be addressed. Learn to deliver difficult conversations easier by using tools like the situation-behavior-impact technique.

6) **Not preparing for high-stakes communications**. Poorly prepared presentations, reports or emails can frustrate your audience. Prepare and plan your communications carefully.

7) **Not practicing for high-stakes communications**. Practicing is different than preparing. Make sure you practice for important events like public presentations and board of directors' presentations.

8) **Not actively listening**. All communication is two-way. Not paying careful attention will result in lost opportunities to connect on a deeper level and better understand opposing sides of a disagreement.

9) **Assuming you have been understood**. You want to make sure that people have understood what you're trying to communicate. Make sure that people have the chance to ask questions and that they've understood what you're saying. Always look for that feedback loop.

10) **Ignoring nonverbal communication.** Again, most communication is actually nonverbal. Watch your body language and the body language of others to make sure you're communicating effectively.

# SAMPLE SECURITY CHARTER

**Information Security Program Charter**

Approved on: XXXX

Owner: Chief Information Security Officer

## TABLE OF CONTENTS

# 1 SCOPE

## 1.1 Applicability

This Information Security Program Charter ("Charter") applies XXX Company ("THE COMPANY" or the "Company)".

## 1.2 Effective Date

This document shall take effect as of XXX date.

# 2 PURPOSE

## 2.1 Purpose and Objective

The Company Risk Committee has approved this Charter as of [Date], in recognition of the Company's commitment to reasonably ensuring the security, confidentiality, and integrity of sensitive business and customer information.

# 3 OVERVIEW

## 3.1 Information Security Program

THE COMPANY management ("Management") shall develop and implement a written Information Security Program ("Program"). The Program shall be designed to:

- Reasonably ensure the security and confidentiality of sensitive business and customer information;
- Reasonably protect against any anticipated threats or hazards to the security or integrity of such information;
- Reasonably protect against unauthorized access to or use of such information that could result in substantial harm or inconvenience to THE COMPANY or any customer; and
- Reasonably ensure the proper disposal of customer digital information.

This Program shall be applicable across THE COMPANY. As such, the COMPANY Chief Information Officer ("CIO") shall have all appropriate authority to carry out all necessary activities in furtherance of the Program.

# 4 ROLES AND RESPONSIBILITIES

## 4.1 Risk Committee

The Risk Committee shall approve this Charter and shall monitor implementation and maintenance of the Program as set forth in this Charter. Such oversight includes the assignment of Management's Program responsibilities through this Charter and reviewing periodic reports from Management as required herein.

## 4.2 THE COMPANY Board of Directors

THE COMPANY Board of Directors shall oversee the Program, subject to direction and oversight provided by the Risk Committee and the requirements of this Charter. Such oversight includes monitoring the performance of Management's Program responsibilities established through this Charter and reviewing periodic reports from Management as required herein. The COMPANY Board of Directors shall also approve this Charter.

## 4.3 Enterprise Risk Management Committee ("ERMC")

The COMPANY ERMC shall recommend this Charter for approval by the COMPANY Board of Directors. The COMPANY ERMC shall also review periodic reports from Management as required herein.

## 4.4 Management

The CIO shall oversee the integration of the Information Security Program with the COMPANY's processes, people, and technology to mitigate risk in accordance with this Charter. The CIO may, in its discretion, create and fill such roles as the CIO deems fit to further the objectives of this Charter.

The CIO shall appoint the COMPANY IT Compliance and Security Leader ("Leader") to oversee the implementation of the Program. The Chief Information Security Officer ("CISO"), as appointed by the Leader, will serve as Chair of the COMPANY Information Security Council ("ISC") and develop and implement the Program established through this Charter.

The CIO shall at least annually report to the Risk Committee and the COMPANY Board of Directors, describing the overall status of the Program and the COMPANY's compliance with [insert any applicable regulatory standards here] as published by our regulatory authorities. This report should also include other matters related to the Program, when determined to be material in the judgment of the CIO. Examples of material matters may include risk assessments, Business Units' risk management and control decisions, security breaches and Management's responses, violations of policies established under the authority of this Charter, information regarding the oversight of service provider arrangements, testing results, and recommended modifications to this Charter. The CIO or its designee shall also periodically report such information to the ERMC, ISC, and other similarly situated bodies as deemed appropriate by the CIO.

The CIO and its designees shall have unrestricted access, as permitted by law, to all COMPANY platforms, functions, records, property, and personnel deemed necessary by the CIO to accomplish the responsibilities of the CIO, Leader, CISO, or ISC under this Charter or policies or guidelines established hereunder.

# 4.5 The COMPANY Information Security Council

The ISC shall assist the CISO with the ongoing development and implementation of the Program.

## Membership

The CIO shall approve a charter governing the membership and rules of the ISC. The CISO shall organize, and serve as Chair of, the ISC. The members of the ISC shall be the CISO, the COMPANY Chief Privacy Officer, one representative for each Business Unit, as selected by the Business Unit's CIO (or, to the extent not so titled, such similarly situated member of the Business Unit's management), and any other persons as determined by the CIO necessary to ensure comprehensive representation on the ISC reflecting the management structure of the COMPANY.

The aforementioned members will serve as the voting members of the ISC. The Chair or voting members of the ISC may also request that others participate as non-voting members.

## Approval of Policies in Support of the Program

The ISC shall issue information security policy proposals to the CIO for approval. Policies approved by the CIO shall be binding across the COMPANY and its Business Units unless otherwise expressly stated therein.

The ISC must endeavor to ensure that policies approved by the CIO include appropriate administrative, physical, and technical safeguards in light of the objectives of the Program. Such policies should be uniform across the COMPANY and the Business Units to the extent practicable and advisable. The ISC must also endeavor to ensure that all such policies and the Program elements are coordinated and follow a common framework.

The ISC may recommend to the CIO adjustments to the Program for reasons such as changes in technology, laws, the sensitivity of customer information, or internal or external threats to information.

### Creation of Subcommittees

To assist the ISC in accomplishing its objectives, the ISC or ISC Chair may establish subcommittees. Any such subcommittee shall be subject to this Charter.

## 4.6  Business Units

"Business Unit" in the context of this Charter refers to the major operating units defined by the COMPANY. Business Units are responsible for reasonably ensuring the security, confidentiality, and integrity of sensitive customer or business information they create, transmit, or process. Business Units are also responsible for implementing appropriate risk management and control decisions, if any, in light of the risk assessment and control testing performed pursuant to this Charter.

The CIO shall issue policies in support of this Charter, and the Business Units' practices and procedures shall be consistent with such policies. Business Units shall ensure that their records management program is coordinated with [the records management program of] the ISC to ensure appropriate safeguards for sensitive customer and business information in all of its forms. Moreover, Business Units shall fully and timely execute their responsibilities identified under this Charter and policies or guidelines issued hereunder in support of the Program.

# 5  SPECIFIC PROGRAM ELEMENTS

The policies issued by the CIO under this Charter shall include at least the following elements in support of the Program, in addition to those otherwise deemed advisable by the CIO.

## 5.1  Risk Assessments

The CISO shall establish a framework designed to assess the sufficiency of the policies and processes in place to control reasonably foreseeable risks to sensitive customer and business information. All Business Units shall ensure appropriate and timely completion of their responsibilities under this framework, including reporting requirements.

## 5.2 Incident Response Program

As part of the Program, the CISO shall establish an incident response program that specifies the actions to be taken by the Company when they detect that unauthorized individuals have gained access to information systems or that other security incidents have occurred.

## 5.3 Testing Key Controls

Policies established in support of the Program shall require that the COMPANY periodically test key controls, systems, and procedures of the Program. The frequency and nature of such tests should be informed by the risk assessment framework in section 5.1 above. The ISC shall assist the CISO in identifying the relevant key controls and determining the frequency and nature of such tests, subject to the final approval of the CISO.

These tests should be conducted, or reviewed, by independent third parties or staff independent of those that develop or maintain the subject controls. Such parties or staff shall provide the results of this testing in the manner required by the CISO. Testing of controls performed by Internal Audit can satisfy the independent testing requirements of this Charter.

## 5.4 Mitigation Response

Utilizing a risk-based approach in light of the objectives of this Charter, the CISO and such Business Unit personnel the CISO deems appropriate shall ensure timely mitigation of information security risks identified through the Program including, but not limited to, risks identified through sections 5.1 and 5.3 above or security risks introduced by relevant changes in technology, internal or external threats, or business arrangements such as acquisitions or alliances. The CIO may, in its discretion, require that a Business Unit implement certain mitigation efforts in the manner and timeframe necessary to satisfy the objectives of this Charter.

## 5.5 Additional Requirements for Jurisdictions Inside and Outside of the United States

The CISO, in conjunction with counsel as appropriate, shall ensure that the policies or guidelines established through this Charter satisfy any information security legal or regulatory expectations of jurisdictions inside and outside of the United States, or sub-jurisdictions thereof, at a minimum as such expectations are applicable to particular Business Units or subparts thereof. While all policies and guidance developed under this section 5.5 need not be uniform across the COMPANY, their design and implementation shall be coordinated by the Business Units in conjunction with the Leader, CISO, and the ISC.

# 6 TRAINING

The CISO, in coordination with the Business Units, will develop materials to train employees on the Program to support the COMPANY policies, to the extent appropriate. Business Units are responsible for ensuring the timely completion of such training by their employees. All training materials developed regarding information security shall be consistent with policies issued pursuant to this Charter.

# 7 PRE-EXISTING POLICIES AND SUPERSEDING EFFECT OF POLICIES ISSUED UNDER THIS CHARTER

THE COMPANY and its Business Units have long placed a high priority on protecting sensitive business and customer information through policies providing for administrative, physical, and technical safeguards. Unless in conflict with this Charter, any policy pre-existing this Charter shall remain in force until superseded through the authority of this Charter. The CIO has the final authority to resolve any question on such superseding effect.

# 8 MODIFICATIONS

This Charter may need to be modified from time to time to adapt to changing business arrangements, practices, or regulatory requirements. As such, the CIO shall review annually this Charter and recommend to the Risk Committee, the COMPANY Board of Directors, or ERMC any Charter modifications for the final approval of the Risk Committee.

## DOCUMENT CHANGE HISTORY

| | APPROVAL | | | | |
|---|---|---|---|---|---|
| VERSION | RISK COMMITTEE | THE COMPANY BOARD | ERMC | CIO | CHANGES |
| | | | | | |
| | | | | | |
| | | | | | |
| | | | | | |

## DRAFT DOCUMENT CHANGE HISTORY

| | DRAFT AND REVIEW | | | | |
|---|---|---|---|---|---|
| VERSION | | | | | DATES |
| | | | | | |
| | | | | | |
| | | | | | |
| | | | | | |

# SAMPLE PROGRAM STATUS REPORT

## Summary

### Progress Since Prior Report

- Tasks
- Deliverables
- Milestones
- Meetings
- Communications
- Decisions
- Risks and issues

### What's Happening This Week

- Tasks
- Deliverables
- Milestones

- Meetings
- Communications
- Decisions
- Risks and issues

## Overall Budget Spent

00% spent; 00% remaining

## Upcoming Tasks and Milestones

| TASK/MILESTONE | TARGET DATE | NOTES |
|---|---|---|
| | | |
| | | |
| | | |
| | | |

## Action Items

| ACTION ITEM | OWNER | DUE DATE | NOTES |
|---|---|---|---|
| | | | |
| | | | |
| | | | |
| | | | |

## Project Risks, Issues, and Mitigation Plans

| PROJECT ISSUE | RISK | MITIGATION | NOTES |
|---|---|---|---|
| | | | |
| | | | |
| | | | |
| | | | |

# SAMPLE RISK ACCEPTANCE FORM

Reference # _____

**Department:** _____
**Next Review Date:** _____

**Risk Acceptance Form**

**Name and title of Originator:** _____

---

**Summary of Request:**
*(Detail the specifics of the risk to be accepted, including what policy exceptions are required)*

**Overview of the Business Service Impacted:**
*(Discuss what specific business processes are in scope for this exception)*

**Business Benefit of Accepting This Risk:**
*(Add the perceived benefit of not following security policy in business terms)*

**CISO Recommendation:**
*(Add the CISO recommendation & disposition and highlight if you are in agreement or not)*

**Alternatives Evaluated:**
*(Discuss any alternatives proposed as a way to eliminate or reduce this risk)*

**Summary of the Risk being accepted:**

**Summary of Compensating Controls:**
*(Describe the technical and procedural controls implemented to address the vulnerabilities and risks above. If you are not putting any controls in place, simply write "None")*

**Estimated Probability of this Risk Occurring (To be completed by the security team):**
*(Low, medium, high with brief justification or scenario description)*

I understand that compliance with information security policies and standards is expected for all organizational units, information systems, and communication systems. I believe that the control(s) required by our information security policies and guidance from the IT security group cannot be complied with due to the reasons documented above. I understand and accept the risks documented in this form and certify that my department will be responsible for any direct and indirect costs incurred due to incidents related to these identified risks. I also understand that this exception may be revoked by the Chief Information Security Officer at any time and may be subject to alternate opinions from Internal Audit.

**Signature of responsible person**

**Printed name of responsible person**


**CISO/CTO/CIO acknowledgement**                                      **Date**


**Data Owner Signature**                                               **Date**


**Data Owner Name and Title**


Date received by Office of Information Security: _____


Date accepted by the Security Advisory Committee: _____

---

# INCIDENT RESPONSE SAMPLE COMMUNICATIONS

The following template is intended to use as a starting point when communicating about an incident. There are four templates here, one for a low-severity incident, one for a moderate-high security incident, a third-party notification template and an internal message template. Communicating incidents is not something you should do alone. You should involve other groups and areas, including senior management and corporate communications. Use these only as a starting point; please tailor them to your individual business requirements.

## Low-Severity Message Template

One of our clients has informed us of a security incident involving unauthorized access to their systems through the use of stolen credentials. Please note that this system does not involve any unauthorized access or compromise to [your company name] systems and that the intrusion is linked specifically to a company within our own client network.

At this stage, we are unable to provide any further details, and information will remain limited until our client's investigation is complete. Our client is committed to sharing details with us that we can pass along to you. We intend to provide full transparency regarding this incident. We are working with our client to enable a quick and effective resolution to this issue.

[add company contact information]

## Moderate-to-High-Severity Message Template

[Your company name] is informing you of a situation where unauthorized access to our computer systems has been confirmed. We have notified federal and local authorities and are working closely with them to respond to this incident. We are working diligently to minimize the impact to our clients and restore all security measures to our systems.

It is too early to understand what client information, if any, has been compromised. We are taking all precautions in investigating this situation, and at this point we have no reason to believe that client information is at risk. [Your company name] leads with integrity and is committed to working through this investigation and addressing any concerns that our clients might have. The trust that our clients place with us is important and something we do not take lightly.

We will continue to keep you updated on this situation. In the meantime, should you have any additional questions, please contact [name] at [phone or email].

## Third-Party Message Template

[Your company name] was made aware of a security vulnerability by one of our third-party vendors, [vendor/third-party name]. Currently, [vendor/third-party name] is working to understand the source and impact of this problem and come to a quick resolution. We believe that our clients will experience minimal impact during this investigation.

We are monitoring this situation closely and will continue to keep you informed as we learn more. We want to stress that this vulnerability was not a result of any direct compromise to our systems. We are committed to helping [vendor/third-party name] address this issue and cascade any/all important information to you once available.

We take security issues very seriously and will keep you updated on further developments. We will provide an email communication with further details as soon as they are available. In the meantime, should you have any additional questions, please contact [name] at [phone or email].

## Internal Message Template

We can confirm that on [day/date], our network was compromised as a result of a cyberattack. This situation remains under investigation, and at this time, we believe we have contained this incident and that no sensitive data was exposed.

The safety of our employees, customers and business operations is paramount. We will continue to provide updates when we have more information to share.

Should you be contacted by any outside channels for information about this incident, please direct those individuals to [name/department/number], who is the only person authorized to speak on behalf of our company about this issue.

# SAMPLE JOB SEARCH COMMUNICATIONS

## Sample Interview Follow-up Email

Dear [first name],

I wanted to drop you a note to thank you for meeting with me regarding the CISO role at ABC Corporation. I really appreciated hearing your perspective on what areas of the business are growing and where you need the most focus from the security program. I feel like I have a much stronger understanding of ABC Corporation's needs and believe I can take the security program to the next level and help support the business.

I especially enjoyed our discussion around this position and your organization's goals. In particular, I enjoyed hearing about your journey to the cloud and some of the new regulatory and privacy challenges that come along with that. I believe there's a lot I can offer in these and other areas and look forward to continuing our conversation.

Thank you again for your time, and I look forward to the next steps in this process.

Sincerely,

*[your name and contact information]*

# EXECUTIVE JOB SEARCH CHECKLIST

This checklist is intended to help you plan and organize a job search. Use this as a framework to help focus your efforts.

## Start with Your Mindset

☐ If you have been laid off, you will need to spend some time on your mindset. Do what you need to do to get over your last role and start focusing on the future. Reframe the situation as an opportunity to try something new and a path to better opportunities.

☐ If you are employed and simply looking for your next role, you need to get ready to dedicate time and attention to the process. Looking for a new job is a job by itself. If you are still actively working in a role, you want to make sure you're getting your work done but also making room for all the new activities that come along with a job search.

☐ Make sure you set reasonable expectations. Landing an executive role can take six months or longer from start to finish. Yes, we are essentially in a zero-unemployment industry, but you need to find the right opportunity in the right location with the right salary. And you must beat out your competitors. Sometimes, you might be the most qualified candidate in the mix, but there are other "good enough" alternatives that are a better fit or, frankly, cheaper than you. Strive not to take any of this personally. It's just business. Keep positive and keep looking.

## Focus Your Efforts

☐ Focus on your ideal industry, role, and geographic location. If you're going to make a change, maybe it's time for a big change. You'll want to discuss things like relocation with your family before going too far with roles that require moving.

☐ Once you have your criteria set, select some target companies that might fit these criteria.

☐ Develop your value proposition and elevator pitch.

☐ Define your professional branding documents, including your résumé and LinkedIn profile.

☐ Tailor your cover letters and résumés for different roles based on job descriptions, the company or other criteria.

☐ Do some salary research that factors in the role, the industry and the geographic location. Do not expect to be paid like a Financial Services CISO in New York City if you are applying to a deputy CISO role in Kansas. What you made before will not always translate into a bump in pay. In many states, it is now illegal to ask about your previous salary. Instead, be prepared to answer questions about your "salary expectations." Be careful being too aggressive in the early stages and try to either respond about finding the right role first or give a range that you have researched, and think is reasonable for your demographic and experience.

## Plan and Organize

☐ Build a pipeline of potential opportunities. Do not depend on a single company and opportunity no matter how promising it seems at first. A lot can and will happen that will derail that "sure thing" role.

☐ Use all methods of networking at your disposal: people you already know, people you would like to meet, recruiters and applying to posted job leads.

☐ Try to find a referral to any companies where you submit your résumé. Look through first and second-level contacts on LinkedIn and see if you can find any connections that might be able to help from the inside.

# Networking

☐ Organize your connections and let them know you're looking for new opportunities.
☐ Get involved with industry events and networking events.
☐ Consider volunteer work if you are not actively employed. You'll want to be able to explain any gaps in your résumé with value-add activities that are keeping you current.
☐ Reach out to carefully selected recruiters. If you don't have any, reach out to your network and see if you can get some recommendations.

# Plan Your Time

☐ Organize your job search. You'll want to log what roles you've either discussed or applied for and where they all stand. If you've tailored cover letters or résumés to specific companies or industries, keep copies of them so you can leverage them for similar companies or industries you apply for in the future.
☐ If you are between jobs, spend more than half of your time on networking opportunities and reaching out to contacts. Do not spend the majority of your time browsing online postings; this is the least likely method to work out for you in the long run.
☐ Celebrate your successes and keep positive.

# Interview Like a Pro

☐ Are you confident in researching companies and interviewers and feel prepared for interviews?
☐ Have you practiced the STAR method, and do you have specific examples of past accomplishments to present during conversations?
☐ Do you understand how to approach interviews like a consultant, focusing on offering solutions to challenges?
☐ Make sure you consistently follow up after all interviews.

# Negotiate Your Best Possible Offer

☐ Do you understand the compensation negotiation process and strategy that includes more than just salary?
☐ Do you know what your "must have's" are and where you can compromise?

## Transition into Your Next Opportunity

- ☐ Once you have accepted a new position, begin immediately putting together your 100-day plan that includes meeting senior business leaders 1-1.
- ☐ Update your résumé and LinkedIn profile with your new role.
- ☐ Inform and thank your contacts and those who assisted you throughout your search.
- ☐ Set some new career development goals for continued networking and professional development. You should never consider yourself done with this process.

# Index

Printed in the United States
by Baker & Taylor Publisher Services